TEACHER'S PLANNING GUIDE

Project-Based Inquiry Science™

ANIMALS IN ACTION

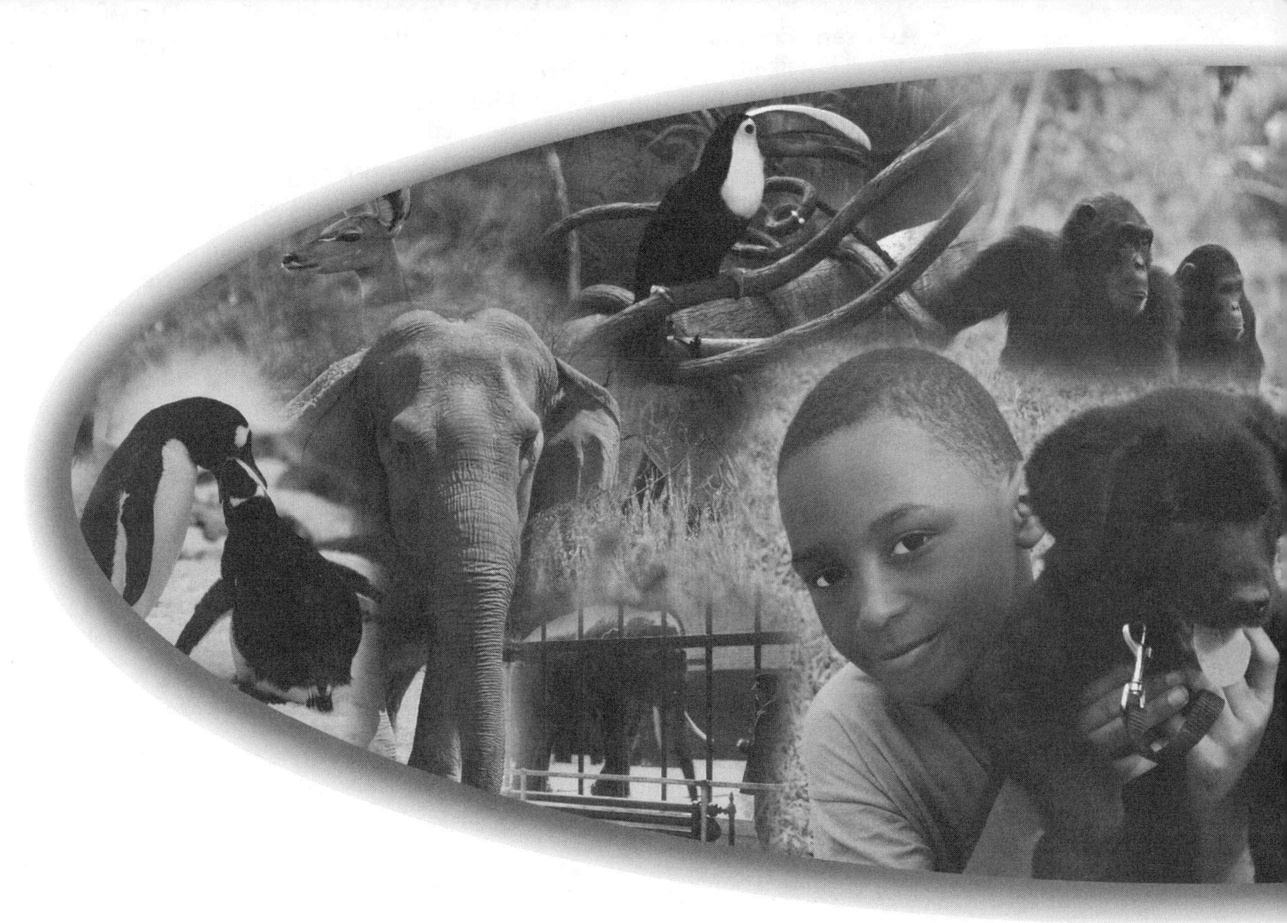

84 Business Park Drive, Armonk, NY 10504
Phone (914) 273-2233 Fax (914) 273-2227
www.its-about-time.com

Publishing Team

President
Tom Laster

Director of Product Development
Barbara Zahm, Ph.D.

Managing Editor
Maureen Grassi

Project Development Editor
Ruta Demery

Project Manager
Sarah V. Gruber

Assistant Editors
Gail Foreman
Susan Gibian
Nomi Schwartz

Assistant Editors, Teacher's Planning Guide
Danielle Bouchat-Friedman
Kelly Crowley
Edward Denecke
Heide M. Doss
Jake Gillis
Rhonda Gordon

Safety and Content Reviewer
Edward Robeck
Barbara Speziale

Creative Director
John Nordland

Production/Studio Manager
Robert Schwalb

Layout and Production
Sean Campbell
Doreen Flaherty

Illustrator
Dennis Falcon

Technical Art/ Photo Research
Sean Campbell
Michael Hortens
Marie Killoran

Equipment Kit Developers
Dana Turner
Joseph DeMarco

Printed and bound in the United States of America.

ISBN-13: 978-1-58591-632-0
1 2 3 4 5 VH 12 11 10 09 08

This project was supported, in part, by the ***National Science Foundation*** under grant nos. 0137807, 0527341, 0639978.

Opinions expressed are those of the authors and not necessarily those of the National Science Foundation.

PBIS Principal Investigators

Janet L. Kolodner is a Regents' Professor in the School of Interactive Computing in the Georgia Institute of Technology's College of Computing. Since 1978, her research has focused on learning from experience, both in computers and in people. She pioneered the Artificial Intelligence method called *case-based reasoning*, providing a way for computers to solve new problems based on their past experiences. Her book, *Case-Based Reasoning*, synthesizes work across the case-based reasoning research community from its inception to 1993.

Since 1994, Dr. Kolodner has focused on the applications and implications of case-based reasoning for education. In her approach to science education, called Learning by Design™ (LBD), students learn science while pursuing design challenges. Dr. Kolodner has investigated how to create a culture of collaboration and rigorous science talk in classrooms, how to use a project challenge to promote focus on science content, and how students learn and develop when classrooms function as learning communities. Currently, Dr. Kolodner is investigating how to help young people come to think of themselves as scientific reasoners. Dr. Kolodner's research results have been widely published, including in *Cognitive Science, Design Studies,* and the *Journal of the Learning Sciences.*

Dr. Kolodner was founding Director of Georgia Tech's EduTech Institute, served as coordinator of Georgia Tech's Cognitive Science program for many years, and is founding Editor in Chief of the *Journal of the Learning Sciences*. She is a founder of the International Society for the Learning Sciences, and she served as its first Executive Officer. She is a fellow of the American Association of Artificial Intelligence.

Joseph S. Krajcik is a Professor of Science Education and Associate Dean for Research in the School of Education at the University of Michigan. He works with teachers in science classrooms to bring about sustained change by creating classroom environments in which students find solutions to important intellectual questions that subsume essential curriculum standards and use learning technologies as productivity tools. He seeks to discover what students learn in such environments, as well as to explore and find solutions to challenges that teachers face in enacting such complex instruction.

Dr. Krajcik has authored and co-authored over 100 manuscripts and makes frequent presentations at international, national, and regional conferences that focus on his research, as well as presentations that translate research findings into classroom practice. He is a fellow of the American Association for the Advancement of Science and served as president of the National Association for Research in Science Teaching. Dr. Krajcik co-directs the Center for Highly Interactive Classrooms, Curriculum and Computing in Education at the University of Michigan and is a co-principal investigator in the Center for Curriculum Materials in Science and The National Center for Learning and Teaching Nanoscale Science and Engineering. In 2002, Dr. Krajcik was honored to receive a Guest Professorship from Beijing Normal University in Beijing, China. In winter 2005, he was the Weston Visiting Professor of Science Education at the Weizmann Institute of Science in Rehovot, Israel.

Daniel C. Edelson is Vice President for Education and Children's Programs at the National Geographic Society. Previously, he was the director of the Geographic Data in Education (GEODE) Initiative at Northwestern University, where he led the development of Planetary Forecaster and Earth Systems and Processes. Since 1992, Dr. Edelson has directed a series of projects exploring the use of technology as a catalyst for reform in science education and has led the development of a number of software environments for education. These include My World GIS, a geographic information system for inquiry-based learning, and WorldWatcher, a data visualization and analysis system for gridded geographic data. Dr. Edelson is the author of the high school environmental science text, *Investigations in Environmental Science: A Case-Based Approach to the Study of Environmental Systems*. His research has been widely published, including in the *Journal of the Learning Sciences*, the *Journal of Research on Science Teaching*, *Science Educator*, and *Science Teacher*.

Brian J. Reiser is a Professor of Learning Sciences in the School of Education and Social Policy at Northwestern University. Professor Reiser served as chair of Northwestern's Learning Sciences Ph.D. program from 1993, shortly after its inception, until 2001. His research focuses on the design and enactment of learning environments that support students' inquiry in science, including both science curriculum materials and scaffolded software tools. His research investigates the design of learning environments that scaffold scientific practices, including investigation, argumentation, and explanation; design principles for technology-infused curricula that engage students in inquiry projects; and the teaching practices that support student inquiry. Professor Reiser also directed BGuILE (Biology Guided Inquiry Learning Environments) to develop software tools for supporting middle school and high school students in analyzing data and constructing explanations with biological data. Reiser is a co-principal investigator in the NSF Center for Curriculum Materials in Science. He served as a member of the NRC panel authoring the report Taking Science to School.

Mary L. Starr is a Research Specialist in Science Education in the School of Education at the University of Michigan. She collaborates with teachers and students in elementary and middle school science classrooms around the United States who are implementing *Project-Based Inquiry Science*. Before joining the PBIS team, Dr. Starr created professional learning experiences in science, math, and technology, designed to assist teachers in successfully changing their classroom practices to promote student learning from coherent inquiry experiences. She has developed instructional materials in several STEM areas, including nanoscale science education, has presented at national and regional teacher education and educational research meetings, and has served in a leadership role in the Michigan Science Education Leadership Association. Dr. Starr has authored articles and book chapters, and has worked to improve elementary science teacher preparation through teaching science courses for pre-service teachers and acting as a consultant in elementary science teacher preparation. As part of the PBIS team, Dr. Starr has played a lead role in making units cohere as a curriculum, in developing the framework for PBIS Teachers Planning Guides, and in developing teacher professional development experiences and materials.

Acknowledgements

Three research teams contributed to the development of *Project-Based Inquiry Science* (PBIS): a team at the Georgia Institute of Technology headed by Janet L. Kolodner, a team at Northwestern University headed by Daniel Edelson and Brian Reiser, and a team at the University of Michigan headed by Joseph Krajcik and Ron Marx. Each of the PBIS units was originally developed by one of these teams and then later revised and edited to be a part of the full three-year middle-school curriculum that became PBIS.

PBIS has its roots in two educational approaches, Project-Based Science and Learning by Design™. Project-Based Science suggests that students should learn science through engaging in the same kinds of inquiry practices scientists use, in the context of scientific problems relevant to their lives and using tools authentic to science. Project-Based Science was originally conceived in the hi-ce Center at the University of Michigan, with funding from the National Science Foundation. Learning by Design™ derives from Problem-Based Learning and suggests sequencing, social practices, and reflective activities for promoting learning. It engages students in design practices, including the use of iteration and deliberate reflection. LBD was conceived at the Georgia Institute of Technology, with funding from the National Science Foundation, DARPA, and the McDonnell Foundation.

The development of the integrated PBIS curriculum was supported by the National Science Foundation under grants no. 0137807, 0527341, and 0639978. Any opinions, findings and conclusions, or recommendations expressed in this material are those of the authors and do not necessarily reflect the views of the National Science Foundation.

PBIS Team

Principal Investigator
Janet L. Kolodner

Co-Principal Investigators
Daniel C. Edelson
Joseph S. Krajcik
Brian J. Reiser

NSF Program Officer
Gerhard Salinger

Curriculum Developers
Michael T. Ryan
Mary L. Starr

Teacher's Planning Guide Developers
Rebecca M. Schneider
Mary L. Starr

Literacy Specialist
LeeAnn M. Sutherland

NSF Program Reviewer
Arthur Eisenkraft

Project Coordinator
Juliana Lancaster

External Evaluators
The Learning Partnership
Steven M. McGee
Jennifer Witers

The Georgia Institute of Technology Team

Project Director:
Janet L. Kolodner

Development of PBIS units at the Georgia Institute of Technology was conducted in conjunction with the Learning by Design™ Research group (LBD), Janet L. Kolodner, PI.

Lead Developers, Physical Science:
David Crismond
Michael T. Ryan

Lead Developer, Earth Science:
Paul J. Camp

Assessment and Evaluation:
Barbara Fasse
Daniel Hickey
Jackie Gray
Laura Vandewiele
Jennifer Holbrook

Project Pioneers:
JoAnne Collins
David Crismond
Joanna Fox
Alice Gertzman
Mark Guzdial
Cindy Hmelo-Silver
Douglas Holton
Roland Hubscher
N. Hari Narayanan
Wendy Newstetter
Valery Petrushin
Kathy Politis
Sadhana Puntambekar
David Rector
Janice Young

The Northwestern University Team

Project Directors:
Daniel Edelson
Brian Reiser

Lead Developer, Biology:
David Kanter

Lead Developers, Earth Science:
Jennifer Mundt Leimberer
Darlene Slusher

Development of PBIS units at Northwestern was conducted in conjunction with:

The Center for Learning Technologies in Urban Schools (LeTUS) at Northwestern, and the Chicago Public Schools
Louis Gomez, PI;
Clifton Burgess, PI
for Chicago Public Schools.

The BioQ Collaborative
David Kanter, PI.

The Biology Guided Inquiry Learning Environments (BGuILE) Project
Brian Reiser, PI.

The Geographic Data in Education (GEODE) Initiative
Daniel Edelson, Director

The Center for Curriculum Materials in Science at Northwestern
Brian Reiser,
Daniel Edelson,
Bruce Sherin, PIs.

The University of Michigan Team

Project Directors:
Joseph Krajcik
Ron Marx

Literacy Specialist:
LeeAnn M. Sutherland

Project Coordinator:
Mary L. Starr

Development of PBIS units at the University of Michigan was conducted in conjunction with:

The Center for Learning Technologies in Urban Schools (LeTUS)
Ron Marx, Phyllis Blumenfeld,
Barry Fishman,
Joseph Krajcik,
Elliot Soloway, PIs.

The Detroit Public Schools
Juanita Clay-Chambers
Deborah Peek-Brown

The Center for Highly Interactive Computing in Education (hi-ce)
Ron Marx,
Phyllis Blumenfeld,
Barry Fishman,
Joseph Krajcik,
Elliot Soloway,
Elizabeth Moje,
LeeAnn Sutherland, PIs.

Field-Test Teachers

National Field Test
Tamica Andrew
Leslie Baker
Jeanne Bayer
Gretchen Bryant
Boris Consuegra
Daun D'Aversa
Candi DiMauro
Kristie L. Divinski
Donna M. Dowd
Jason Fiorito
Lara Fish
Christine Gleason
Christine Hallerman
Terri L. Hart-Parker
Jennifer Hunn
Rhonda K. Hunter
Jessica Jones
Dawn Kuppersmith
Anthony F. Lawrence
Ann Novak
Rise Orsini
Tracy E. Parham
Cheryl Sgro-Ellis
Debra Tenenbaum
Sarah B. Topper
Becky Watts
Debra A. Williams
Ingrid M. Woolfolk
Ping-Jade Yang

New York City Field Test

Several sequences of PBIS units have been field-tested in New York City under the leadership of Whitney Lukens, Staff Developer for Region 9, and Greg Borman, Science Instructional Specialist, New York City Department of Education

6th Grade
Norman Agard
Tazinmudin Ali
Heather Guthartz Aniba
Asher Arzonane
Asli Aydin
Shareese Blakely
John J. Blaylock
Joshua Blum
Tsedey Bogale
Filomena Borrero
Zachary Brachio
Thelma Brown
Alicia Browne-Jones
Scott Bullis
Maximo Cabral
Lionel Callender
Matthew Carpenter
Ana Maria Castro
Diane Castro
Anne Chan
Ligia Chiorean
Boris Consuegra
Careen Halton Cooper
Cinnamon Czarnecki
Kristin Decker
Nancy Dejean
Gina DiCicco
Donna Dowd
Lizanne Espina
Joan Ferrato
Matt Finnerty
Jacqueline Flicker
Helen Fludd
Leigh Summers Frey
Helene Friedman-Hager
Diana Gering
Matthew Giles
Lucy Gill
Steven Gladden
Greg Grambo
Carrie Grodin-Vehling
Stephan Joanides
Kathryn Kadei
Paraskevi Karangunis
Cynthia Kerns
Martine Lalanne
Erin Lalor
Jennifer Lerman
Sara Lugert
Whitney Lukens
Dana Martorella
Christine Mazurek
Janine McGeown
Chevelle McKeever
Kevin Meyer
Jennifer Miller
Nicholas Miller
Diana Neligan
Caitlin Van Ness
Marlyn Orque
Eloisa Gelo Ortiz
Gina Papadopoulos
Tim Perez
Albertha Petrochilos
Christopher Poli
Kristina Rodriguez
Nadiesta Sanchez
Annette Schavez
Hilary Sedgwitch
Elissa Seto
Laura Shectman
Audrey Shmuel
Katherine Silva
Ragini Singhal
C. Nicole Smith
Gitangali Sohit
Justin Stein
Thomas Tapia
Eilish Walsh-Lennon
Lisa Wong
Brian Yanek
Cesar Yarleque
David Zaretsky
Colleen Zarinsky

7th Grade
Mayra Amaro
Emmanuel Anastasiou
Cheryl Barnhill
Bryce Cahn
Ligia Chiorean
Ben Colella
Boris Consuegra
Careen Halton Cooper
Elizabeth Derse
Urmilla Dhanraj
Gina DiCicco
Lydia Doubleday
Lizanne Espina
Matt Finnerty
Steven Gladden
Stephanie Goldberg
Nicholas Graham
Robert Hunter
Charlene Joseph
Ketlynne Joseph
Kimberly Kavazanjian
Christine Kennedy
Bakwah Kotung
Lisa Kraker
Anthony Lett
Herb Lippe
Jennifer Lopez
Jill Mastromarino
Kerry McKie
Christie Morgado
Patrick O'Connor
Agnes Ochiagha
Tim Perez
Nadia Piltser
Chris Poli
Carmelo Ruiz
Kim Sanders
Leslie Schiavone
Ileana Solla
Jacqueline Taylor
Purvi Vora
Ester Wiltz
Carla Yuille
Marcy Sexauer Zacchea
Lidan Zhou

8th Grade
Emmanuel Anastasio
Jennifer Applebaum
Marsha Armstrong
Jenine Barunas
Vito Cipolla
Kathy Critharis
Patrecia Davis
Alison Earle
Lizanne Espina
Matt Finnerty
Ursula Fokine
Kirsis Genao
Steven Gladden
Stephanie Goldberg
Peter Gooding
Matthew Herschfeld
Mike Horowitz
Charlene Jenkins
Ruben Jimenez
Ketlynne Joseph
Kimberly Kavazanjian
Lisa Kraker
Dora Kravitz
Anthony Lett
Emilie Lubis
George McCarthy
David Mckinney
Michael McMahon
Paul Melhado
Jen Miller
Christie Morgado
Ms. Oporto
Maria Jenny Pineda
Anastasia Plaunova
Carmelo Ruiz
Riza Sanchez
Kim Sanders
Maureen Stefanides
Dave Thompson
Matthew Ulmann
Maria Verosa
Tony Yaskulski

Animals in Action

The PBIS version of *Animals in Action* is an adaptation of the unit *Animals in Action* developed by a team at University of Toledo. The unit was inspired by and includes activities adapted from *Behavior Matters*, developed jointly by the Brookfield Zoo in Chicago and the Center for Learning Technologies in Urban Schools at Northwestern University.

Animals in Action

PBIS Editorial Team:

Francesca Casella
Mary L. Starr

University of Toledo Team

Lead Developer

Rebecca M. Schneider

Other Developers

Cara Aschliman
Jodie VonSeggern

Pilot Teachers

Lara Fish
Cheri Sgro-Ellis
Debra Tennenbaum

Behavior Matters

Project Directors

Brian J. Reiser
Susan Margulis

Writers/Editors

Marilyn Havlik
Debora Ward

Contributors and Pilot Teachers

Steve Arnam
Jerry Balin
Cindy Cho
Robin Dembeck
Michele Fleming
Vanessa Go
Ravit Golan
Carl Koch
Lanis Petrik
Brian Reiser
Keith Winsten
Angela Wrobel
Sherry Yarema

The development of *Animals in Action* was supported by the National Science Foundation under grants no. 0137807, 0527341, 0639978. The development of *Behavior Matters* was funded by the National Science Foundation under grant nos. 9809636 and 9720383. We are grateful for the recommendations of Whitney Lukens of the NYC Public Schools during development of the PBIS Unit. Any opinions, findings, and conclusions or recommendations expressed in this material are those of the authors and do not necessarily reflect the views of the National Science Foundation or the Brookfield Zoo.

Animals In Action Teacher's Planning Guide

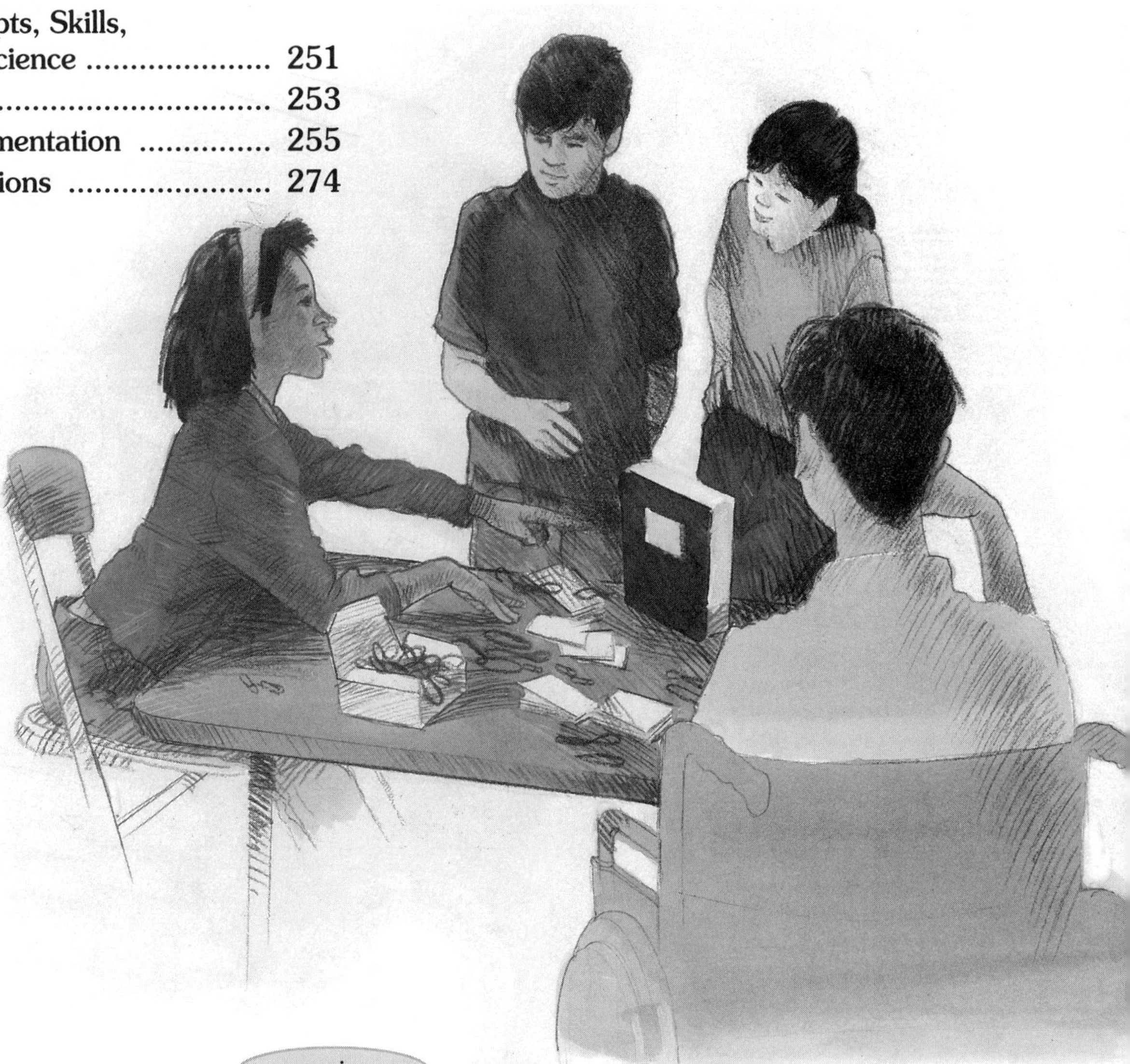

Learning Set 3

What Affects How Animals Communicate?

Science Concepts: *Verbal and non-verbal communication, structure and function, effects of habitat, adaptation, the waggle dance, echolocation, sonar, sound and sound waves, how humans hear, careful observation, keeping good records, finding trends in data, reliable data, interpretation, using evidence to support claims, explanation, collaboration, building on the work of others, criteria and constraints.*

3.0 Learning Set Introduction

3.1 Understand the Question

Thinking About What Affects How Animals Communicate?

3.2 Investigate

How Do Humans Communicate?

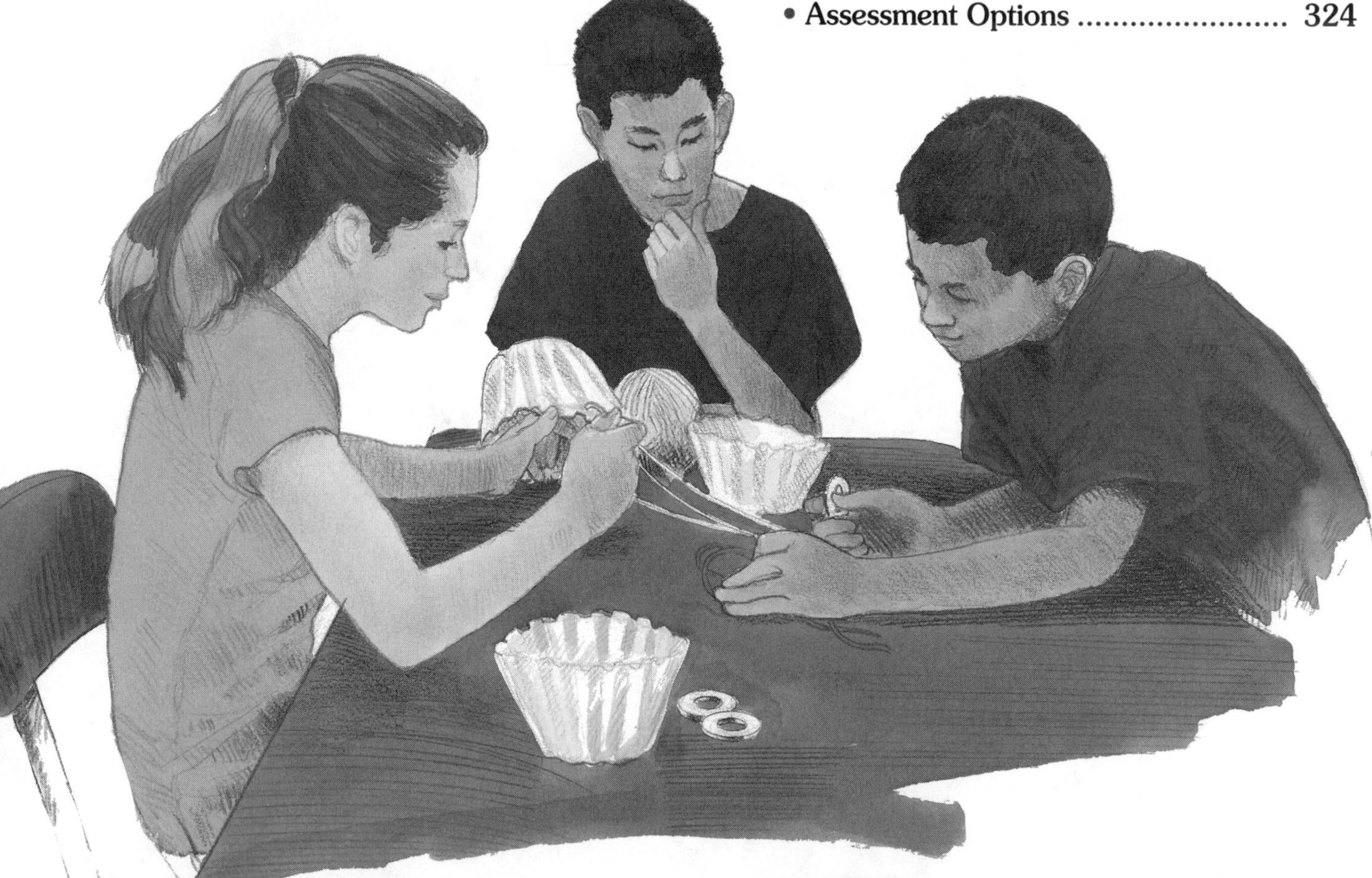

Welcome to Project-Based Inquiry Science!

Welcome to Project-Based Inquiry Science (PBIS): A Middle-School Science Curriculum!

This year, your students will be learning the way scientists learn, exploring interesting questions and challenges, reading about what other scientists have discovered, investigating, experimenting, gathering evidence, and forming explanations. They will learn to collaborate with others to find answers and to share their learning in a variety of ways. In the process, they will come to see science in a whole new, exciting way that will motivate them throughout their educational experiences and beyond.

What is PBIS?

In project-based inquiry learning, students investigate scientific content and learn science practices in the context of attempting to address challenges in or answer questions about the world around them. Early activities introducing students to a challenge help them to generate issues that need to be investigated, making inquiry a student-driven endeavor. Students investigate as scientists would, through observations, designing and running experiments, designing, building, and running models, reading written material, and so on, as appropriate. Throughout each project, students might make use of technology and computer tools that support their efforts in observation, experimentation, modeling, analysis, and reflection. Teachers support and guide the student inquiries by framing the guiding challenge or question, presenting crucial lessons, managing the sequencing of activities, and eliciting and steering discussion and collaboration among the students. At the completion of a project, students publicly exhibit what they have learned along with their solutions to the specific challenge. Personal reflection to help students learn from the experience is embedded in student activities, as are opportunities for assessment.

The curriculum will provide three years of piloted project-based inquiry materials for middle-school science. Individual curriculum units have been defined that cover the scope of the national content and process standards for the middle-school grades. Each Unit focuses on helping students acquire qualitative understanding of targeted science principles and move toward quantitative understanding, is infused with technology, and provides a foundation in reasoning skills, science content, and science process that will ready them for more advanced science. The curriculum as a whole introduces students to a wide range of investigative approaches in science (e.g., experimentation, modeling) and is designed to help them develop scientific reasoning skills that span those investigative approaches.

Technology can be used in project-based inquiry to make available to students some of the same kinds of tools and aids used by scientists in the field. These range from pencil-and-paper tools for organized data recording, collection, and management to software tools for analysis, simulation, modeling, and other tasks. Such infusion provides a platform for providing prompts, hints, examples, and other kinds of aids to students as they are engaging in scientific reasoning. The learning technologies and tools that are integrated into the curriculum offer essential scaffolding to students as they are developing their scientific reasoning skills, and are seamlessly infused into the overall completion of project activities and investigations.

Standards-Based Development

Development of each curriculum Unit begins by identifying the specific relevant national standards to be addressed. Each Unit has been designed to cover a specific portion of the national standards. This phase of development also includes an analysis of curriculum requirements across multiple states. Our intent is to deliver a product that will provide coverage of the content deemed essential on the widest practical scope and that will be easily adaptable to the needs of teachers across the country.

Once the appropriate standards have been identified, the development team works to define specific learning goals built from those standards, and takes into account conceptions and misunderstandings common among middle-school students. An orienting design challenge or driving question for investigation is chosen that motivates achieving those learning goals, and the team then sequences activities and the presentation of specific concepts so that students can construct an accurate understanding of the subject matter.

Inquiry-Based Design

The individual curriculum Units present two types of projects: engineering-design challenges and driving-question investigations. Design-challenge Units begin by presenting students with a scenario and problem and challenging them to design a device or plan that will solve the problem. Driving-question investigations begin by presenting students with a complex question with real-world implications. Students are challenged to develop answers to the questions. The scenario and problem in the design Units and the driving question in the investigation Units are carefully selected to lead the students into investigation of specific science concepts, and the solution processes are carefully structured to require use of specific scientific reasoning skills.

Pedagogical Rationale

Research shows that individual project-based learning units promote excitement and deep learning of the targeted concepts. However, achieving deep, flexible, transferable learning of cross-disciplinary content (e.g., the notion of a model, time scale, variable, experiment) and science practice requires a learning environment that consistently, persistently, and pervasively encourages the use of such content and practices over an extended period of time. By developing project-based inquiry materials that cover the spectrum of middle-school science content in a coherent framework, we provide this extended exposure to the type of learning environment most likely to produce competent scientific thinkers who are well grounded in their understanding of both basic science concepts and the standards and practices of science in general.

Evidence of Effectiveness

There is compelling evidence showing that a project-based inquiry approach meets this goal. Working at Georgia Tech, the University of Michigan, and Northwestern University, we have developed, piloted, and/or field-tested many individual project-based units. Our evaluation evidence shows that these materials engage students well and are manageable by teachers, and that students learn both content and process skills. In every summative evaluation, student performance on post-tests improved significantly from pretest performance (Krajcik, et al., 2000; Holbrook, et al., 2001; Gray et. al. 2001). For example, in the second year in a project-based classroom in Detroit, the average student at post-test scored at about the 95th percentile of the pre-test distribution. Further, we have repeatedly documented significant gains in content knowledge relative to other inquiry-based (but not project-based) instructional methods. In one set of results, performance by a project-based class

in Atlanta doubled on the content test while the matched comparison class (with an excellent teacher) experienced only a 20% gain (significance $p < .001$). Other comparisons have shown more modest differences, but project-based students consistently perform better than their comparisons. Most exciting about the Atlanta results is that results from performance assessments show that, within comparable student populations, project-based students score higher on all categories of problem-solving and analysis and are more sophisticated at science practice and managing a collaborative scientific investigation. Indeed, the performance of average-ability project-based students is often statistically indistinguishable from or better than performance of comparison honors students learning in an inquiry-oriented but not project-based classroom. The Chicago group also has documented significant change in process skills in project-based classrooms. Students become more effective in constructing and critiquing scientific arguments (Sandoval, 1998) and in constructing scientific explanations using discipline-specific knowledge, such as evolutionary explanations for animal behavior (Smith & Reiser, 1998).

Researchers at Northwestern have also investigated the change in classroom practices that are elicited by project-based units. Analyses of the artifacts students produce indicate that students are engaging in ambitious learning practices, requiring weighing and synthesizing many results from complex analyses of data, and constructing scientific arguments that require synthesizing results from multiple complex analyses of data (Edelson et al, 1998; Reiser et al, 2001). Students are engaged in planning, performing, monitoring and revising their investigations, and reporting on their investigation processes as well as their results (Loh et al. 1998). In general, the classrooms engaging in project-based activities reveal substantial moves toward a scientific discourse community in which students focus on arguing from evidence, critiquing ideas, and conjecturing, rather than simply reporting on what they have read or been told (Tabak & Reiser, 1997).

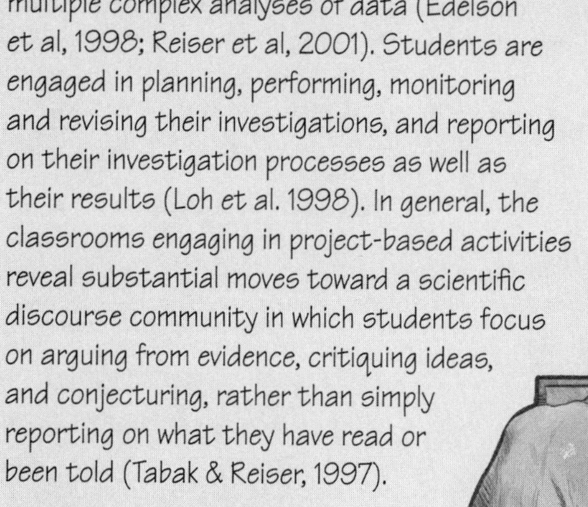

ntroducing PBIS

Vhat Do Scientists Do?

) Scientists...address big challenges and big questions.

tudents will find many different kinds of *Big Challenges* and *uestions* in *PBIS* Units. Some ask them to think about why omething is a certain way. Some ask them to think about hat causes something to change. Some challenge them to esign a solution to a problem. Most are about things that can nd do happen in the real world.

nderstand the Big Challenge or Question

s students get started with each Unit, they will do ctivities that help them understand the *Big Question* or *hallenge* for that Unit. They will think about what they ready know that might help them, and they will identify ome of the new things they will need to learn.

roject Board

he *Project Board* helps you and your students eep track of their learning. For each challenge r question, they will use a *Project Board* to eep track of what they know, what they need learn, and what they are learning. As they arn and gather evidence, they will record hat on the *Project Board*. After they have nswered each small question or challenge, ney will return to the *Project Board* to record ow what they have learned helps them nswer the *Big Question* or *Challenge*.

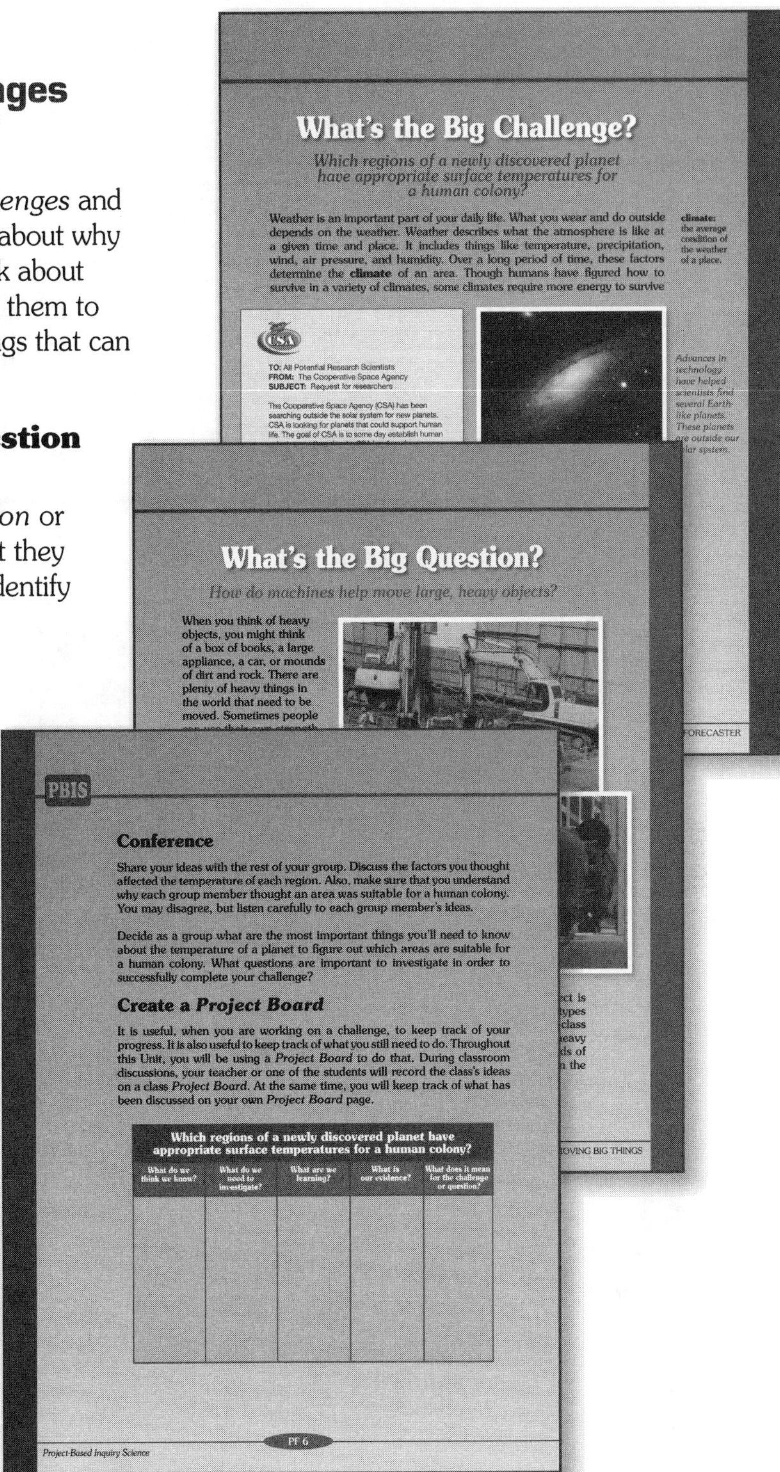

What's the Big Challenge?

Which regions of a newly discovered planet have appropriate surface temperatures for a human colony?

Weather is an important part of your daily life. What you wear and do outside depends on the weather. Weather describes what the atmosphere is like at a given time and place. It includes things like temperature, precipitation, wind, air pressure, and humidity. Over a long period of time, these factors determine the **climate** of an area. Though humans have figured how to survive in a variety of climates, some climates require more energy to survive

climate: the average condition of the weather of a place.

CSA

TO: All Potential Research Scientists
FROM: The Cooperative Space Agency
SUBJECT: Request for researchers

The Cooperative Space Agency (CSA) has been searching outside the solar system for new planets. CSA is looking for planets that could support human life. The goal of CSA is to some day establish human

Advances in technology have helped scientists find several Earth-like planets. These planets are outside our solar system.

What's the Big Question?

How do machines help move large, heavy objects?

When you think of heavy objects, you might think of a box of books, a large appliance, a car, or mounds of dirt and rock. There are plenty of heavy things in the world that need to be moved. Sometimes people

PBIS

Conference

Share your ideas with the rest of your group. Discuss the factors you thought affected the temperature of each region. Also, make sure that you understand why each group member thought an area was suitable for a human colony. You may disagree, but listen carefully to each group member's ideas.

Decide as a group what are the most important things you'll need to know about the temperature of a planet to figure out which areas are suitable for a human colony. What questions are important to investigate in order to successfully complete your challenge?

Create a *Project Board*

It is useful, when you are working on a challenge, to keep track of your progress. It is also useful to keep track of what you still need to do. Throughout this Unit, you will be using a *Project Board* to do that. During classroom discussions, your teacher or one of the students will record the class's ideas on a class *Project Board*. At the same time, you will keep track of what has been discussed on your own *Project Board* page.

Which regions of a newly discovered planet have appropriate surface temperatures for a human colony?				
What do we think we know?	What do we need to investigate?	What are we learning?	What is our evidence?	What does it mean for the challenge or question?

Project-Based Inquiry Science PF 6

Learning Set 1

How Do Flowing Water and Land Interact in a Community?

The big question for this unit is *How does water quality affect the ecology of a community?* So far you have considered what you already know about what water quality is. Now you may be wondering where the water you use comes from. If you live in a city or town, the water you use may come from a river. You would want to know the quality of the water you are using. To do so, it is important to know how the water gets into the river. You also need to know what happens to the water as the river flows across the land.

You may have seen rivers or other water bodies near your home, your school, or in your city. Think about the river closest to where you live. Consider from where the water in the river comes. If you have traveled along the river, think about what the land around the river looks like. Try to figure out what human activities occur in the area. Speculate as to whether these activities affect the quality of water in the river.

To answer the big question, you need to break it down into smaller questions. In this *Learning Set*, you will investigate two smaller questions. As you will discover, these questions are very closely related and very hard to separate. The smaller questions are *How does water affect the land as it moves through the community?* and *How does land use affect water*

Address the Big Challenge

How Do Scientists Work Together to Solve Problems?

You began this unit with the question, *how do scientists work together to solve problems?* You did several small challenges. As you worked on those challenges you learned about how scientists solve problems. You will now watch a video about real-life designers. You will see what the people in the video are doing that is like what you have been doing. Then you will think about all the different things you have been doing during this unit. Lastly, you will write about what you have learned about doing science and being a scientist.

Watch

IDEO Video

The video you will watch follows a group of designers at IDEO. IDEO is an innovation and design firm. In the video, they face the challenge of designing and building a new kind of shopping cart. These designers are doing many of the same things that you did. They also use other practices that you did not use. As you watch the video, record the interesting things you see.

After watching the video, answer the questions on the next page. You might want to look at them before you watch the video. Answering these questions should help you answer the big question of this unit: *How do scientists work together to solve problems?*

Learning Sets

Each Unit is composed of a group of *Learning Sets*, one for each of the smaller questions that needs to be answered to address the *Big Question* or *Challenge*. In each *Learning Set*, students will investigate and read to find answers to the *Learning Set's* question. They will also have a chance to share the results of their investigations with their classmates and work together to make sense of what they are learning. As students come to understand answers to the questions on the *Project Board*, you will record those answers and the evidence they collected. At the end of each *Learning Set*, they will apply their knowledgeto the *Big Question* or *Challenge*.

Answer the Big Question/ Address the Big Challenge

At the end of each Unit, students will put everything they have learned together to tackle the *Big Question* or *Challenge*.

2) Scientists...address smaller questions and challenges.

What Students Do in a Learning Set

Understanding the Question or Challenge

At the start of each *Learning Set*, students will usually do activities that will help them understand the *Learning Set's* question or challenge and recognize what they already know that can help them answer the question or achieve the challenge. Usually, they will visit the *Project Board* after these activities and record on it the even smaller questions that they need to investigate to answer a *Learning Set's* question.

Investigate/Explore

There are many different kinds of investigations students might do to find answers to questions. In the *Learning Sets,* they might

- design and run experiments;
- design and run simulations;
- design and build models;
- examine large sets of data.

Don't worry if your students haven't done these things before. The text will provide them with lots of help in designing their investigations and in analyzing thier data.

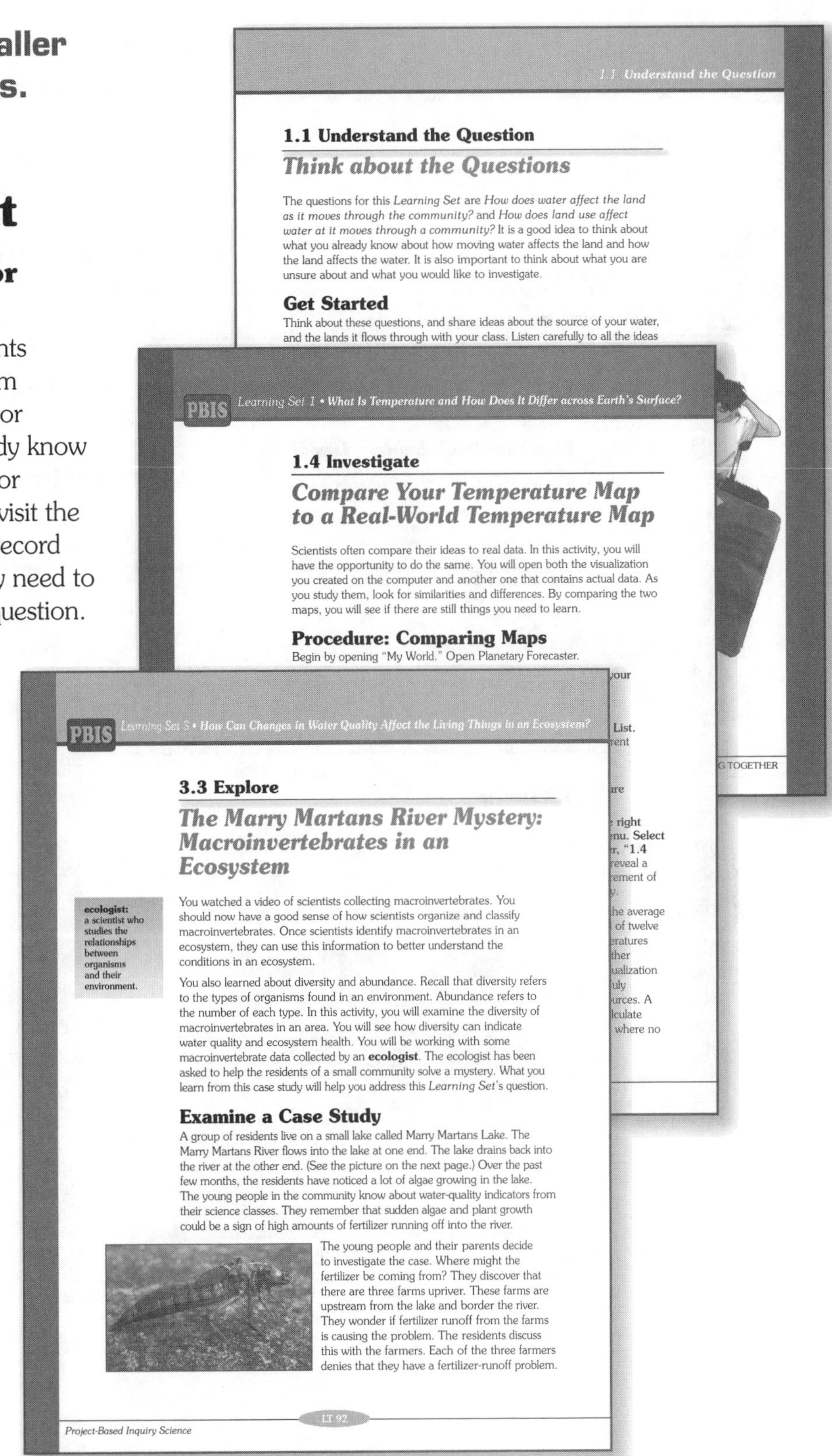

1.1 Understand the Question

1.1 Understand the Question

Think about the Questions

The questions for this *Learning Set* are *How does water affect the land as it moves through the community?* and *How does land use affect water at it moves through a community?* It is a good idea to think about what you already know about how moving water affects the land and how the land affects the water. It is also important to think about what you are unsure about and what you would like to investigate.

Get Started

Think about these questions, and share ideas about the source of your water, and the lands it flows through with your class. Listen carefully to all the ideas

PBIS Learning Set 1 • What Is Temperature and How Does It Differ across Earth's Surface?

1.4 Investigate

Compare Your Temperature Map to a Real-World Temperature Map

Scientists often compare their ideas to real data. In this activity, you will have the opportunity to do the same. You will open both the visualization you created on the computer and another one that contains actual data. As you study them, look for similarities and differences. By comparing the two maps, you will see if there are still things you need to learn.

Procedure: Comparing Maps

Begin by opening "My World." Open Planetary Forecaster.

PBIS Learning Set 3 • How Can Changes in Water Quality Affect the Living Things in an Ecosystem?

3.3 Explore

The Marry Martans River Mystery: Macroinvertebrates in an Ecosystem

ecologist: a scientist who studies the relationships between organisms and their environment.

You watched a video of scientists collecting macroinvertebrates. You should now have a good sense of how scientists organize and classify macroinvertebrates. Once scientists identify macroinvertebrates in an ecosystem, they can use this information to better understand the conditions in an ecosystem.

You also learned about diversity and abundance. Recall that diversity refers to the types of organisms found in an environment. Abundance refers to the number of each type. In this activity, you will examine the diversity of macroinvertebrates in an area. You will see how diversity can indicate water quality and ecosystem health. You will be working with some macroinvertebrate data collected by an **ecologist**. The ecologist has been asked to help the residents of a small community solve a mystery. What you learn from this case study will help you address this *Learning Set's* question.

Examine a Case Study

A group of residents live on a small lake called Marry Martans Lake. The Marry Martans River flows into the lake at one end. The lake drains back into the river at the other end. (See the picture on the next page.) Over the past few months, the residents have noticed a lot of algae growing in the lake. The young people in the community know about water-quality indicators from their science classes. They remember that sudden algae and plant growth could be a sign of high amounts of fertilizer running off into the river.

The young people and their parents decide to investigate the case. Where might the fertilizer be coming from? They discover that there are three farms upriver. These farms are upstream from the lake and border the river. They wonder if fertilizer runoff from the farms is causing the problem. The residents discuss this with the farmers. Each of the three farmers denies that they have a fertilizer-runoff problem.

Project-Based Inquiry Science LT 92

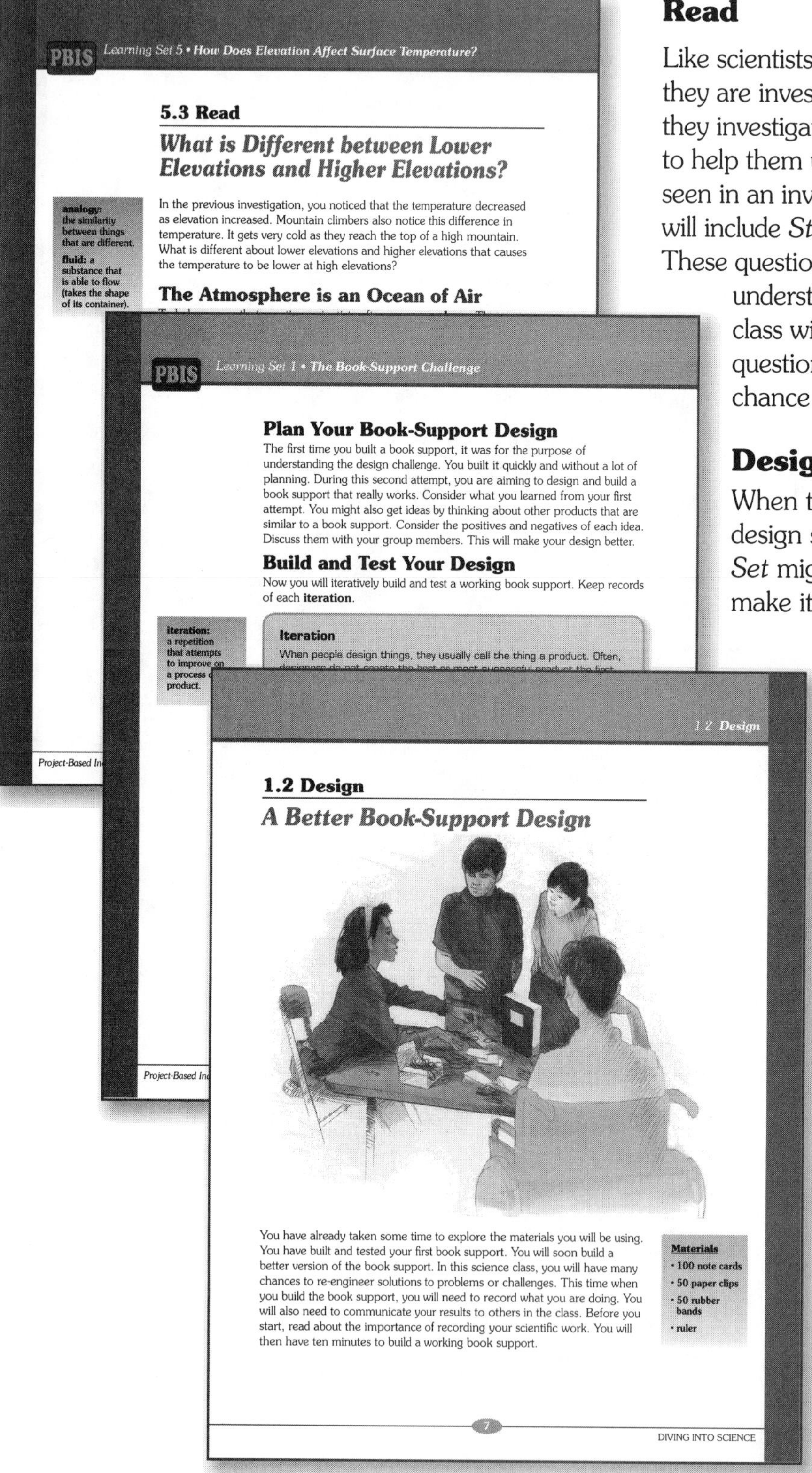

PBIS Learning Set 5 • How Does Elevation Affect Surface Temperature?

5.3 Read

What is Different between Lower Elevations and Higher Elevations?

analogy: the similarity between things that are different.

fluid: a substance that is able to flow (takes the shape of its container).

In the previous investigation, you noticed that the temperature decreased as elevation increased. Mountain climbers also notice this difference in temperature. It gets very cold as they reach the top of a high mountain. What is different about lower elevations and higher elevations that causes the temperature to be lower at high elevations?

The Atmosphere is an Ocean of Air

PBIS Learning Set 1 • The Book-Support Challenge

Plan Your Book-Support Design

The first time you built a book support, it was for the purpose of understanding the design challenge. You built it quickly and without a lot of planning. During this second attempt, you are aiming to design and build a book support that really works. Consider what you learned from your first attempt. You might also get ideas by thinking about other products that are similar to a book support. Consider the positives and negatives of each idea. Discuss them with your group members. This will make your design better.

Build and Test Your Design

Now you will iteratively build and test a working book support. Keep records of each **iteration**.

iteration: a repetition that attempts to improve on a process or product.

Iteration

When people design things, they usually call the thing a product. Often,

1.2 Design

1.2 Design

A Better Book-Support Design

You have already taken some time to explore the materials you will be using. You have built and tested your first book support. You will soon build a better version of the book support. In this science class, you will have many chances to re-engineer solutions to problems or challenges. This time when you build the book support, you will need to record what you are doing. You will also need to communicate your results to others in the class. Before you start, read about the importance of recording your scientific work. You will then have ten minutes to build a working book support.

Materials
- 100 note cards
- 50 paper clips
- 50 rubber bands
- ruler

7 DIVING INTO SCIENCE

Read

Like scientists, students will also read about the science they are investigating. They will read a little bit before they investigate, but most of the reading they do will be to help them understand what they have experienced or seen in an investigation. Each time they read, the text will include *Stop and Think* questions after the reading. These questions will help students gauge how well they understand what they have read. Usually, the class will discuss the answers to *Stop and Think* questions before going on so that everybody has a chance to make sense of the reading.

Design and Build

When the *Big Challenge* for a Unit asks them to design something, the challenge in a *Learning Set* might also ask them to design something and make it work. Often students will design a part of the thing they will design and build for the *Big Challenge*. When a *Learning Set* challenges students to design and build something, they will do several things:

- identify what questions they need to answer to be successful
- investigate to find answers to those questions
- use those answers to plan a good design solution
- build and test their design

Because designs don't always work the way one wants them to, students will usually do a design challenge more than once. Each time through, they will test their design. If their design doesn't work as well as they would like, they will determine why it is not working and identify other things they need to investigate to make it work better. Then, they will learn those things and try again.

Explain and Recommend

A big part of what scientists do is explain, or try to make sense of why things happen the way they do. An explanation describes why something is the way it is or behaves the way it does. An explanation is a statement one makes built from claims (what you think you know), evidence (from an investigation) that supports the claim, and science knowledge. As they learn, scientists get better at explaining. You will see that students get better, too, as they work through the *Learning Sets*.

A recommendation is a special kind of claim—one where you advise somebody about what to do. Students will make recommendations and support them with evidence, science knowledge, and explanations.

3.5 Explain

Create an Explanation

After scientists get results from an investigation, they try to make a claim. They base their claim on what their evidence shows. They also use what they already know to make their claim. They explain why their claim is valid. The purpose of a science explanation is to help others understand the following:

- what was learned from a set of investigations
- why the scientists reached this conclusion

Later, other scientists will use these explanations to help them explain other phenomena. The explanations will also help them predict what will happen in other situations.

You will do the same thing now. Your claim will be the trend you found in your experiment. You will use data you collected and science knowledge you have read to create a good explanation. This will help you decide whether your claim is valid. You will be reporting the results of the investigation to your classmates. With a good explanation that matches your claim, you can convince them that your claim is valid.

Because your understanding of the science of forces is not complete, you may not be able to fully explain your results. But you will use what you have read to come up with your best explanation. Scientists finding out about new things do the same thing. When they only partly understand something, it is impossible for them to form a "perfect" explanation. They do the best they can based on what they understand. As they learn more, they make their explanations better. This is what you will do now and what you will be doing throughout PBIS. You will explain your results the best you can based

4.3 Explain and Recommend

Explanations and Recommendations about Parachutes

As you did after your whirligig experiments, you will spend some time now explaining your results. You will also try to come up with recommendations. Remember that explanations include your claims, the evidence for your claims, and the science you know that can help you understand the claim. A recommendation is a statement about what someone should do. The best recommendations also have evidence, science, and an explanation associated with them. In the *Whirligig Challenge*, you created explanations and recommendations separately from each other. This time you will work on both at the same time.

Create and Share Your Recommendation and Explanation

Work with your group. Use the hints on the *Create Your Explanation* pages to make your first attempt at explaining your results. You'll read about parachute science later. After that, you will probably want to revise your explanations. Right now, use the science you learned during the *Whirligig Challenge* for your first attempt.

Write your recommendation. It should be about designing a slow-falling parachute. Remember that it should be written so that it will help someone else. They should be able to apply what you have learned about the effects of your variable. If you are having trouble, review the example in *Learning Set 3*.

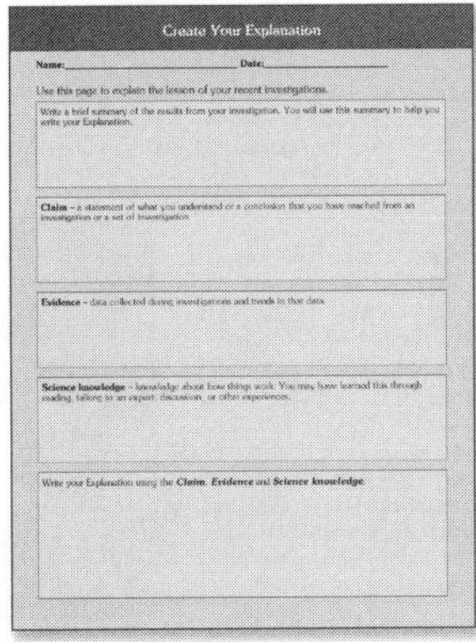

Create Your Explanation

Name:______ Date:______

Use this page to explain the lesson of your recent investigations.

Write a brief summary of the results from your investigation. You will use this summary to help you write your Explanation.

Claim – a statement of what you understand or a conclusion that you have reached from an investigation or a set of investigation.

Evidence – data collected during investigations and trends in that data.

Science knowledge – knowledge about how things work. You may have learned this through reading, talking to an expert, discussion, or other experiences.

Write your Explanation using the **Claim**, **Evidence** and **Science knowledge**.

PBIS Learning Set 1 • How Do Flowing Water and Land Interact in a Community?

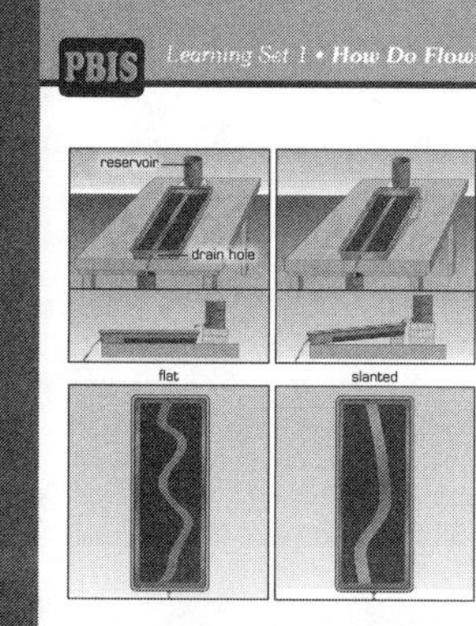

Your teacher will set up the stream table in four different ways, as shown in the diagrams.

Sketch the different models. As you watch the water flow through the model, pay very close attention to the way the land on both sides of the river changes. Pay attention to

- how the soil moves,
- where along the bank the soil moves, and
- where the soil ends up.

Make notes about what you observe for each of these situations. You might want to mark your sketches based on what you observed.

Stop and Think

Look at your sketches and the notes you took about the river models you observed. What did you notice about how the soil was moved by the river? Answer these questions. Be prepared to discuss your answers with your group and the class.

1. When the river was straight and the pan was level, how did the soil move along the river?
2. When your teacher made the pan more slanted by lifting the water end of the pan, how did the water move compared to the level pan? How did that change affect the soil that the river moved?
3. Your teacher also made rivers that were more curved. How did that change the way the soil moved along the river?

Project-Ba

1.3 Read

Reflect

Think about the book support you designed and built so far. Try to think about the science concepts you have read about and discussed as a class. Answer the following questions. Be prepared to discuss your answers with the class.

1. Was your structure strong? If not, did it collapse because of folding, compression, or both?
2. How could you make the structure stronger to resist folding or compression?
3. Was your book support stable? That is, did it provide support so that the book did not tip over? Did it provide this support well? Draw a picture of your book support showing the center of mass of the book and the places in your book support that resist the load of your book.
4. How could you make your book support more stable?
5. How successful were the book supports that used columns in their design?
6. How could you make your book support work more effectively by including columns into the design?
7. Explain how the pull on the book could better be resisted by the use of columns in your design. Be sure to discuss both the strength and the stability of the columns in your design. You might find it easier to draw a sketch and label it to explain how the columns do this.
8. Think about some of the structures that supported the book well. What designs and building decisions were used?

You are going to get another chance to design a book support. You will use the same materials. Think about how your group could design your next book support to better meet the challenge. Consider what you now know about the science that explains how structures support objects.

19 DIVING INTO SCIENCE

3) Scientists...reflect in many different ways.

PBIS provides guidance to help students think about what they are doing and to recognize what they are learning. Doing this often as they are working will help students be successful student scientists.

Tools for Making Sense

Stop and Think

Stop and Think sections help students make sense of what they have been doing in the section they are working on. *Stop and Think* sections include a set of questions to help students understand what they have just read or done. Sometimes the questions will remind them of something they need to pay more attention to. Sometimes they will help students connect what they have just read to things they already know. When there is a *Stop and Think* in the text, students will work individually or with a partner to answer the questions, and then the whole class will discuss the answers.

Reflect

Reflect sections help students connect what they have just done with other things they have read or done earlier in the Unit (or in another Unit). When there is a *Reflect* in the text, students will work individually or with a partner or small group to answer the questions. Then, the whole class will discuss the answers. You may want to ask students to answer *Reflect* questions for homework.

Analyze Your Data

Whenever students have to analyze data, the text will provide hints about how to do that and what to look for.

Mess About

"Messing about" is a term that comes from design. It means exploring the materials to be used for designing or building something or examining something that works like what is to be designed. Messing about helps students discover new ideas—and it can be a lot of fun. The text will usually give them ideas about things to notice as they are messing about.

What's the Point?

At the end of each *Learning Set*, students will find a summary, called *What's the Point?*, of the important information from the *Learning Set*. These summaries can help students remember how what they did and learned is connected to the *Big Question* or *Challenge* they are working on.

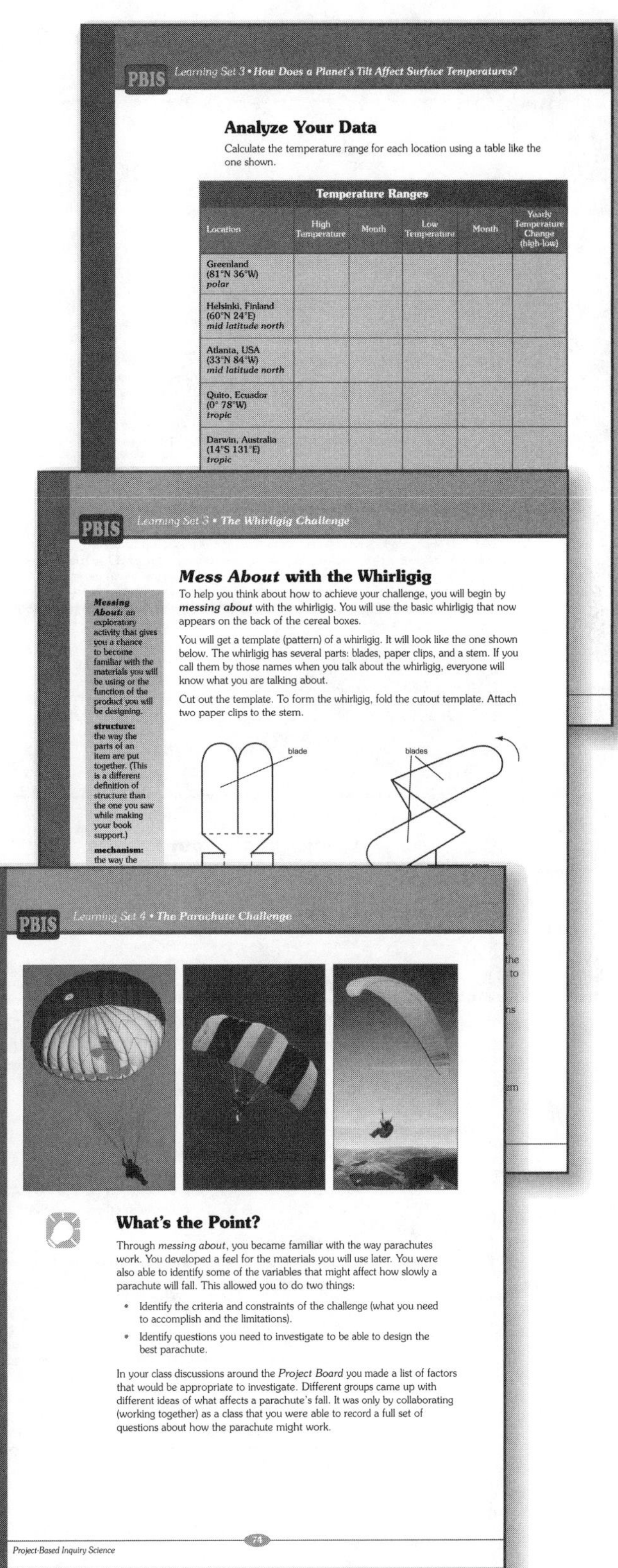

PBIS Learning Set 3 • How Does a Planet's Tilt Affect Surface Temperatures?

Analyze Your Data

Calculate the temperature range for each location using a table like the one shown.

Temperature Ranges					
Location	High Temperature	Month	Low Temperature	Month	Yearly Temperature Change (high-low)
Greenland (81°N 36°W) *polar*					
Helsinki, Finland (60°N 24°E) *mid latitude north*					
Atlanta, USA (33°N 84°W) *mid latitude north*					
Quito, Ecuador (0° 78°W) *tropic*					
Darwin, Australia (14°S 131°E) *tropic*					

PBIS Learning Set 3 • The Whirligig Challenge

Mess About with the Whirligig

To help you think about how to achieve your challenge, you will begin by ***messing about*** with the whirligig. You will use the basic whirligig that now appears on the back of the cereal boxes.

You will get a template (pattern) of a whirligig. It will look like the one shown below. The whirligig has several parts: blades, paper clips, and a stem. If you call them by those names when you talk about the whirligig, everyone will know what you are talking about.

Cut out the template. To form the whirligig, fold the cutout template. Attach two paper clips to the stem.

Messing About: an exploratory activity that gives you a chance to become familiar with the materials you will be using or the function of the product you will be designing.

structure: the way the parts of an item are put together. (This is a different definition of structure than the one you saw while making your book support.)

mechanism: the way the

PBIS Learning Set 4 • The Parachute Challenge

What's the Point?

Through *messing about*, you became familiar with the way parachutes work. You developed a feel for the materials you will use later. You were also able to identify some of the variables that might affect how slowly a parachute will fall. This allowed you to do two things:

- Identify the criteria and constraints of the challenge (what you need to accomplish and the limitations).
- Identify questions you need to investigate to be able to design the best parachute.

In your class discussions around the *Project Board* you made a list of factors that would be appropriate to investigate. Different groups came up with different ideas of what affects a parachute's fall. It was only by collaborating (working together) as a class that you were able to record a full set of questions about how the parachute might work.

Project-Based Inquiry Science 74

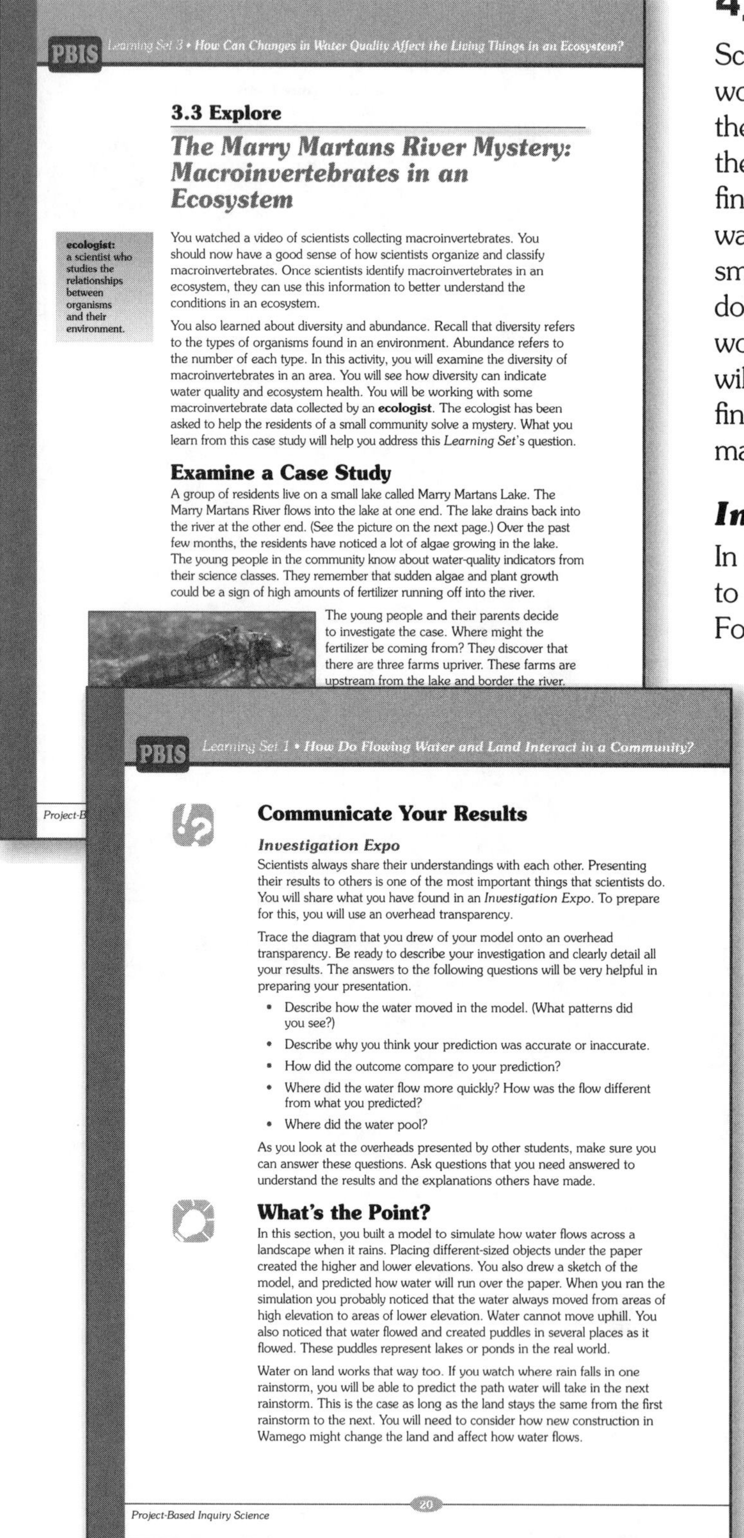

3.3 Explore

The Marry Martans River Mystery: Macroinvertebrates in an Ecosystem

ecologist: a scientist who studies the relationships between organisms and their environment.

You watched a video of scientists collecting macroinvertebrates. You should now have a good sense of how scientists organize and classify macroinvertebrates. Once scientists identify macroinvertebrates in an ecosystem, they can use this information to better understand the conditions in an ecosystem.

You also learned about diversity and abundance. Recall that diversity refers to the types of organisms found in an environment. Abundance refers to the number of each type. In this activity, you will examine the diversity of macroinvertebrates in an area. You will see how diversity can indicate water quality and ecosystem health. You will be working with some macroinvertebrate data collected by an **ecologist**. The ecologist has been asked to help the residents of a small community solve a mystery. What you learn from this case study will help you address this *Learning Set*'s question.

Examine a Case Study

A group of residents live on a small lake called Marry Martans Lake. The Marry Martans River flows into the lake at one end. The lake drains back into the river at the other end. (See the picture on the next page.) Over the past few months, the residents have noticed a lot of algae growing in the lake. The young people in the community know about water-quality indicators from their science classes. They remember that sudden algae and plant growth could be a sign of high amounts of fertilizer running off into the river.

The young people and their parents decide to investigate the case. Where might the fertilizer be coming from? They discover that there are three farms upriver. These farms are upstream from the lake and border the river.

Communicate Your Results

Investigation Expo

Scientists always share their understandings with each other. Presenting their results to others is one of the most important things that scientists do. You will share what you have found in an *Investigation Expo*. To prepare for this, you will use an overhead transparency.

Trace the diagram that you drew of your model onto an overhead transparency. Be ready to describe your investigation and clearly detail all your results. The answers to the following questions will be very helpful in preparing your presentation.

- Describe how the water moved in the model. (What patterns did you see?)
- Describe why you think your prediction was accurate or inaccurate.
- How did the outcome compare to your prediction?
- Where did the water flow more quickly? How was the flow different from what you predicted?
- Where did the water pool?

As you look at the overheads presented by other students, make sure you can answer these questions. Ask questions that you need answered to understand the results and the explanations others have made.

What's the Point?

In this section, you built a model to simulate how water flows across a landscape when it rains. Placing different-sized objects under the paper created the higher and lower elevations. You also drew a sketch of the model, and predicted how water will run over the paper. When you ran the simulation you probably noticed that the water always moved from areas of high elevation to areas of lower elevation. Water cannot move uphill. You also noticed that water flowed and created puddles in several places as it flowed. These puddles represent lakes or ponds in the real world.

Water on land works that way too. If you watch where rain falls in one rainstorm, you will be able to predict the path water will take in the next rainstorm. This is the case as long as the land stays the same from the first rainstorm to the next. You will need to consider how new construction in Wamego might change the land and affect how water flows.

4) Scientists...collaborate.

Scientists never do all their work alone. They work with other scientists (collaborate) and share their knowledge. *PBIS* helps students by giving them lots of opportunities for sharing their findings, ideas, and discoveries with others (the way scientists do). Students will work together in small groups to investigate, design, explain, and do other science activities. Sometimes they will work in pairs to figure out things together. They will also have lots of opportunities to share their findings with the rest of their classmates and make sense together of what they are learning.

Investigation Expo

In an *Investigation Expo*, small groups report to the class about an investigation they've done. For each *Investigation Expo*, students will make a poster detailing what they were trying to learn from their investigation, what they did, their data, and their interpretation of the data. The text gives them hints about what to present and what to look for in other groups' presentations. *Investigation Expos* are always followed by discussions about the investigations and about how to do science well. You may want to ask students to write a lab report following an investigation.

Plan Briefing/Solution Briefing/ Idea Briefing

Briefings are presentations of work in progress. They give students a chance to get advice from their classmates that can help them move forward. During a *Plan Briefing*, students present their plans to the class. They might be plans for an experiment for solving a problem or achieving a challenge. During a *Solution Briefing*, students present their solutions in progress and ask the class to help them make their solutions better. During an *Idea Briefing*, students present their ideas, including their evidence in support of their plans, solutions, or ideas. Often, they will prepare posters to help them make their presentation. Briefings are almost always followed by discussions of their investigations and how they will move forward.

Solution Showcase

Solution Showcases usually happen near the end of a Unit. During a *Solution Showcase*, students show their classmates their finished products—either their answer to a question or solution to a challenge. Students will also tell the class why they think it is a good answer or solution, what evidence and science they used to get to their solution, and what they tried along the way before getting to their answers or solutions. Sometimes a *Solution Showcase* is followed by a competition. It is almost always followed by a discussion comparing and contrasting the different answers and solutions groups have come up with. You may want to ask students to write a report or paper following a *Solution Showcase*.

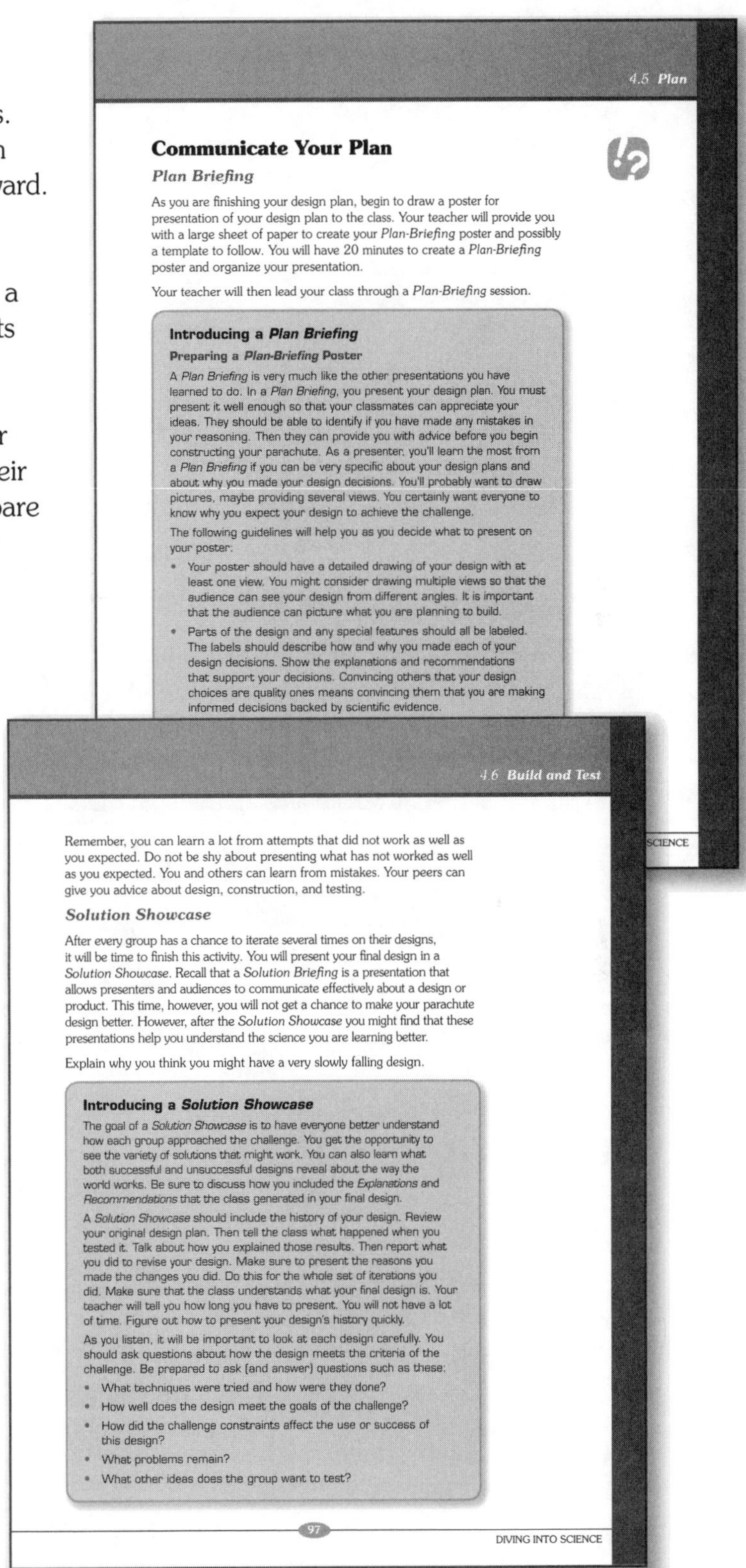

4.5 Plan

Communicate Your Plan

Plan Briefing

As you are finishing your design plan, begin to draw a poster for presentation of your design plan to the class. Your teacher will provide you with a large sheet of paper to create your *Plan-Briefing* poster and possibly a template to follow. You will have 20 minutes to create a *Plan-Briefing* poster and organize your presentation.

Your teacher will then lead your class through a *Plan-Briefing* session.

Introducing a *Plan Briefing*

Preparing a *Plan-Briefing* Poster

A *Plan Briefing* is very much like the other presentations you have learned to do. In a *Plan Briefing*, you present your design plan. You must present it well enough so that your classmates can appreciate your ideas. They should be able to identify if you have made any mistakes in your reasoning. Then they can provide you with advice before you begin constructing your parachute. As a presenter, you'll learn the most from a *Plan Briefing* if you can be very specific about your design plans and about why you made your design decisions. You'll probably want to draw pictures, maybe providing several views. You certainly want everyone to know why you expect your design to achieve the challenge.

The following guidelines will help you as you decide what to present on your poster:

- Your poster should have a detailed drawing of your design with at least one view. You might consider drawing multiple views so that the audience can see your design from different angles. It is important that the audience can picture what you are planning to build.
- Parts of the design and any special features should all be labeled. The labels should describe how and why you made each of your design decisions. Show the explanations and recommendations that support your decisions. Convincing others that your design choices are quality ones means convincing them that you are making informed decisions backed by scientific evidence.

4.6 Build and Test

Remember, you can learn a lot from attempts that did not work as well as you expected. Do not be shy about presenting what has not worked as well as you expected. You and others can learn from mistakes. Your peers can give you advice about design, construction, and testing.

Solution Showcase

After every group has a chance to iterate several times on their designs, it will be time to finish this activity. You will present your final design in a *Solution Showcase*. Recall that a *Solution Briefing* is a presentation that allows presenters and audiences to communicate effectively about a design or product. This time, however, you will not get a chance to make your parachute design better. However, after the *Solution Showcase* you might find that these presentations help you understand the science you are learning better.

Explain why you think you might have a very slowly falling design.

Introducing a *Solution Showcase*

The goal of a *Solution Showcase* is to have everyone better understand how each group approached the challenge. You get the opportunity to see the variety of solutions that might work. You can also learn what both successful and unsuccessful designs reveal about the way the world works. Be sure to discuss how you included the *Explanations* and *Recommendations* that the class generated in your final design.

A *Solution Showcase* should include the history of your design. Review your original design plan. Then tell the class what happened when you tested it. Talk about how you explained those results. Then report what you did to revise your design. Make sure to present the reasons you made the changes you did. Do this for the whole set of iterations you did. Make sure that the class understands what your final design is. Your teacher will tell you how long you have to present. You will not have a lot of time. Figure out how to present your design's history quickly.

As you listen, it will be important to look at each design carefully. You should ask questions about how the design meets the criteria of the challenge. Be prepared to ask (and answer) questions such as these:

- What techniques were tried and how were they done?
- How well does the design meet the goals of the challenge?
- How did the challenge constraints affect the use or success of this design?
- What problems remain?
- What other ideas does the group want to test?

97 DIVING INTO SCIENCE

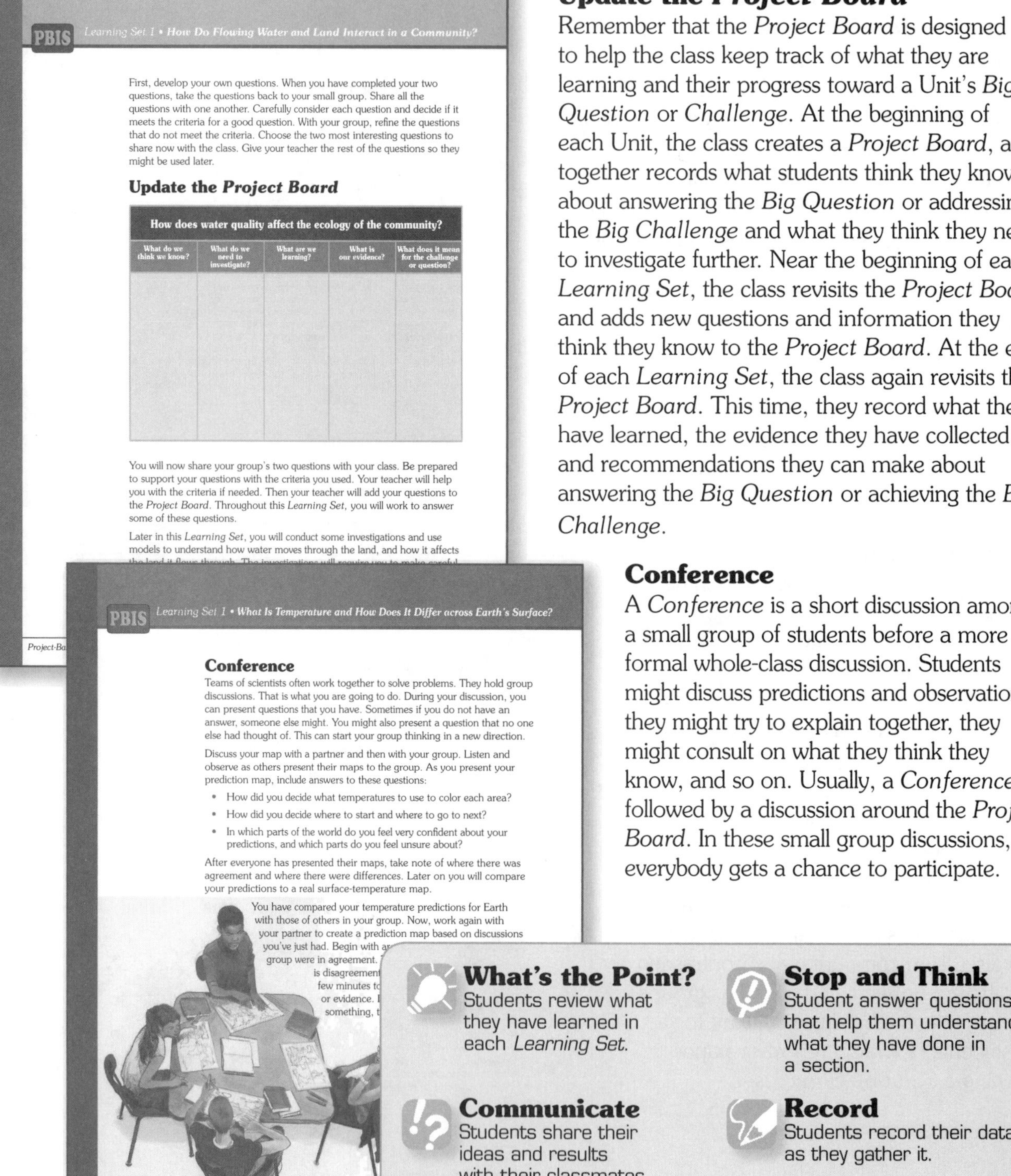

First, develop your own questions. When you have completed your two questions, take the questions back to your small group. Share all the questions with one another. Carefully consider each question and decide if it meets the criteria for a good question. With your group, refine the questions that do not meet the criteria. Choose the two most interesting questions to share now with the class. Give your teacher the rest of the questions so they might be used later.

Update the *Project Board*

How does water quality affect the ecology of the community?				
What do we think we know?	What do we need to investigate?	What are we learning?	What is our evidence?	What does it mean for the challenge or question?

You will now share your group's two questions with your class. Be prepared to support your questions with the criteria you used. Your teacher will help you with the criteria if needed. Then your teacher will add your questions to the *Project Board*. Throughout this *Learning Set*, you will work to answer some of these questions.

Later in this *Learning Set*, you will conduct some investigations and use models to understand how water moves through the land, and how it affects

Conference

Teams of scientists often work together to solve problems. They hold group discussions. That is what you are going to do. During your discussion, you can present questions that you have. Sometimes if you do not have an answer, someone else might. You might also present a question that no one else had thought of. This can start your group thinking in a new direction.

Discuss your map with a partner and then with your group. Listen and observe as others present their maps to the group. As you present your prediction map, include answers to these questions:

- How did you decide what temperatures to use to color each area?
- How did you decide where to start and where to go to next?
- In which parts of the world do you feel very confident about your predictions, and which parts do you feel unsure about?

After everyone has presented their maps, take note of where there was agreement and where there were differences. Later on you will compare your predictions to a real surface-temperature map.

What's the Point?
Students review what they have learned in each *Learning Set*.
Stop and Think
Student answer questions that help them understand what they have done in a section.
Communicate
Students share their ideas and results with their classmates.
Record
Students record their data as they gather it.

Update the *Project Board*

Remember that the *Project Board* is designed to help the class keep track of what they are learning and their progress toward a Unit's *Big Question* or *Challenge*. At the beginning of each Unit, the class creates a *Project Board*, and together records what students think they know about answering the *Big Question* or addressing the *Big Challenge* and what they think they need to investigate further. Near the beginning of each *Learning Set*, the class revisits the *Project Board* and adds new questions and information they think they know to the *Project Board*. At the end of each *Learning Set*, the class again revisits the *Project Board*. This time, they record what they have learned, the evidence they have collected, and recommendations they can make about answering the *Big Question* or achieving the *Big Challenge*.

Conference

A *Conference* is a short discussion among a small group of students before a more formal whole-class discussion. Students might discuss predictions and observations, they might try to explain together, they might consult on what they think they know, and so on. Usually, a *Conference* is followed by a discussion around the *Project Board*. In these small group discussions, everybody gets a chance to participate.

NOTES

NOTES

NOTES

NOTES

Teacher's Planning Guide

ANIMALS IN ACTION

As a student scientist, you will...

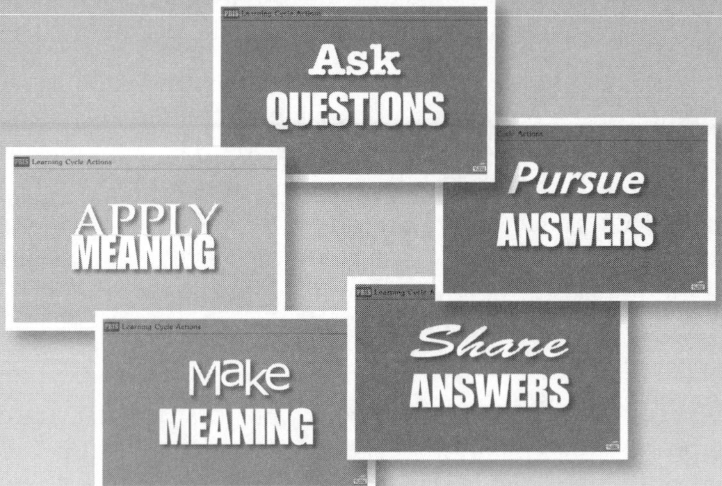

UNIT OVERVIEW

Content

Animals in Action is a Life-Sciences Unit. Students begin this Unit by learning how to design an observation plan, implement it, record the data, analyze the data, and create an explanation based on the data. Students use these principles and procedures to answer the *Big Question, How do scientists answer big questions and solve big problems?* and to work toward addressing the *Big Challenge*.

For the *Big Challenge,* students will design an animal enclosure based on two criteria: it must allow the animal to feed and to communicate as it would in its natural habitat, and it also must allow biologists to make accurate observations.

During this Unit, students use observations to investigate two animal behaviors: feeding and communication. Students will consider why animals engage in these two behaviors, and what affects how they feed and communicate. These two behaviors are common to all animals. Students design observational methods for data collection, recording, and analysis. They observe a variety of species and identify several principles that guide biology: form and function, adaptations, and the impact of the environment on behavior. Students implement reliable procedures, and use evidence to construct explanations and make recommendations.

The goals of *Learning Set 1* focus on the social practices of scientists and on the investigative process, including observation, planning, record keeping, criteria and constraints, and reliable data. In *Learning Sets 2* and *3,* students use the scientific methods they have practiced to observe animals and create explanations of their behavior. Feeding is the behavior studied in *Learning Set 2*. Students observe several animals feeding, determine the needs of animals for feeding, create methods to help make their observations efficient and reliable, and read about scientists' work in observing feeding behaviors. Ultimately, students understand that an animal's body structure and environment affects how it feeds. In *Learning Set 3,* the students investigate communication methods used by different species. Through observation, students determine what affects how animals communicate. To conclude the Unit, students *Address the Big Challenge*. They design an animal enclosure that will allow an animal to feed or to communicate as it would in its natural habitat, and to be observed by biologists.

LOOKING AHEAD

Students will observe several animal behaviors using videos shown throughout this Unit. Each video will require some way to project the video so that all students can view it and hear its sounds.

32 class periods*

*A class period is considered to be one 40 to 50 minute class.

Investigations

In *Learning Set 1,* students work together in small groups to make observations, and to collect and analyze data. They will also develop questions and new observational plans based on their experiences. Students learn about the importance of criteria and constraints in scientific work, and the value of an iterative scientific process, as they redesign their observational plans and repeat their observations. In *Learning Set 2,* students use the observational procedures they have practiced to observe animal feeding behavior. They begin to create explanations for why animals feed as they do. Students revise these explanations as they learn more about different animals. Students continue to practice collection, organization, and analysis of data. Students engage in scientific social practices as their groups present results to the class and discuss them. Animal communication is observed and analyzed in *Learning Set 3*. Small groups apply their developing knowledge of the factors that affect an animal's behavior to the new behavior of communication. Students complete the Unit by addressing the *Big Challenge,* in which they design an enclosure that allows an animal to feed or communicate as it would in its natural habitat. Students also include their understanding of the work of ethologists as they make the enclosure more usable for scientific investigation by humans.

LOOKING AHEAD

In the first *Learning Set*, a small group of students will model for the class a student snacking situation. The students will need to have a snack available, and some space in the classroom where all the other students can observe them.

***Learning Set 2's More to Learn Section* is a flower-dissection lab, for which you will need to acquire fresh flowers such as azaleas, lilies, gladioli, tulips, daffodils, geraniums, snapdragons, or sweet peas. The flowers should be large, deep flowers with closed corollas, in which the parts are easily recognizable.**

Nature of Science

Animals in Action is a Life-Sciences Unit focused on animal behavior. The scientific work of ethologists, biologists who study animal behavior, is different from that of physicists or geologists. In this Unit, students learn and practice scientific procedures of constructing investigative questions and planning, collecting, analyzing, and explaining observational data. The observational data they collect is the type of data ethologists collect.

Throughout *Animals in Action,* students work as ethologists when they use an iterative process to refine their observational plans, data collection, and analysis procedures. Students further mirror scientific behavior as they display and share their procedures, recordings, analyses, and explanations with their own scientific community (their classmates). Through their discussions and presentations, students realize the value of sharing and building on each other's ideas.

In the *Unit Introduction,* students read about the evolution of zoos, from places of entertainment to organizations focused on conservation and deepening awareness. They will use this perspective as they address the *Big Challenge* at the end of the Unit.

In *Learning Set 1,* students are introduced to the four large questions ethologists attempt to answer. They begin the process of finding some answers to these questions through their animal observations. As they are making observations, they learn to identify interpretations, separate those from the observations, and use them with their evidence to support their explanations.

In *Learning Sets 2* and *3,* students engage in further observations of animals by watching a video. They practice their observational skills, and share in analysis and explanation tasks with their classmates. They are introduced to other scientists who have done work as ethologists. Students consider the role of scientific fieldwork in learning about animal behavior, by reading about several scientists who have done this work. To reinforce the idea that scientific knowledge is tenuous, students read about how human understanding of animal behavior has changed. They also read about debates in the study of animal behavior.

The *Project Board* also mirrors scientific practice. Students record their developing understanding, discuss their ideas in a public forum, keep a visual record of what they think they know and what they are learning, support their learning with evidence, connect what they are learning to the *Big Question* or *Challenge* they are addressing, and track additional investigation ideas.

LOOKING AHEAD

The final challenge at the end of this Unit requires students to refer back to the *Observing and Interpreting Animal Behavior* pages from throughout the Unit. Students will benefit if they have some means of keeping these pages organized.

Decide on how you want to create student groups for this Unit. Usually, groups of three or four work best. If possible, create eight groups per class, because some sections will require eight groups. It would be most effective to maintain these small groups for the entire Unit.

Artifacts

Throughout *Animals in Action,* students develop artifacts of their learning. The *Project Board* is one example of a class artifact that is displayed and which changes throughout the Unit. Students use the *Observing and Interpreting Animal Behavior* pages to record their observations and interpretations. These pages are organizational tools for students, as well as artifacts of their learning. Posters that synthesize their data collection procedures, data analysis, and explanations, often accompany students' presentations. These posters are artifacts of their learning, and provide a scientific record of what the students have learned.

Students use a *Create Your Explanation* page to think about and justify their recommendations for the town of Wamego. The Unit culminates with a presentation to the town council. Final recommendations from each group are in the form of a poster, an electronic presentation, or a skit.

Targeted Concepts, Skills, and Nature of Science	Section
Scientists often work together and then share their findings. Sharing findings makes new information available, and helps scientists refine their ideas and build on others' ideas. When another person's or group's idea is used, credit needs to be given.	All *Learning Sets*
Scientists must keep clear, accurate, and descriptive records of what they do so they can share their work with others, and consider what they did, why they did it, and what they want to do next.	All *Learning Sets*
Tables are an effective way to communicate results of a scientific investigation.	All *Learning Sets*
Observations and measurements are considered reliable if the results are repeatable by other scientists using the same procedures.	*LS 1, LS 2*
Scientists differentiate between observations and interpretations. They use their observations and interpretations to explain animal behavior.	All *Learning Sets*
Studying the work of different scientists provides understanding of scientific inquiry and reminds students that science is a human endeavor.	*LS 2, LS 3*
Scientific knowledge is developed through observations, recording and analysis of data, and development of explanations based on evidence.	All *Learning Sets*
Scientists make claims (conclusions) based on evidence obtained (trends in data) from reliable investigations.	All *Learning Sets*
Explanations are claims supported by evidence. Evidence can be experimental results, observational data, and other accepted scientific knowledge.	All *Learning Sets*
Scientists use models to simulate processes that happen too fast, too slow, on a scale that cannot be observed directly (either too small or too large), or are too dangerous.	*LS 2*
Criteria and constraints are important in determining effective scientific procedures and in answering scientific questions.	All *Learning Sets*

Targeted Concepts, Skills, and Nature of Science	Section
Behavior is a type of response to internal or external stimulus. Behavior is determined by experience, physical characteristics, and environment.	All *Learning Sets*
The structure and function of animals' bodies are complementary and affect animal behavior.	All *Learning Sets*
Organisms need food to grow, reproduce, and maintain their bodies. They must be able to live in a changing environment.	*LS 1, LS 2*
Animals' sense of sight is adapted to their environment. Some animals see things that humans cannot see.	*LS 2, LS 3*
To see an object, light reflected from the object must enter the eye. Eyes of various animal species work differently to make them effective in their environment.	*LS 2*
White light is composed of all the colors of the rainbow. The sun emits electromagnetic radiation, of which humans can only detect a small portion with their eyes.	*LS 2*
Flowers are the reproductive organs of plants, as well as food suppliers for many animals.	*LS 2*
Animals' bodies have similarities to simple machines. These machines help the animals survive.	*LS 2*
Animals communicate with other animals using sound. The sounds they can make, and the sounds they can hear, are adaptations to their environment.	*LS 3*
Vibrations of molecules produce sound. Sound is compression waves that can be described by amplitude, frequency, and wavelength. Sound moves differently through different matter.	*LS 3*
Animals' ears are adapted to hearing sounds in their environment. Some animals use sound that is out of the range of human hearing.	*LS 3*

Unit Material List

Quantities for groups of 4-6 students.		
Unit Durable Group Items	**Section**	**Quantity**
Flower Cards, set of 24	2.4, 2.5	1
UV spy pen light	2.5	1
Hand lens	2.6	2
Tweezers	2.6	2
Small tray	2.6	1
Small puzzle, 12–24 pieces	3.2	2

Quantities for 5 classes of 8 groups.		
Unit Durable Classroom Items	**Section**	**Quantity**
Project Board, laminated	ABQ, 2.1, 2.2, 2.3, 2.5, 2.7, 3.1, 3.2, 3.3, 3.5, 3.7	5
Project Board transparency	ABQ, 2.1, 2.2, 2.3, 2.5, 2.7, 3.1, 3.2, 3.3, 3.5, 3.7	1
Animals In Action DVD	2.1, 2.2, 2.7, 3.3, 3.4, 3.5, 3.6, 3.7	1

Quantities for groups of 4-6 students.		
Unit Consumable Group Items	**Section**	**Quantity**
Battery, AA type	2.5	2
Invisible tape	2.6	1

Unit Material List

Quantities for 5 classes of 8 groups.		
Consumable Classroom Items	**Section**	**Quantity**
Index cards, pkg. of 100	1.2, ABQ, 2.2, 2.7, 3.1, 3.4, 3.6	4
Restickable easel pad	2.2, 2.6, 3.2, 3.4, 3.6, ABQ	4
Set of colored pencils	2.2, 2.6, 3.2, 3.4, 3.6, ABQ	6

Additional Items Needed Not Supplied	Section	Quantity
Snacks for a group of 3	1.1, 1.2, BBQ	1 per classroom
Sticky notepad	1.2, ABQ, 2.2, 2.7, 3.1, 3.4, 3.6	1 per group
Newspaper or paper towels	2.6	1 per group
Flowers such as azalea, lily, gladiolus, tulip, daffodil, geranium, snapdragon or sweet pea	2.6	2 per group
Safety scissors	2.6	1 per group
Tuning fork (optional)	3.5	1 per classroom

UNIT INTRODUCTION

What's the Big Question?

How Do Scientists Answer Big Questions and Solve Big Problems?

◀ *1 class period**

*A class period is considered to be one 40 to 50 minute class.

Overview

Students are introduced to the *Big Question* and the *Big Challenge* of the Unit. Students learn that, to answer the *Big Question,* they will be answering smaller questions and engaging in the scientific endeavor. Students gain experience acting as scientists as they proceed through the Unit to solve the *Big Challenge,* designing a zoo enclosure that allows scientists to observe an animal's feeding and communication behavior in an environment similar to the animal's natural habitat. Students are also introduced to the ideas of scientific observation, and criteria and constraints, by considering how they will address the *Big Challenge*. The class is then introduced to the *Project Board*. This is an organizational tool they will use to keep track of their ideas, investigative questions, what they are learning, the evidence, and how it helps them to solve the challenge. The class decides through discussion on what to record for what they think they know about animal behavior, and the investigative questions they think they need to pursue.

Targeted Concepts, Skills, and Nature of Science	Performance Expectations
Scientists often work together and then share their findings. Sharing findings makes new information available, and helps scientists refine their ideas and build on others' ideas.	Students should realize that, during the class discussions, they are sharing and refining their ideas, and building on others' ideas.
Scientists must keep clear, accurate, and descriptive records of what they do so they can share their work with others and consider what they did, why they did it, and what they want to do next.	Students should record accurate and descriptive information concerning the criteria and constraints of the *Big Challenge* and their ideas in their *Project Board* pages.

Targeted Concepts, Skills, and Nature of Science	Performance Expectations
Criteria and constraints are important in determining effective scientific procedures and in answering scientific questions.	Students should describe what criteria and constraints are, and should give specific examples of these for the zoo enclosure.

Materials	
1 per class	Class *Project Board*
1 per student	*Project Board* page

Activity Setup and Preparations

Project Board

Project Boards are described more completely in the *Teacher's Resource Guide,* the teacher implementation notes on introducing the *Project Board,* and the student text. The *Project Board* is a tool used to organize information and ideas when working on a project. It consists of five columns labeled: *What do we think we know? What do we need to investigate? What are we learning? What is our evidence?* and *What does it mean for the challenge or question?*

- Each class makes their own *Project Board,* and will be revisiting it and updating it through the end of the Unit. You will need to create a *Project Board* for each class and update it throughout every Unit.
- To set up the *Project Board,* decide if you will use butcher block paper, a projected computer file, or some other medium. You will record the class's ideas and questions on the *Project Board* during class discussions. Each student will also have their own record of the *Project Board* that they will keep with them. Add the *Big Challenge* to the header of the *Project Board. (How can you develop a zoo enclosure to support natural animal behavior?)*

Homework Options

Reflection

- **Science Process:** What do you think the usefulness of the *Project Board* is? *(Students should recognize that the* Project Board *helps to organize ideas and information.)*
- **Science Process:** Why are criteria and constraints important in design? *(Criteria and constraints are important in design because they describe what needs to be accomplished and the limitations to do that.)*
- **Science Process:** Imagine it is 3 P.M. and you just got home from school. You have a pet dog that needs to go out for a walk and be fed before it gets dark at 6 P.M. You have estimated that you have three hours of homework to do, and it must be done by 7 P.M. Your challenge is to get it all done. List the criteria and constraints of this challenge and your plan for meeting this challenge. *(Criteria = walk dog, feed dog, get 3 hours of homework done. Constraints = time. You have only 4 hours to complete everything, until 6 P.M. to walk and feed the dog, and homework must be done by 7 P.M.)*

Preparation for 1.1

- **Science Process:** What are things you would look for to gain insights into an animal's behavior? Give examples using an animal you are familiar with, such as a cat, dog, or fish. *(The purpose of this is to get students to think about making observations.)*

NOTES

UNIT INTRODUCTION IMPLEMENTATION

◀ *1 class period**

What's the Big Question?

How do scientists answer big questions and solve big problems?

Imagine that on your way to school one morning you see a bird. It swoops down to the ground then flies back into the tree. You stop to watch and wonder what the bird is doing. Why is it going back and forth between the ground and the tree? You see a small pile of crumbs on the ground. Then, you notice a nest built in the tree branches. All these things help you realize that the bird is feeding its young in the nest. When you take the time to watch the world in this way, you are acting much like **biologists** do.

biologist: a scientist who studies living things.

observe: to use one of the five senses to gather information about an object or phenomenon.

Biologists are a group of scientists who study living things. One thing they study is the behavior of animals. Studying animal behavior helps them better understand the living world. The results of their studies can help you to see animals in a new and different way. Through the study of animals, you may even better understand your own behavior or the behavior of your friends.

In this Unit, you will **observe** and study the behavior of several animals. You will learn about how animals feed and communicate with each other and what conditions affect those things. You will also develop some tools to help you collect and organize your observations. These tools will help you analyze your data as you try to answer the *Big Question: How do scientists answer big questions and solve big problems?* You will answer this question in the context of a science question: *Why do animals behave the way they do?*

Welcome to Animals In Action!
Enjoy being a student scientist.

What's the Big Question?

How do scientists answer big questions and solve big problems?

5 min.

Introduce PBIS, *the* Big Question, *and the* Big Challenge.

Engage

Help students know that in this class they will be doing science the way scientists do. They will need help knowing what this means, but there is no need to overwhelm them with details. It will become clearer as they engage in *PBIS* activities. The introduction for this Unit can help, as can the *Letter to the Students.*

*A class period is considered to be one 40 to 50 minute class.

TEACHER TALK

"When you think about the work of biologists, what picture comes to your mind? Biologists study living things, like plants and animals. How do you think they might do their work? What kind of people become biologists? What are some characteristics of biologists? What do people do when they study animals?"

META NOTES

Many students think scientists are men with white, unkempt hair, wearing white lab coats and glasses, surrounded by flasks of smoking liquids. They may think their work setting is dark and usually windowless. These ideas are deeply held by middle-school students, and may take many experiences to change. Through *PBIS*, students learn that science is about answering big questions and applying what's known to addressing real-world challenges.

Record students' initial ideas so the class can refer back to them at the end of the Unit.

△ Guide

Let students know that, in this Unit, they will be learning the methods biologists use to study animal behavior. They will also learn about what affects animal behavior.

○ Engage

Ask students what they know about animals, what animals live around them, and what some of the animals' behaviors are.

TEACHER TALK

"In this Unit, you will be studying animals and how biologists study animals. What animals are you familiar with? How would you describe their behaviors? What animals live around your neighborhood? How would you describe them? How would you describe their behavior?"

After eliciting animal behaviors from the class, ask students to categorize these behaviors into different types, such as communication, feeding, movement, protecting their young, etc.

TEACHER TALK

"Can we categorize any of these behaviors? For example, are there behaviors that are related to feeding? What other categories are there?""

△ Guide

Then point out the difference between describing physical characteristics (what the animal looks like) and behavioral characteristics (what the animal is doing). Let students know that biology is the study of living things, and that biologists study living things such as plants and animals.

Use the example in the student text to describe what biologists do.

Next, introduce and emphasize the *Big Question: How do scientists answer big questions and solve big problems?*

PBIS

Think about the Big Question

In this Unit, you will respond to a challenge to answer the *Big Question*. Before you start to think about what you already know about the *Big Question*, read about the challenge you will address.

Your Challenge

Look at the pictures on this page. They show animal **enclosures** found in zoos early in the 20th century. Some of these zoos were built over 100 years ago. Observe the pictures closely. The animals in these enclosures are all in cages. The zoos kept very large animals and smaller mammals, birds, and reptiles in similar enclosures. Zoos built these kinds of enclosures for animals at a time when zoos were designed for the display of animals and as places where people could have fun. In addition to animals, the zoo might also have had an amusement park, a playground, or a dance hall. The animal cages were usually very primitive. They were made of steel with cement floors and only sometimes included trees or water.

Zoos do not have amusement–park rides anymore. Today, zoos are concerned with **conservation** and education rather than the display of unusual animals. As the purposes of zoos have changed, so have spaces built for the animals. Nowadays, many zoos build animal spaces in ways that allow the animals to live more like they do in their natural habitat and allow people to learn about animals by watching them. For example, in the 1920s, the Detroit Zoo built a new home for birds. The large domed building held many cages for the birds to live in. The building was a bright and comfortable place for zoo visitors to watch the birds.

In 1996, the Detroit Zoo renovated the old bird house transforming it into a new butterfly house and interpretive center. Now birds, as well as butterflies, are free

AIA 4

Project-Based Inquiry Science

Think about the *Big Question*

5 min.

Inform students how they will be answering the Big Question.

△ Guide

Let students know that, to answer the *Big Question,* they will need to break it up into smaller questions. As a group, they will be answering these smaller questions, and they will also be working on a challenge like scientists do that will help them answer the *Big Question.*

Your Challenge

15 min.

Introduce the Big Challenge *to the class.*

d about the cha... address.

Your Challenge

Look at the pictures on this page. They show animal **enclosures** found in zoos early in the 20th century. Some of these zoos were built over 100 years ago. Observe the pictures closely. The animals in these enclosures are all in cages. The zoos kept very large animals and smaller mammals, birds, and reptiles in similar enclosures. Zoos built these kinds of enclosures for animals at a time when zoos were designed for the display of animals and as places where people could have fun. In addition to animals, the zoo might also have had an amusement park, a playground, or a dance hall. The animal cages were usually very primitive. They were made of steel with cement floors and only sometimes included trees or water.

Zoos do not have amusement-park rides anymore. Today, zoos are concerned with **conservation** and education rather than the display of unusual animals. As the purposes of zoos have changed, so have spaces built for the animals. Nowadays, many zoos build animal spaces in ways that allow the animals to live more like they do in their natural habitat and allow people to learn about animals by watching them. For example, in the 1920s, the Detroit Zoo built a new home for birds. The large domed building held many cages for the birds to live in. The building was a bright and comfortable place for zoo visitors to watch the birds.

In 1996, the Detroit Zoo renovated the old bird house transforming it into a new butterfly house and interpretive center. Now birds, as well as butterflies, are free

AIA 4

Project-Based Inquiry Science

○ Engage

Ask students what they know about animal enclosures at the zoo.

TEACHER TALK

"Have any of you seen old animal enclosures at the zoo? New animal enclosures? What are they like? How are they different from each other? What else have you noticed about animal enclosures at the zoo?"

META NOTES

Eliciting prior knowledge will help ground the discussion of the animal enclosures, both new and old.

△ Guide

Ask students to look at the pictures of zoo enclosures in the student text, and to share their experiences with similar or different types of enclosures. Describe the differences in zoo goals over the last one hundred years.

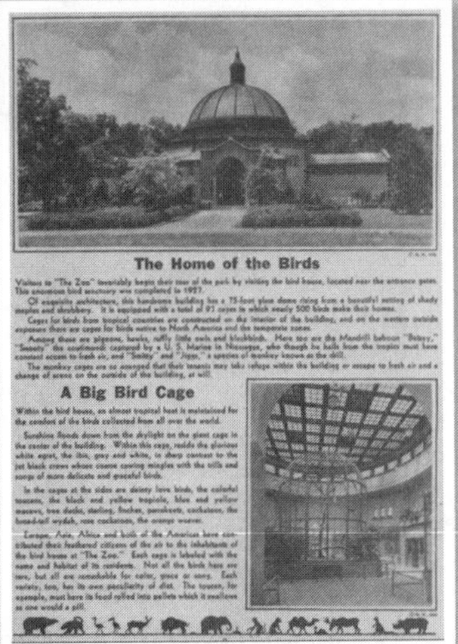

The Home of the Birds

A Big Bird Cage

The Detroit Zoo's Bird exhibit, built in the 1920s.

Toucan at the Central Park Zoo in New York City.

This panda's exhibit at the National Zoo, in Washington, DC, is based on the needs of the panda.

to fly within the building. There are many plants for shelter and water in the renovated exhibit. The space created for the animals has changed, and so has what people can learn by watching the animals.

One goal of zoos is to make the zoo **habitat** as close to the animal's natural environment as possible. When animals live in areas that look more like their natural surroundings, they are more likely to act naturally. This way, biologists can find out more about animal behavior, and zoo visitors can better see how animals behave in their natural habitats.

Recently, the panda enclosure at the National Zoo in Washington, DC was updated. The National Zoo is committed to making the captive animals' lives as similar to their natural life as possible. The design of the new panda area was based on scientists' observations of pandas in the field. To determine what important features the new environment would need, scientists watched as the pandas ate, played, slept, and interacted with one another. The pandas are thriving in the new habitat, and zoo visitors and scientists are learning more about these animals.

enclosure: an area that is surrounded by something like a fence or a wall.

conservation: the preservation, management, and care of natural and cultural resources.

habitat: a place where animals (including people) live.

AIA 5

ANIMALS IN ACTION

TEACHER TALK

"Zoos used to be more like amusement parks, with rides and unusual looking animals. Now zoos are more concerned about studying animals, and helping to preserve species and their habitats. This is called conservation. Many zoos try to help repopulate endangered species. Biologists and zoo visitors can learn more about animals and how they live in their natural habitats by observing them in their enclosures, but only if the enclosures are close to their natural habitats."

PBIS

WHAT'S NEW IN ZOOS

All zoos are putting the needs of the animals first. In 2004, Michigan's Detroit Zoo was one of the first zoos to permanently close its elephant exhibit and retire two female Asian elephants, Winky and Wanda, to a sanctuary. The Detroit Zoo chose to do this for ethical reasons. Detroit Zoo Director Ron Kagan stated, "Now we understand how much more is needed to be able to meet all the physical and psychological needs of elephants in captivity, especially in a cold climate."

Winky, now age 51, and Wanda, now age 46, were captured in the wild as babies and have been companions at the Detroit Zoo since 1994. Thanks to the Detroit Zoo's humane decision, Winky and Wanda will enjoy full, enriching years of retirement, roaming through hundreds of acres of natural habitat in the company of many other elephants.

Your challenge for this Unit is to design a new enclosure that will accommodate the feeding or communication of one of the animals you study in this Unit. Your goal will be to design the zoo environment so it is similar enough to the natural environment of the animal to allow the animal to feed or communicate effectively. The enclosure will also have to allow visitors and scientists to observe the animals clearly.

Discuss with students the example provided in the student text about the elephant enclosure constructed in the Detroit Zoo in Michigan.

Introduce the *Big Challenge* to students: to design a new enclosure that will encourage the feeding or communication of one of the animals they study in this Unit. Let students know that it is important that the enclosure model the animal's natural habitat, taking into account factors that affect how the animal feeds or communicates, and that it allows zoo visitors and scientists to observe and study the animals. Let students know they will be working in groups to solve this challenge.

Identify Criteria and Constraints

Before getting started on a challenge, it is important to make sure you understand the challenge. Design challenges have two parts: **criteria** and **constraints**.

Criteria are goals that must be satisfied to achieve the challenge. For the zoo enclosure challenge, this will include designing the enclosure so that the animals it will hold can communicate or feed as they would in their natural habitat. It will have to allow scientists and zoo visitors to observe the animals effectively.

Constraints are factors that limit how you can address a challenge. Your biggest constraint will be that the enclosure be built close to where you live. You will be able to assume that space can be found, but you will have to think about the weather where you live and how it will affect your enclosure design. You can probably think of other constraints.

With your class, identify the full set of criteria and constraints for this challenge, and put them on a chart like the one below so that you will remember them as you move through the Unit.

criteria: (singular, criterion) goals that must be satisfied to successfully achieve a challenge.

constraints: factors that limit how you can achieve a challenge.

The Zoo Enclosure Challenge

Criteria	Constraints
The animal you choose has to be able to feed or communicate as it would in its natural habitat.	The weather where we live will require...
It has to be easy for zoo visitors to observe the feeding or communication.	

AIA 7 ANIMALS IN ACTION

Identify Criteria and Constraints

10 min.

Describe the criteria and constraints for the Zoo-Enclosure Challenge, and use these as examples to introduce the terms "criteria" and "constraints."

△ Guide

Lead students in thinking more about the challenge, and what criteria and constraints are, before using the words criteria and constraints.

TEACHER TALK

"What must be accomplished?

What are the limitations in how we can do this?"

META NOTES

At the end of the Unit, each animal enclosure design will be assessed according to how it meets the criteria and constraints.

As students identify criteria and constraints, record them on a class list. When there are several items for each, introduce the terms criteria and constraints. Criteria are goals that must be satisfied, and constraints are limitations. Then separate the list into two columns, one for criteria and one for constraints.

◇ Evaluate

Make sure the following criteria and constraints are listed:

Criteria: The animal's enclosure must encourage feeding and communication. The environment inside the enclosure must be like the animal's natural habitat. Zoo visitors and scientists must be able to observe and study the animal inside the enclosure.

Constraint: The animal chosen must be one studied in this Unit. The enclosure must be built close to where you live, so that local factors such as the weather can be taken into consideration.

NOTES

PBIS

Create a *Project Board*

Project Board: a chart for keeping track of progress as you work on a project over a long period of time.

In this Unit, you will be working toward achieving a *Big Challenge*. This Unit, like other Units in *Project-Based Inquiry Science (PBIS)*, is broken into *Learning Sets*. Each *Learning Set* helps you learn a different set of concepts and skills. At the end of each *Learning Set*, you will work toward applying what you have just learned to the *Big Challenge*. Then at the end of the whole Unit, you will return to the *Big Challenge* again to pull everything together and create a solution.

When you work on a big project, it is useful to keep track of your progress and what you still need to do. You will use a *Project Board* to do that.

Be a Scientist

Introducing the *Project Board*

When you work on a project, it is useful to keep track of your progress and what you still need to do. A *Project Board* gives you a place to keep track of your scientific understanding as you make your way through a Unit. It is designed to help your class organize its questions, investigations, results, and conclusions. The *Project Board* will also help you decide what to do next. During classroom discussions, you will record the class's ideas on a class *Project Board*. At the same time, you will also keep your own *Project Board* page.

The *Project Board* has space for answering five guiding questions:

- What do we think we know?
- What do we need to investigate?
- What are we learning?
- What is our evidence?
- What does it mean for the challenge or question?

Each time you use the *Project Board*, you will record as much as you can in each column. As you work through a Unit, you will return over and over again to the *Project Board*. You will add more information and revise what you have recorded. Everything you write in the columns will be based on what you know or what you have learned. In addition to text, you will sometimes want to put pictures or data on the board.

Create a Project Board

20 min.

Introduce the Project Board *to the class, and include the class's ideas on what they think they know and what they need to find out to meet the challenge.*

Engage

Remind students of the enclosures they described earlier, and ask them to think about what ideas they already have about designing an enclosure. Help students understand that they will need a way to keep track of all their ideas, so they can monitor their progress and figure out what they know and what they still need to find out. Then, let them know they will be using the *Project Board* for this purpose.

To get started on this *Project Board*, review the questions you are answering and the challenge you are addressing. Your challenge is to design an enclosure for an animal that will allow the animal to behave as it would in its natural habitat and that will allow visitors and scientists to observe and study the animal. This will help you answer the questions: *How do scientists answer big questions and solve big problems?* and *Why do animals behave the way they do?* Record these questions in the top area of the *Project Board as shown below.*

How do scientists answer big questions and solve big problems? Why do animals behave the way they do				
What do we think we know?	What do we need to investigate?	What are we learning?	What is our evidence?	What does it mean for the challenge or question?

As you create your *Project Board* for this Unit, you will focus on animal behavior and on making good observations. Think about what you know that would help you address the challenge. You might have experiences with animal behavior, the jobs of biologists, or making observations and inferences that will be important for addressing this challenge. You may also have some questions about zoos, zoo animals, or studying animals.

You will begin by focusing on the first two columns: *What do we think we know?* and *What do we need to investigate?*

TEACHER TALK

"Remember we are working on designing an enclosure for a zoo. You've already described a number of zoo enclosures you have seen such as ..., and you've been introduced to the challenge. You probably already have some ideas about designing an enclosure. There will be many ideas to consider and many things to think about. You will need to keep track of things you know, what you need to figure out, and things to test before you get to a good design for your enclosure. You will need a way to keep track of your ideas and progress, just as biologists do as they are addressing challenges. We'll do this using a *Project Board*."

PBIS

What do we think we know?

In this column, you will record what you think you know that is important to the challenge. This might be what you know about animal behavior or about studying animal behavior. You might also want to record what you know about animal enclosures or the behavior of animals in captivity. You probably think you know a lot about animal behavior. Some things may not be completely accurate. It is important to record those things anyway, for two reasons:

- When you look at the *Project Board* later, you will be able to see how much you have learned.
- Discussion with your class will help you figure out what you need to investigate.

What do we need to investigate?

In this column, you will record the things you need to learn more about to address the challenge. You probably have many ideas now about what you need to investigate. Work with your class to get these ideas on the *Project Board*. Later in this Unit, you will add other questions. Later, you may find things you are confused about. You and your classmates might disagree about some ideas. You will be recording in this column what you do not understand well or what you disagree about.

Sometimes you will be unsure about how to word your idea as a question. One of the things your class will do together around the *Project Board* is turn the things you are curious about into questions you can investigate.

You will return to the *Project Board* many times in this Unit. You will continue to add information to the board. You will record many of the ideas you have and things you are learning. You will then see how your ideas change. By the end of the Unit, you will fill in all of the columns.

The *Project Board* is a great place to start discussions. You may find that you disagree with other classmates about what you have learned and the evidence for it. This is a part of what scientists do. Such discussions help participants identify what they or others do not understand well and what else they need to learn or investigate. The class will fill in the large *Project Board*. Make sure to record the same information on your own *Project Board* page.

AIA 10

Project-Based Inquiry Science

△ Guide

Introduce the *Project Board* and how it is used.

Show students the class *Project Board*. Explain that the *Project Board* is a tool that will be used throughout their science class this year. The class will use the *Project Board* to organize their questions, their learning and evidence, and the relationship of all of this to the challenge of designing an enclosure for a zoo.

Let students know that, while they are working toward a design solution, they will also be learning different concepts, skills, and practices that biologists use. The *Project Board* helps keep track of it all. Help students understand that the *Project Board* provides a way for the class to develop questions and answers to the *Big Questions* and the *Big Challenges* by answering a series of smaller questions, and by the sharing of ideas and information.

TEACHER TALK

"Over here I have set up our class *Project Board* for our zoo enclosure challenge. All of you have ideas about enclosures in zoos, how animals behave, and how you might design your animal's enclosure. You may have experiences with seeing enclosures at a zoo or you may have experiences with pets or animals in your neighborhood. These experiences may help you begin thinking about the challenge.

The *Project Board* will help organize our ideas as we work on this challenge. We will need to answer many questions as we work on the *Big Question* of the Unit. Our *Project Board* will help us record what we know, what we need to know, and what we've learned when we are working on a project over time. We will be returning to our class *Project Board* and updating it many times, sharing our ideas and information, while we work on answering our *Big Question: How do biologists answer big questions and solve big problems?* and while we are designing our zoo enclosure."

META NOTES

Students may be surprised that the first column is called "What do we think we know?" instead of "What do we know?" These are students' initial ideas and may or may not be correct. The intent is to identify their thoughts. Ideas they disagree upon should be turned into investigative questions for the second column.

Describe the five columns: *What do we think we know? What do we need to investigate? What we are learning? What is our evidence? What does it mean for the big challenge or question?*

Emphasize that the class will be filling out the first two columns *(What do we think we know?* and *What do we need to investigate?)* today, and that they will add to all the columns as they work on the challenge. Explain that the third and fourth columns go hand in hand (column three lists claims, and column four lists the evidence that backs up those claims based on observations and information from experts). The fifth column explains how the information they are learning is connected with the challenge or a bigger question.

META NOTES

Students do not think about science or engineering as dynamic processes that change as new information becomes available. They may be surprised to hear that they will be returning to the *Project Board* and updating it over time.

You may want to distribute the student *Project Board* pages at this time. Let students know that they should be keeping a personal copy of the class *Project Board* that they can refer to as needed, and that you will be keeping the class *Project Board* and recording the class's ideas and questions on it until the end of the Unit.

⬡ Get Going

Begin students on the *Project Board* by starting with the question: *How do scientists answer big questions and solve big problems?* Then, tell students that they will be answering this as they work together to solve the *Big Challenge.* Write the question for the challenge in the *Project Board: What do we need to know about an animal to design a zoo enclosure that allows the animals to behave as in their natural habitat and be observed by biologists?*

NOTE: you will later want to write the *Learning Set* questions across the top of the *Project Board* or within the *Project Board* in Column 2.

△ Guide

Now is time to help students post ideas and questions in the first two columns of the *Project Board*. This will require helping them share what they have discussed about the zoo enclosures so far, and reminding them of the behaviors they have already observed and the categories they came up with. Guide students by referring them to the list they created at the start of this section. Further guide students in this discussion by asking questions such as:

TEACHER TALK

"What did you list as things you think you know?

- **What do you think you know so far about zoo enclosures, based on our discussion of old and new zoo enclosures?**
- **What do you think you know about zoo enclosures based on your experiences?**
- **What do we need to investigate?**
- **What aren't you sure about yet?**
- **What would you need to investigate or find out more about before designing the enclosure?"**

As students share their ideas and questions with the class, record their ideas on the *Project Board*. Use the first column to record what they think they know. There is no need now to figure out if these ideas are right or wrong.

Use the second column for their questions. This includes questions about ideas students do not agree upon, or have a hard time accepting. Students will disagree about some of the items in the first column. These are indications of questions that need to go in the second column. Things students are surprised by should go in the second column.

META NOTES

You should date items posted on the Project Board so you and the class can monitor changes in the class's ideas and their progress on the challenge.

META NOTES

Linking items with arrows in different columns that are related to each other helps students see how ideas and questions are connected.

META NOTES

During the discussion, listen for students' reasoning behind their ideas, and encourage them to provide their reasons.

META NOTES

At this point do not expect students to be experts at the *Project Board.* They will be using this *Project Board* over the entire Unit, and will understand its purpose and process better as they gain more experience with it.

Sample *Project Board*

How do scientists answer big questions and solve big problems? Why do animals behave the way they do? How can you develop a zoo enclosure to support natural animal behavior?				
What do we think we know?	**What do we need to investigate?**	**What are we learning?**	**What is our evidence?**	**What does it mean for the challenge or question?**
Animals eat and drink. *March 30* Animals communicate. *March 30* Animals groom themselves. *March 30* Animals reproduce. *March 30* Animals grow. *March 30*	What do specific animals eat? *March 30* How do specific animals eat? *March 30* What parts of their body do they use to gather food and eat it? *March 30* How do specific animals communicate? *March 30* How do specific animals care for their young? *March 30* How do specific animals sleep? *March 30* How do specific animals communicate? *March 30*			

◇ Evaluate

With students, look over the ideas and questions on the *Project Board*. Make sure everyone has had the opportunity to contribute and that their ideas are represented. Also, make sure to include ideas or questions about communication and feeding.

Assessment Options

Targeted Concepts, Skills, and Nature of Science	How do I know if students got it?
Scientists often work together and then share their findings. Sharing findings makes new information available, and helps scientists refine their ideas and build on others' ideas.	**ASK:** How did your ideas change based on the class discussions? **LISTEN:** Students should be able to describe how sharing their ideas during the class discussion changed their ideas.
Scientists must keep clear, accurate, and descriptive records of what they do so they can share their work with others and consider what they did, why they did it, and what they want to do next.	**ASK:** Why do you think the *Project Board* is useful? Is it useful to record items on it clearly and accurately? **LISTEN:** Students should be able to describe how the *Project Board* will help them organize their ideas, and the importance of having clear and accurate records to refer to when sharing or remembering information.
Criteria and constraints are important in determining effective scientific procedures and, answering scientific questions.	**ASK:** What are criteria and constraints? Provide some examples for the zoo enclosure you will design. **LISTEN:** Students should describe what criteria and constraints are, and should give specific examples of these for the zoo enclosure.

eacher Reflection Questions

- What difficulties did students have understanding the challenge, and how can you guide their understanding?
- How can you further guide your students to understand the importance of the *Project Board,* which will be utilized throughout the *PBIS* curriculum?
- How can you assist students to lead and participate in the class discussions?

NOTES

LEARNING SET 1 INTRODUCTION

Learning Set 1

How Do Biologists Study Animal Behavior?

◀ ***9 class periods****

*A class period is considered to be one 40 to 50 minute class.

Student groups practice ethology as they make observations and interpretations based on images, and construct explanations.

Overview

Student groups are introduced to the *Learning Set's Big Question, How Do Biologists Study Animal Behavior?* Students then practice being ethologists as they learn about how to make observations, interpretations, and scientific explanations. They begin by making observations of middle-school students. During discussions, students realize that good observations need to be descriptive enough so that someone else might be able to draw an image of what was observed. Through categorizing behaviors, and through group and class discussions, students become aware of the difference between an observation and an interpretation. The students then learn what a scientific explanation is, and begin constructing explanations and revising them as they learn more about animal behavior. Students learn the importance of the iterative process, and the importance of engaging in the social practices of scientists as they build on each other's ideas and refine their own.

Targeted Concepts, Skills, and Nature of Science	Section
Scientists often work together and then share their findings. Sharing findings makes new information available, and helps scientists refine their ideas and build on others' ideas. When another person's or group's idea is used, credit must be given.	1.1, 1.2, 1.3, 1.4
Scientists must keep clear, accurate and descriptive records of what they do, so they can share their work with others, and consider what they did, why they did it, and what they want to do next.	1.1, 1.2, 1.3
Tables are an effective way to communicate results of a scientific investigation.	1.2, 1.3, 1.4

Targeted Concepts, Skills, and Nature of Science	Section
Observations and measurements are considered reliable if the results are repeatable by other scientists using the same procedures.	1.2
Scientists differentiate between observations and interpretations. They use their observations and interpretations to explain animal behavior.	1.3, BBQ
Scientific knowledge is developed through observations, recording and analysis of data, and development of explanations based on evidence.	1.4, 1.5
Scientists make claims (conclusions) based on evidence obtained (trends in data) from reliable investigations.	1.3, 1.4, 1.5, BBQ
Explanations are claims supported by evidence. Evidence can be experimental results, observational data, and other accepted scientific knowledge.	1.4, 1.5, BBQ
Criteria and constraints are important in determining effective scientific procedures and in answering scientific questions.	*Unit Introduction*
Behavior is a type of response to internal or external stimulus. Behavior is determined by experience, physical characteristics, and environment.	1.3, 1.4, 1.5
The structure and function of animal's bodies are complementary, and affect animal behavior.	1.5
Organisms need food to grow, reproduce and maintain their bodies. They must be able to live in a changing environment.	*Unit Introduction,* 1.5
Animals communicate with other animals using sound. The sounds they can make and the sounds they can hear are adaptations to their environment.	*Unit Introduction,* 1.1, 1.2, 1.5

Students' Initial Conceptions and Capabilities

- Students may have difficulties understanding the difference between observations and inferences (Allen, Statkiewitz, & Donovan, 1983; Kuhn 1991, 1992; Roseberry, Warrant, & Conant, 1992).
- Students are familiar with many animals, but may be unaware of the animals' behaviors, and of how to observe those behaviors.

Students' Initial Conceptions and Capabilities

- Students may anthropomorphize (attribute human characteristics to animal behavior observed), as we all do. It is important to be careful when interpreting animal behavior by using knowledge about human behavior. Other animals do not have the same goals and capabilities as humans, nor do they engage in the same activities as humans. Sometimes, describing animal behavior in human terms helps us understand animals better. When we do this we realize how similar, and dissimilar, humans can be from other animals. If students begin to use their knowledge of humans to describe other animals, they should be asked to support their claims with additional knowledge. Usually, when pressed for more information, or when pressed to describe why they think animal behavior resembles human behavior, students realize that this anthropomorphized animal behavior is less human-like than they previously thought.

Understanding for Teachers

Project Based Inquiry Science Units always revolve around answering a Unit question. The *Big Question* is *How do scientists answer big questions and solve big problems?* In *Animals In Action,* students also address a *Big Challenge.* They must use their knowledge to design an animal enclosure in which the animal will be able to feed or communicate in ways that approximate the animals' behavior in its natural environment. Students learn observational skills, and then use those skills to investigate feeding and communication.

Scientific Practice

Animals In Action is a life-science-based Launcher Unit and serves two goals. First, it introduces students to the social practices common throughout the *PBIS* curriculum. Second, it emphasizes the methods used by ethologists when studying animal behavior: planning, recording, analyzing, and explaining observations of animal behavior.

Project Based Inquiry Science emphasizes students creating knowledge as individuals, in small groups, and in large classes. *PBIS* uses many different social practices to assist students in learning science. The overall goal of these practices is to increase students' opportunities to become student scientists. Students will spend significant time planning, enacting, and reflecting on their observations. They will use an iterative process to develop explanations of animal behavior. These explanations include a scientific claim, evidence, and scientific knowledge to support the claim. They explain what they see, then learn more and revise their explanation. They will share their developing understanding with other students, and will listen to their feedback. The feedback will often be incorporated into a new plan. This type of sharing mirrors the methods of scientists who share their work with others.

Scientific Observation

In *Animals In Action,* students will engage in the scientific practice of observation. They will observe animal behavior and learn that carefully made, detailed observations allow for the development of explanations. Students identify the difference between observation and interpretation, and attempt to differentiate between them as they make their observations of animals. They also identify characteristics of the animals' bodies and of their environment that affect animal behavior.

Animal Behavior, Feeding, and Communication

Animals In Action is a Life-Science Unit about animal behavior. In this Unit students engage in the work of ethologists (biologists who study animal behavior). Ethologists consider questions about why animals behave as they do and about what affects animal behavior. In this Unit, students focus on two different behaviors: feeding and communicating. Each behavior is required for the survival of individual organisms and the species.

Animal behavior is affected by many different factors. In *Animals In Action,* students consider animals' body structure and their environment as influences on their feeding and communication. Throughout the Unit, students use these two factors to develop explanations for feeding and communication.

Animals feed to survive. All animals feed. Some have complex feeding behaviors, and others have very simple feeding behaviors. When animals feed, they are constrained by their body structure. Animals use the resource of their bodies to obtain food. Bees, chimpanzees, and elephants are all foragers, although they forage for different foods. They each use their bodies in specific ways to make sure they have enough to eat. Animals' adaptations make them efficient feeders.

Animals communicate in specific ways depending on their environment, their communication needs, and their body structure. Animals have adapted to their environment to become effective communicators.

LEARNING SET 1 IMPLEMENTATION

◀ *1 class period**

Learning Set 1

How Do Biologists Study Animal Behavior?

Animals are interesting to people. Humans work hard to understand why animals behave the way they do. People have been observing animals for hundreds of years. The *Big Question* for this Unit is: *How do scientists answer big questions and solve big problems?* You will answer this question in the context of the science question: *Why do animals behave the way they do?* You will apply what you are learning to designing a zoo enclosure. Before you start your observations, you need to break the *Big Question* into smaller questions. The smaller question you will answer in this *Learning Set* is: *How do biologists study animal behavior?* To answer this question, it will be important to work as biologists work, by making careful observations. You will be thinking about how to gather, record, and analyze data about animals in a way biologists do.

Notice how the animals in these pictures behave differently. The biologists who study animal behavior make careful observations and gather, record, and analyze data to determine why animals behave the way they do.

AIA 11 ANIMALS IN ACTION

Learning Set 1

How Do Biologists Study Animal Behavior?

5 min.

Students are introduced to the Learning Set.

△ Guide

Let students know that during this *Learning Set* they will be focusing on a smaller question, *How do biologists study animal behavior?*, as a first step to answering the Unit's *Big Question, How do scientists answer big questions and solve big problems?* and to addressing the *Big Challenge* of designing a zoo enclosure.

*A class period is considered to be one 40 to 50 minute class.

TEACHER TALK

"Our goal is to answer the *Big Question: How do scientists answer big questions and solve big problems?* Our first step is to answer the smaller question: *How Do Biologists Study Animal Behavior?*

We'll need to know about animal behavior in order to design our zoo enclosure, and as we learn about animal behavior we'll be learning what scientists do and how they do it.

How do you think scientists learn about animal behavior?"

Record students' ideas.

Let students know how scientists observe animals.

Then, let students know that they will be focusing on making and recording careful observations, analyzing those observations, interpreting their observations to find trends, make claims, and construct explanations on how animals behave and why they behave as they do.

TEACHER TALK

"When a scientist is trying to learn about an animal's behavior, they watch the animal for a long time, and record what they see, hear, smell, and other measurements they may make. These are their observations. After they collect many observations, they have to refer back to them to find repeated behaviors, and to figure out what those behaviors represent and why. You will be making and recording observations during this *Learning Set,* and you will practice making claims about those observations and constructing explanations about animal behavior."

SECTION 1.1 INTRODUCTION

1.1 Understand the Question

Observing Animal Behavior

◀ ***1 class period****

*A class period is considered to be one 40 to 50 minute class.

Overview

Student groups act as ethologists as they observe middle-school students' behavior while feeding. Students engage in the social practices of scientists, share their observations with the class, and realize the importance of keeping good records.

Targeted Concepts, Skills, and Nature of Science	Performance Expectations
Scientists often work together and then share their findings. Sharing findings makes new information available, and helps scientists refine their ideas and build on others' ideas. When another person's or group's idea is used, credit needs to be given.	Students should realize that, by sharing their ideas, they may get new ideas, revising their old ones and improving them.
Scientists must keep clear, accurate, and descriptive records of what they do so they can share their work with others, consider what they did, why they did it, and what they want to do next.	Students should be able to record accurate and descriptive observations of other students' feeding and communicating behaviors.
Animals communicate with other animals using sounds. The sounds they make and the sounds they can hear are adaptations to their environment.	Students should observe that communication between middle-school students occurs both verbally and nonverbally.

Materials	
1 per class	Snacks for a group of three students

Activity Setup and Preparation

Small groups will collaborate to make observations of other students. Decide how you want to create student groups of three or four. If possible, it would be best if there were eight groups per class, since some sections will require eight groups. It would be most effective to maintain these small groups for the entire Unit.

Prepare snacks for one group of students. They will eat this snack while the rest of the class observes. Also, pick a topic of discussion for the snacking group such as, What did you do over the weekend?

Consider having an assignment on the board for students to write their observations of you as the class gets started.

Homework Options

Reflection

- **Science Process:** Observe and record feeding, socializing, or communicating behavior outside the middle-school environment. How are the behaviors you observed similar to and different from those in the middle school? *(Students should recognize similarities in the feeding and communicating behaviors of other students, and differences due to a change of environment.)*
- **Science Process:** Describe how an animal (other than a human) might behave during feeding. What are the similarities and differences between humans and animals feeding? *(Students might include differences in how animals eat, e.g, not using hands or paws, and they may indicate differences in how animals socialize during feeding.)*

Preparation for 1.2

- **Science Process:** What are some ways ethologists might organize the data they collect? *(Students should discuss different ways of categorizing observations, for example, by gender, by feeding, by interactions with others, etc.)*
- **Science Process:** What do you think the difference is between observations and interpretations? *(Students should discuss the difference between observations — what they actually see, hear, taste, smell, and can measure in some way — and interpretations.)*
- **Science Process:** List some behaviors of animals other than humans that you have observed. *(Students should list observations of other animals.)*

SECTION 1.1 IMPLEMENTATION

1.1 Understand the Question

Observing Animal Behavior

data: (singular, datum) recorded measurements or observations.

ethologist: a biologist who studies the behavior of animals in their natural environment.

Whether you are observing animals, people, machines, or stars, observation is a critical science skill. Without careful observations, scientists could not explain the way the world works. When scientists make observations, they gather a lot of **data**. The data serve as a record of what the scientists observed. By making a record, a scientist can return to the data often and for different purposes. Without a record, the scientist would just be working from memory, and memory is not always reliable.

Ethologists are biologists who study how animals behave. They observe animals to explain why animals behave the way they do in their natural environment. Ethologists usually observe animals to answer the following questions:

- How do different behaviors help an animal survive?
- How do the animal's environment and learning affect different behaviors?
- How do behaviors change as the animal grows?
- How do animals that are similar to each other act in similar or different ways?

To answer these four questions, ethologists observe very carefully and for long periods of time. Ethologists collect a lot of information, or data. To make sense of all the data they collect, ethologists use specific rules to describe and classify what they observe.

You, too, can be an ethologist as you investigate the behavior of animals. In this Unit, you will think about some of these questions. In this *Learning Set*, you will develop some of the tools and observation methods ethologists use when they observe animal behavior. You will work by yourself, with your group, and with your class to record and analyze your observations. You will then use those observations and what you know about the animals to explain why they are behaving the way they are.

Get Started

You will begin your investigations of animal behavior by observing familiar animals: middle-school students. Your teacher has asked a small group of

1.1 Understand the Question

Observing Animal Behavior

10 min.

Introduce the work of ethologists and explain why detailed observations and records are important.

○ Engage

Begin by reviewing the class *Project Board,* and remind students of what they need to investigate. If there are questions on the *Project Board* which are similar to the ethologists' questions, then begin discussing these and move directly to the *Guide* segment below. Otherwise, continue with this *Engage* segment.

Ask students what they observed about you since they entered the classroom. Then, ask students to describe what you were doing the last time they met with you when they walked in the classroom.

META NOTES

Consider having an assignment on the board for students to write their observations of you as the class gets started.

Expect that students will not be able to describe what you did at the beginning of the previous class meeting, as well as they did during this class meeting. Emphasize this, and remind students that they will have to make careful observations of an animal in order to design an enclosure for that animal.

TEACHER TALK

"What have you observed about me while we were getting ready for class?

What did you observe about me yesterday while we were getting ready for class?

Do you remember as much about yesterday as today? Do you remember what shoes I wore? What things did I do the same? What things did I do differently?

In order for you to meet your challenge of designing a zoo enclosure you will have to make observations of animals living in the wild. How will you make these observations, and what will you look for?"

△ Guide

Describe how observations are the data biologists collect in order to learn more about animal behavior. Discuss the importance of recording observations so that scientists can review their data and not work from memory alone.

Then, introduce students to what ethologists do.

META NOTES

The four questions of ethology are traditional questions that ethologists have used since the 1920s. In this Unit, only the first two questions will be investigated.

TEACHER TALK

"An ethologist is a biologist who studies animal behavior. Ethologists observe animals in their natural habitats, and try to figure out why they do the things they do. They usually try to determine how behaviors help animals survive, how an animal's environment and learning affect its behaviors, how its behaviors change as it grows, and how animals that are similar to each other act in similar or different ways. These four things, bulleted in the student text, comprise the four questions that ethologists try to answer."

> **Get Started**
>
> You will begin your investigations of animal behavior by observing familiar animals: middle-school students. Your teacher has asked a small group of
>
> AIA 12
>
> *Project-Based Inquiry Science*

Get Started

10 min.

Students observe a small group of students.

△ Guide

Tell students that they will begin their work as ethologists by observing other middle-school students. They will watch as the group of students you selected share a snack and talk for three minutes. The rest of the class should record their observations as carefully and specifically as they can.

TEACHER TALK

"Today you will be doing the work of ethologists. You will be observing the animal you are most familiar with, the middle-school student. I will select a group of students to sit in front of the class and enjoy a snack and discuss a certain topic. The rest of you will make observations of their behavior."

META NOTES

NOTE: Students may have trouble talking to one another. Suggest a topic, such as what they did over the weekend, movies, sports, or homework.

⬡ Get Going

Have the group of students you selected share a snack and talk for three minutes while the other groups observe and record their observations. Consider assigning a topic of discussion. The group should talk as if they were not being watched.

META NOTES

This is the students' first opportunity to make observations. They will be familiar with the scene because they have experienced it many times. They will make many observations, but may miss a lot of detail because they lack an observational plan.

Communicate

15 min.

Lead a class discussion of students' observations.

your classmates to act as middle-school students usually act. They will show you about three minutes of behavior. They will try to act as they would in their natural environment. Your goal is to accurately observe and record the details of the students' behavior. Be sure to watch carefully and record your observations as specifically as you can.

Communicate

Three minutes of behavior might have seemed short when you first started watching it, but it is amazing how much there is to see in three minutes. Share your observations with your class. Describe what you saw as carefully as you can. Listen carefully so you can decide if all of you saw the same things or if you saw different things. Listen to the details of each observation so you can decide if the observation was accurate.

Discuss with your class the challenges of making observations. Describe the way you made your observations. Did you focus on one person or the whole group? Perhaps you listened more than you watched. Tell your class how you made your observations. Tell your class how confident you are of your observations.

Reflect

You probably saw many of the same things as your classmates. But you might have described them a little differently. Details are very important. Details help you tell the difference between accurate observations and inaccurate ones. Answer these questions to help you think about making observations and how you might improve on making observations.

1. After you listened to others' descriptions, how confident were you about your observations?
2. The next time you make observations, what will you do to make your observations as detailed as possible?
3. How important was the amount of detail people included in their observations? How did the level of detail help you know what happened in the middle-school scene?
4. Look back at your observations. If someone else read them, could they sketch a picture of what you saw? What could you do to help someone be able to sketch a picture?

META NOTES

Students may disagree with each other about their observations. Students may have observed different parts of the scene. They may also disagree with those who were being observed. Inconsistencies in the observations are expected. This should lead students to begin improving their observational plan, and should help students make better observations. They will have other opportunities to make their observations with a plan.

△ Guide and Assess

Have students share their observations with each other to decide if they all saw the same things. Record students' observations. Students should describe what they saw as carefully as they can. They should include how they made their observations, and how confident they are of their observations.

For a few presentations, have students draw a sketch of one of the observations described. This is an indicator of how descriptive the observations were. This should not take a long time, and will be used as an example later in the section.

Notice what students are observing and how they are communicating with each other. Keep in mind any issues that they are having trouble with to help them in later presentations.

△ Guide

Transition the discussion by asking students how confident they were in their observations and what they would do differently. Students should begin to realize that they need a plan for observing.

> **META NOTES**
>
> The students' observations should become more detailed and accurate as their observational skills improve. Keep their current abilities in mind as they write an observation plan and implement it.

your ob

Reflect

You probably saw many of the same things as your classmates. But you might have described them a little differently. Details are very important. Details help you tell the difference between accurate observations and inaccurate ones. Answer these questions to help you think about making observations and how you might improve on making observations.

1. After you listened to others' descriptions, how confident were you about your observations?
2. The next time you make observations, what will you do to make your observations as detailed as possible?
3. How important was the amount of detail people included in their observations? How did the level of detail help you know what happened in the middle-school scene?
4. Look back at your observations. If someone else read them, could they sketch a picture of what you saw? What could you do to help someone be able to sketch a picture?

Reflect

10 min.

Lead a class discussion of the differences in students' observations.

> **META NOTES**
>
> This is a good opportunity for formative assessment.

⬡ Get Going

Let students know how much time they have to answer the questions and to get prepared for a class discussion.

△ Guide and Assess

Begin with a discussion of how confident students were of their results and what they might do differently next time. Then discuss each of the *Reflect* questions with the class.

Listen for the following responses to the questions in the student text:

1. Students should base their answers on the disagreements that came up, and on how their observations compared to the observations of the rest of the class.
2. Students should think about ways to make their observations more accurate. This is a good place to begin talking about planning how

to make observations and the many different things there are to observe, such as what people are saying or how they are eating, etc. Tie this in to the need for a plan for their observations.

3. Students should recognize that details are necessary to reconstruct what happened.
4. Students should be aware of the level of detail in their observations. Even if they say that someone could have drawn a picture from their observation, the second question will help them identify the importance of detail and careful records. You should remind students of the few sketches they drew during the presentation of others' observations. Emphasize that their observations should be detailed enough for someone to be able to draw a picture of it.

NOTES

What's the Point?

Scientists make observations to help them find out more about the world. When scientists make observations, they include as much detail as possible. When they include details, other scientists can read their descriptions and understand them. Then other scientists are able to decide if the observations are accurate. An ethologist is a scientist who studies how animals behave.

Eastern bongo antelope calves in their natural habitat. Their coats are reddish brown with bright white strips, which help provide camouflage from their enemies.

What's the Point?

5 min.

This summary provides the main ideas and goals of the section.

META NOTES

A *What's the Point?* summary ends most sections. In this summary, the main ideas and goals of the section are provided. It gives students a sense of what they have accomplished, and it also provides you with an opportunity to evaluate students' conceptions and capabilities. Sometimes a discussion will be in order; other times, the most recent discussion will have just covered the content of a *What's the Point?* summary.

◇ Evaluate

Make sure students see the importance of accurate and detailed observations, and that it is helpful to have a plan. Refer back to the *Reflect* questions if students still need guidance.

Assessment Options

Targeted Concepts, Skills, and Nature of Science	How do I know if students got it?
Scientists must keep clear, accurate, and descriptive records of what they do so they can share their work with others and consider what they did, why they did it, and what they want to do next.	**ASK:** Why is it important to make observations very detailed? **LISTEN:** Students should include information about being able to accurately describe what they saw, and the importance of this when studying animal behavior. They may refer to the differences in their observations of the same thing, and note how these may be resolved with more accurate and detailed observations.
Scientists often work together and then share their findings. Sharing findings makes new information available, and helps scientists refine their ideas and build on others' ideas. When another person's or group's idea is used, credit needs to be given.	**ASK:** How did sharing your observations help you to understand the need for accurate observations? **LISTEN:** Students' responses should include information about how hearing others' observations and ideas led them to think about ways to revise and improve their method of observations, recording of observations, and ideas.
Animals communicate with other animals using sounds. The sounds they make and the sounds they can hear are adaptations to their environment.	**ASK:** What did you observe about how middle-school students communicate? **LISTEN:** Students should include descriptions of verbal (making sounds) and nonverbal (gestures and facial expressions) communication.

Teacher Reflection Questions

- How can you tell if students understand the importance of detailed observations in learning about animal behavior? What ideas do you have to help students make accurate and detailed observations as they continue in this Unit?
- How were you able to focus students' attention on the way they were making observations? What can you do to help students pay attention to how they are observing as they are observing?
- Students may have seen different things during their observations, and may have disagreed about what they saw. What techniques did you use to keep the discussions productive?

NOTES

NOTES

SECTION 1.2 INTRODUCTION

1.2 Explore

◀ $1\frac{1}{2}$ *class periods**

*A class period is considered to be one 40 to 50 minute class.

How Can You Improve Your Data Collection?

Overview

Groups develop plans for observing the behavior of middle-school students. They decide beforehand how to divide the work between group members, what kinds of things to note, and how to record their observations. Then each group makes observations and analyzes the data, categorizing the data, and looking for trends. They share their categories and discuss their reasoning for them with the class. Students should discover that they saw the same thing, but they may not have labeled it the same. They should discuss which label is most understandable. Students then discuss how they can improve their observational procedures to get more reliable data.

Targeted Concepts, Skills, and Nature of Science	Performance Expectations
Scientists often work together and then share their findings. Sharing findings makes new information available, and helps scientists refine their ideas and build on others' ideas. When another person's or group's idea is used, credit needs to be given.	Students should realize that, by sharing their ideas, they may get new ideas, and may be able to revise and improve their old ones.
Scientists must keep clear, accurate, and descriptive records of what they do so they can share their work with others, consider what they did, why they did it, and what they want to do next.	Students should record accurate and descriptive observations of middle-school students' feeding and communicating behaviors. Students should describe a good observation as one where someone can draw a picture from.
Observations and measurements are considered reliable if the results are repeatable by other scientists using the same procedures.	Students should determine if their observations were reliable, based on whether they were repeatable or if other groups observed the same behavior.

Targeted Concepts, Skills, and Nature of Science	Performance Expectations
Tables are an effective way to communicate results of a scientific investigation.	Students should describe how creating categories helped them to identify trends and communicate their information.
Animals communicate with other animals using sound. The sounds they can make and the sounds they can hear are adaptations to their environment.	Students should observe that communication between middle-school students occurs both verbally (making sounds) and nonverbally.

Materials	
1 per class	Snacks for a group of three students
1 per group	Sticky note pad
10 per group	Index cards

Activity Setup and Preparation

Prepare snacks for one group of students. Also pick a topic of discussion for the snacking group in case they have trouble with selecting their own topic. Consider topic questions such as: What did you do over the weekend? What was the latest movie you have seen? What is your favorite sport? etc.

Select a different group of students to be observed. Every group will be creating an observation plan, but the group being observed will not be observing others in the class. Decide how this group will gain experience observing. You may want to have them observe students analyzing their data.

Homework Options

Reflection

- **Science Process:** What were some differences between the trends your group identified and the trends other groups identified? What do these differences say about the way you made observations or the way you categorized your data? *(Students should recognize how categorizing behaviors differently or selecting different data can lead them to find different trends.)*
- **Science Process:** What were the most important things you did to get reliable data about animal behavior? *(Planning how to observe, and making careful observations, were probably the most important things students did.)*

Preparation for 1.3

- **Science Process:** Did you include in your observations how the students you observed were feeling? What were your reasons for including them or not including them? *(This helps students to begin thinking about the difference between observation and interpretation.)*
- **Science Content:** What behaviors of humans are like behaviors of other animals? *(This engages students in thinking about the different kinds of animal behavior that can be observed.)*

NOTES

SECTION 1.2 IMPLEMENTATION

1.2 Explore

How Can You Improve Your Data Collection?

Plan Your Data Collection

Ethologists need to record what they see in a way that will allow them to remember it later. One way to record behavioral data is by using an **ethogram**, a table that describes behavior. Scientists also need to agree with each other about what they saw. They might want to compare what they saw one time to what they saw another time. They need good records to do that. They also need to record what they saw well enough so other scientists can use their data. In the previous section, you discussed some of the things that make it difficult to make good records of your observations. In this section, you will use a type of ethogram to keep records.

ethogram: a table used to record observations of animal behavior.

You will have another chance to observe and record the behaviors of middle-school students. This time you will use a type of ethogram to keep records. Work together with your group to plan how to keep good records of what you see. Tell each other what you think you saw the first time. Pay attention to where you agree and disagree with each other. Then, think about what you will have to watch and what you will have to record to be able to make detailed observations that you agree about. The questions below can be used to help you think about issues that might affect your observations. To make accurate and detailed observations, you might also think about whether others could draw a picture based on your description.
If they can, your description is very detailed. Consider the following questions:

- Will you watch an individual or all the students in the small group?
- How will you make sure you have observed all the members of the group?
- How will you make sure you have observed all the different behaviors?
- Will you each watch the whole scene or will it help to divide the observation task among all the members of your group?
- How will you record what you see?
- Will you take notes on everything you see?

AIA 15 ANIMALS IN ACTION

◀ $1\frac{1}{2}$ *class periods**

1.2 Explore

How Can You Improve Your Data Collection?

Plan Your Data Collection

15 min.

Guide students in planning how they will make detailed observations and keep good records.

△ Guide

Remind students of their observations of middle-school students. Then ask them what they can do to make their observations more detailed and accurate, or remind them of their discussion during the previous section of how they could better make and record their observations.

*A class period is considered to be one 40 to 50 minute class.

META NOTES

Within the class there should be various plans. The goal is for students to develop their plans and use them for the next observation session. The plans will not be perfect. Some groups will find that their plans do not provide them with accurate or detailed observations. In subsequent sections, they will discuss the strengths and weaknesses of their plans, refine them, and use the refined plans to make observations. For this reason, it is appropriate to retain the small groups throughout the Unit.

META NOTES

Remember you are setting the stage for the rest of the year. If only a couple of groups are not done, you may want to move forward to set the pace of the class, and to emphasize the need to be on task. One way some teachers manage time is to remind students halfway through, and shortly before the end of the time period, how much time they have left.

META NOTES

Some students will suggest that the class could use a video camera instead of recording observations in the field. Remind them that, even with a video camera, they would still have to learn observation skills in order to know what to look for.

Students may also ask about putting a camera in the lunchroom. If so, you should discuss the ethics of collecting data about people.

Next, ask students what might have happened to the accuracy of their observations if they had a plan for making observations before they started. Explain that now students will work in groups to develop a plan for making and recording their observations. To begin, they will discuss what they saw during the previous observations, noting where they agreed and where they disagreed. Then, they should think about what they need to observe, and what they need to record. Let students know that the questions in the student text will help guide their thinking as they discuss these issues. Their finished plans should describe how they will divide up the work, how to record their observations (what terms will they use and what kinds of pictures will they draw), and what they should make sure to include.

⬡ Get Going

Let groups know how much time they have to construct their plan. It should not take more than ten minutes.

☐ Assess

As groups are working on their plans, monitor students' work and discuss each group's plans with them. Look for observation issues that groups are anticipating, questions they have answered, and how detailed the plans are. Some of the discussion will revolve around the effectiveness of the plans, so it is important to note if groups have plans that are very different from one another, and how those differences might affect the final data. Later, you will point out how the differences in the plans affected how the data was collected and analyzed.

- How can you record quickly enough so you don't miss anything? Can you write key terms and not full sentences?
- Will you draw a picture?
- Will you keep track of the amount of time each behavior lasts? If so, how will you do this?

After you have discussed your ideas with your group, develop a plan you will use to make your observations. Use the questions to help you decide on your plan. Make sure all members of your group agree on the plan and know how to follow through on it.

Observe

As you watch your classmates, pay careful attention to the details of the situation. Make sure you record all your observations accurately. Follow the plan you made in your group as closely as possible.

category: a set or class of things with similar characteristics, properties, or attributes.

trend: something that occurs over and over again.

Analyze Your Data

The next step after collecting observational data is to make sense of it. You need to analyze it to understand what was going on. When scientists analyze observational data, they first check to see if their observations are the same as those of others. They discuss anything different and try to come to agreement. Next, they work to organize the data in a way that will let them see **categories** and **trends**. They identify the trends in the data—sequences of behaviors that they see over and over. You will do that with the observations you and your group have made. After you have completed the steps below, you will present your analysis to the class.

1. Determine what everyone saw.

   Allow each group member to describe the behaviors they saw. Listen carefully to the observations of other group members. Decide if all of you saw the same things or if some of the observations were different. Think about why the observations might be different, and share your ideas with your group. Be sure to report any difficulties you might have had using the observation procedures your group decided on. It is important that you describe not only what you observed but also how you made each observation. This will help other students understand the data.

2. Organize your data into categories.

   The first step in analyzing your data is organizing it. You probably have many notes. You cannot understand your data with disorganized

Observe

5 min.

Have the class observe and record a few students snacking, according to their new observation plans.

△ Guide

Emphasize that it is important to follow their group's plan for making and recording observations, and that they should pay close attention to details and record their observations accurately.

META NOTES

Select a different group of students to model snacking behavior, so that the previous group can practice making and recording observations. Provide the same directions, and ask them to perform similarly to the first group.

○ Get Going

Have a student group share a snack and talk for three minutes while the rest of the class observes them. The snacking group should talk. If they have trouble talking, you should suggest a topic such as movies, homework, sports, etc.

Let students know that they will again observe a few students snacking and talking for three minutes.

Analyze Your Data

15 min.

Have groups discuss, organize, and interpret their observations.

charac... properties, attributes.

trend: something that occurs over and over again.

...you made in you... ...as possible...

Analyze Your Data

The next step after collecting observational data is to make sense of it. You need to analyze it to understand what was going on. When scientists analyze observational data, they first check to see if their observations are the same as those of others. They discuss anything different and try to come to agreement. Next, they work to organize the data in a way that will let them see **categories** and **trends**. They identify the trends in the data—sequences of behaviors that they see over and over. You will do that with the observations you and your group have made. After you have completed the steps below, you will present your analysis to the class.

1. Determine what everyone saw.

   Allow each group member to describe the behaviors they saw. Listen carefully to the observations of other group members. Decide if all of you saw the same things or if some of the observations were different. Think about why the observations might be different, and share your ideas with your group. Be sure to report any difficulties you might have had using the observation procedures your group decided on. It is important that you describe not only what you observed but also how you made each observation. This will help other students understand the data.

2. Organize your data into categories.

   The first step in analyzing your data is organizing it. You probably have many notes. You cannot understand your data with disorganized notes. Begin by copying each observation onto a sticky note. Put only one observation on each note. Your entire group will also copy their observations. If some are repeated, you can make just one sticky note for that observation.

   Once each observation is on a separate sticky note, read through them and discuss how they might be grouped into categories by behavior. Combine observations of similar behaviors. If you disagree with the categories, ask others why they think a behavior fits into that category. Sometimes it is necessary to reorganize categories as more observations are read. Once you have each observation in a pile, use index cards to label the categories.

3. Identify trends.

   As a group, prepare a description of the behaviors in each category. The description needs to be specific enough that other students in your class can recognize it. This will help you describe your groupings to the rest of your class.

△ Guide

Briefly, have students share some of their observations. Point out to students that all of their observations are about the same experience, and that if their observations are disorganized, they will be confusing to someone who has not seen the action. Their goal is to do the work of ethologists, so they need to explain what affects behavior. To do that, they need to organize their data differently.

Next, let students know that they will work within their groups to organize their data. They need to make sure that their methods are documented, to create categories for their data, and to identify trends in their data.

To begin, ask group members to describe what they saw to the rest of the group. The group will discuss reasons for any differences in their observations and any difficulties they had with their observation procedures. Then, groups will categorize their observations. First, they should copy each observation onto a sticky note and group these sticky notes into categories by behavior. They should label the categories using index cards. After categorizing and labeling their observations, groups will identify trends by writing descriptions of the behaviors in each category.

⬡ Get Going

Let groups know how much time they have to work. Analyzing the data should take about ten minutes.

△ Guide and Assess

As groups discuss, organize, and interpret their data, assess whether groups are identifying any inconsistencies in their results and problems with their procedures. Also, check to see if students are having trouble categorizing their data and finding trends. A trend can be something as obvious as "all students chewed their food," or as subtle as "students looked at whoever was speaking, except when they were reaching for a snack." If a behavior is observed more than once, it can be called a trend.

META NOTES

The goal of this activity is to explain human behavior. Students are not ready to make claims about how middle-school students behave. To make claims, they will need to analyze, organize, and find trends in their data. They will then use their observations, categorizations, background knowledge, and experiences, to develop explanations for the human behavior.

META NOTES

Students will have developed a list of observations. They may also have included interpretations. The next *Section* will help them differentiate between the two.

META NOTES

To provide the group that was observed with an observing experience, consider having them observe their classmates analyzing the data for a minute or two. Also consider having them present their analysis after all the other groups have presented.

Communicate

15 min.

Have a discussion of presentations, comparing groups' categories.

notes. Begin by copying each observation onto a sticky note. Put only one observation on each note. Your entire group will also copy their observations. If some are repeated, you can make just one sticky note for that observation.

Once each observation is on a separate sticky note, read through them and discuss how they might be grouped into categories by behavior. Combine observations of similar behaviors. If you disagree with the categories, ask others why they think a behavior fits into that category. Sometimes it is necessary to reorganize categories as more observations are read. Once you have each observation in a pile, use index cards to label the categories.

3. Identify trends.

   As a group, prepare a description of the behaviors in each category. The description needs to be specific enough that other students in your class can recognize it. This will help you describe your groupings to the rest of your class.

Communicate

Each group will now share the categories they created and the reasoning behind their categories. Listen carefully to see if others made the same observations you made. Also, listen as each group describes their category labels. Think about whether they used the same labels as your group. Consider whether each group's labels would be understandable by other scientists.

Feeding is one category of behavior common to all animals, including humans.

Reflect

Each group probably did not create exactly the same categories, but each group had reasons for creating its particular categories. You will be learning more about observing animals and creating categories during this *Learning Set*. To prepare for that, answer the following questions and discuss the answers with your class. Listen carefully as other groups share their

META NOTES

Students should see alternative ways of categorizing the data, and should analyze the effectiveness of each.

△ Guide

Let the class know that each group will share their categories with the class, explaining the reasoning behind their categories, and that the rest of the class should listen carefully for the labels groups used for their categories, and the reasons for each categorization. Let students know what they should consider while they listen.

TEACHER TALK

"While other groups are sharing their categories, think about whether they were using the same categories you used. Did they divide things up the same way? Did they call their categories the same things? Also think about what categorization would be most useful for communicating with other biologists."

△ Guide Presentations and Discussions

Begin this discussion by having student groups read their categories to each other. Have each group describe their categories. Students should support their categories by providing one or two observations which belong in that category. Identify similar observations that students may have categorized differently. Have students identify how the differences might change the data analysis and the trends they saw. Have students identify categories that have the same type of behaviors, but are labeled differently. Discuss how the differences in categories and naming may impact how the behavior analysis is completed. After the groups have presented, discuss the labels that groups chose, and determine which labels everyone agrees on as the most understandable.

During the discussion, model discussion skills for the students. Expect students to be active listeners and to ask appropriate questions. Help students to always focus on content-driven questions rather than on more superficial ideas. During the analysis discussion, assist students in their questioning and comments by modeling respectful language such as, "I agree with... because... or, I disagree with... because... or, Could you clarify...?"

META NOTES

The goal of these discussions is for students to discuss ideas with each other. They should be sharing their learning with other biologists (their classmates). Students may not have experience presenting to each other in this way. They will need support to help them focus on presenting to each other, and not to you, the teacher. Therefore, it is important, during this first presentation, to model discussion techniques and assist students in rewording and redirecting their questions as necessary. Consider reading the tips on discussion techniques in the *Teacher's Resource Guide.*

Reflect

20 min.

Have students reflect on their procedures and how to improve them.

Reflect

Each group probably did not create exactly the same categories, but each group had reasons for creating its particular categories. You will be learning more about observing animals and creating categories during this *Learning Set*. To prepare for that, answer the following questions and discuss the answers with your class. Listen carefully as other groups share their

Feeding is one category of behavior common to all animals, including humans.

AIA 17

ANIMALS IN ACTION

○ Engage

First, remind the class that they will need detailed procedures for observing animals. They will need to follow those procedures so that their data will be more reliable. Consistently followed, detailed procedures will allow students to identify trends in the data. Being able to find trends in animal behavior will help them make decisions about designing their animal enclosure when they address the *Big Challenge*.

△ Guide

Let students know that the *Reflect* questions will help them think about how their procedures might have changed what they saw, and how they affected their data analysis. Students should discuss their ideas for each question, and develop their best group response. Then they will share the answers with the class. Emphasize that each student is responsible for contributing to the group discussion and recording the group response on his or her own paper.

⬡ Get Going

Allow students the appropriate time for discussion of these questions. If students seem to get stuck on a question, provide some support for that question. If student groups are taking too long answering the questions or if they do not understand the questions, return to a large group discussion to complete the answers.

META NOTES

One way that students learn through inquiry is by identifying how their ideas change over time. The *Reflect* questions allow students to see their thinking at one point in time. They should be able to return to these questions later in the Unit to refer back to their responses. Therefore, it is important for students to record their group answers, and to track their group's ideas and reasons for those ideas.

ideas. By working together, you may hear ideas that can help you improve your work. You will use the answers to improve your procedures, observations, and analyses.

1. What are the main categories of behaviors you found in your observations of the middle-school students?
2. How difficult was it to stick to the procedure you decided on in your planning?
3. What were the issues in your procedure that affected how you made your observations? Were your observations more complete or less complete because of your procedure?
4. What will you do to improve your procedure the next time you make this type of observation?
5. What will you do to improve your observations and analysis for the next time?
6. Why do you think different groups made different observations and identified different categories?

reliable data: data that is the same when collected many times or by different people.

Be a Scientist

Reliable Data

In this Unit, you will be collecting a lot of information. Most of the information will come from your observations of animals. This information is called **data**. Data can be collected in a lot of different ways. In some Units, you will collect data made up of numbers. You may set up an experiment and measure time, distance, or temperature. In this Unit, you will collect and analyze *observational* data.

Making good scientific decisions requires **reliable data**. Observational data is more reliable when the same procedure is used each time the data is collected. The same procedure also means that another person could use the procedure and collect similar data.

In this Unit, you will use the data from your observations to develop interpretations. Scientists use data to develop their interpretations. If the data are very reliable, then the interpretations are more reliable as well.

△ Guide and Assess

Have a class discussion of the *Reflect* questions. During the discussion, listen for the following responses, and guide students to these responses if they are having difficulty.

1. These should be the categories they shared with the class, and should include feeding and communicating. The categories will not be the same for each group.
2. Students should explain any difficulties they had in following their procedures.

3. Students should evaluate how their procedures affected their observations, and whether or not this made their observations more complete or less complete. This is a good place to review students' understanding of how different procedures may have led to different results. For example, some groups may have decided to have each group member observe only one person. This may give more complete observations of one person, but less complete observations of interactions among people. Other groups may have assigned members to observe only feeding, communication, or body movements. This would provide more complete information in these categories, but less information in other categories.
4. Responses should address difficulties in the type of observations they made, perhaps not being as descriptive as they should have been, or assigning to a given group member too many or too few things to observe.
5. Responses should address inconsistencies in their observations or difficulties in finding trends, for example, due to the categories selected.
6. Students should start to think about the fact that they do not have the same observations, because they are not following a standard set of procedures or using standard categories. As you discuss this question, you should transition into a discussion of what it means for the data to be reliable.

Be a Scientist: Reliable Data

When students plan an observation procedure, their goal is to make observations and collect reliable data: that is, data that everyone can agree on. Since different observational methods can lead to different observations, students will want to find a consistent method for making observations.

Tell students that data (observations or measurements) are reliable when the same information is collected many times or by many different people. This means that the data collected with the same procedures should be very similar. Data collected by a variety of scientists with slightly different procedures should still show the same trends in order to be considered reliable. Also, point out that scientists need reliable data to develop good interpretations.

What's the Point?

Ethologists look for reasons why animals behave the way they do. They try to answer four different questions:

- How do different behaviors help an animal survive?
- How do the animal's environment and learning affect different behaviors?
- How do behaviors change as the animal grows?
- How do animals that are similar to each other act in similar or different ways?

To answer these questions, scientists need to make and analyze observations. Accurate, detailed observations are essential to understanding animal behavior. When you record your data, you can better analyze your data, share your work with others, and answer scientific questions. Ethologists share their observations and analyses with others who can help decide if the observations are accurate and detailed enough to help determine why animals behave as they do. They can do this only if they make careful observations, support their reasoning, and analyze their data to correctly identify categories of behaviors.

The behavior of these ducks could be in the category of parenting.

What's the Point?

5 min.

◇ Evaluate

Review the four questions that ethologists consider:

- How do different behaviors help an animal survive?
- How do the animal's environment and learning affect different behaviors?
- How do behaviors change as the animal grows?
- How do animals that are similar to each other act in similar or different ways?

Emphasize the importance of accurate detailed observations. Review the importance of consistent procedures for making consistent observations and developing explanations.

Assessment Options

Targeted Concepts, Skills, and Nature of Science	How do I know if students got it?
Scientists must keep accurate and descriptive records of what they do so they can share their work with others and consider what they did, why they did it, and what they want to do next.	**ASK:** How did having an observation plan improve your record keeping? How did your record keeping help you to determine trends in your data? **LISTEN:** Students should describe how their improved plans led to more detailed and accurate observations that they could refer back to and find trends in the data. They should also realize that the class's observations were more similar this time.
Scientists often work together and then share their findings. Sharing findings makes new information available, and helps scientists refine their ideas and build on others' ideas.	**ASK:** How did working in a group and sharing your observations and analysis with the class help you? **LISTEN:** Students should describe how working in a group and sharing their ideas helped the group to come up with better observation plans, and how sharing their observations and analysis with the class helped them to come up with ideas for improving their observations.
Tables are an effective way to communicate results of a scientific investigation.	**ASK:** How did categorizing your observations help you? **LISTEN:** Students should describe how creating categories helped them to identify trends and communicate their information.

Targeted Concepts, Skills, and Nature of Science	How do I know if students got it?
Observations and measurements are considered reliable if the results are repeatable by other scientists using the same procedures.	**ASK:** What does it mean for an observation to be reliable? Give an example from your data. **LISTEN:** Students should describe reliable observations as ones that are repeated many times, and observed by different groups. Students should determine if their observations were reliable based on whether or not they were repeatable, or if other groups observed the same behavior.

Teacher Reflection Questions

- What ideas do you have about helping students to develop procedures geared toward increasing the detail of their observations?
- What ideas do you have about how to help students with this providing reasoning for their ideas?
- How did you manage the class time between groups planning their observations and making their observations? What ideas do you have for next time?

NOTES

SECTION 1.3 INTRODUCTION

1.3 Explore

Observing and Interpreting Animal Behavior

◀ *2 class periods**

*A class period is considered to be one 40 to 50 minute class.

Overview

Students explore the difference between observation and interpretation, and the importance of both. Students begin by observing pictures of a familiar animal (humans), and discuss their observations with their groups. By comparing their observations with other group members, and discussing them with the class, they begin to see that they included interpretations of behavior with their observations, and that their interpretations are not all the same. Students should realize that their interpretations need supporting reasons. Students use the *Observing and Interpreting Animal Behavior* pages to help them separate interpretations from observations and provide reasons for interpretations. Students then observe pictures of different animals, and discuss their observations and interpretations with their groups and the class, learning to keep their interpretations separate from their observations, and realizing some of the difficulties of interpreting animal behavior.

Targeted Concepts, Skills, and Nature of Science	Performance Expectations
Scientists often work together and then share their findings. Sharing findings makes new information available, and helps scientists refine their ideas and build on others' ideas. When another person's or group's idea is used, credit needs to be given.	Students should describe how sharing their observations and interpretations with their group members, and with the class, helped them to improve their observations and interpretations.
Scientists must keep clear, accurate, and descriptive records of what they do so they can share their work with others and consider what they did, why they did it, and what they want to do next.	Students should realize the importance of keeping accurate records, and organizing their information using tools such as the *Observing and Interpreting* page.

Targeted Concepts, Skills, and Nature of Science	Performance Expectations
Tables are an effective way to communicate results of a scientific investigation.	Students should describe how the *Observing and Interpreting Animal Behavior* pages helped them to organize their observations, interpretations, and the supporting reasons for their interpretations.
Behavior is a type of response to internal or external stimulus. Behavior is determined by experience, physical characteristics, and environment.	Students should support their interpretations based on some environmental factors and some physical characteristics that they observe in the image, or that they are aware of.
Scientists differentiate between observations and interpretations. They use their observations and interpretations to explain animal behavior.	Students should describe the difference between observations and interpretations, give examples verbally during discussion, and provide written evidence of this on their *Observations and Interpretations* page.
Scientists make claims (conclusions) based on evidence obtained (trends in data) from reliable investigations.	Students should begin to informally make claims (draw conclusions). These are their interpretations.

Materials	
2 per student; 1 additional per group	*Observing and Interpreting Animal Behavior* pages

Activity Setup and Preparation

You may need to regroup your class. At least two groups of students should observe each of the four images, so you should have at least eight groups of students, regardless of how many students are in each group.

Homework Options

Reflection

- **Science Process:** What are some ways that interpreting human behavior is harder than interpreting other animals' behavior? What are some ways that interpreting other animals' behavior is harder than interpreting human behavior? *(It is hard to look at human behavior objectively when we may already have learned incorrect explanations for human behaviors. Animal behaviors may be strange and surprising, or animal behaviors may resemble human behaviors that we understand, without actually serving the same purpose.)*
- **Science Process:** What are some ways you can tell if observations include interpretations? *(If a student's description of the behavior is very different from that of the other students, it may include interpretation. If the description includes things that aren't physical actions, it may include interpretation.)*

Preparation for 1.4

- **Science Process:** How were the reasons for your interpretations similar to others for the same picture? If you changed your mind about any of the reasons that you listed, why did you do so? *(Look to see if students are thinking about the differences between observation and interpretation.)*

NOTES

NOTES

SECTION 1.3 IMPLEMENTATION

1.3 Explore

Observing and Interpreting Animal Behavior

You have made some observations of animal behavior by watching your classmates. While you were making your observations, you might have thought about how difficult it was to make accurate observations of animals. Observing animals' behavior can be difficult. Many things happen in a short time. Not only do you have to watch the animals themselves, but you have to be aware of what is happening around them.

You may have had difficulty sticking with your group's plan. You may have noticed that when you discussed your observations with your group, you had each made different observations. All these experiences show the importance of designing good plans for making observations if you want to understand animal behavior. Ethologists work hard to make accurate observations and use them to support their interpretations of animal behavior.

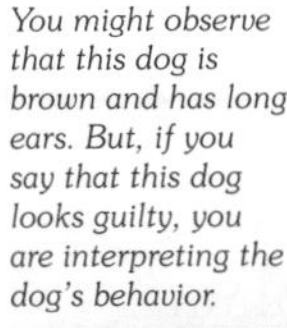

You might observe that this dog is brown and has long ears. But, if you say that this dog looks guilty, you are interpreting the dog's behavior.

In this section, you will apply the observation methods you used in the last section to observe some animal behavior. It would not be possible to bring a lot of animals to your classroom, so you are going to observe pictures of animals to determine what the animals are doing.

As you make your observations, be sure to look at all parts of the picture. Describe what the animal is doing in the picture. Pay attention to whether the animal is with others or alone. Take note of the animal's environment. Keep in mind the questions ethologists investigate, especially how the animal's environment affects its behavior.

◀ *2 class periods**

1.3 Explore

Observing and Interpreting Animal Behavior

5 min.

Prepare students to observe animals in pictures.

○ Engage

Use the example of the dog image in the student text to elicit students' description of what they see. Make two lists of their responses separating observations from interpretations, and have students consider the differences between the lists.

*A class period is considered to be one 40 to 50 minute class.

TEACHER TALK

"How would you describe the picture of the dog below? *(Students should make observations: furry, black nose, etc. and a variety of interpretations, sad, guilty, tired,... Create two lists separating observations from interpretations.)*

I've created two lists here of your responses. What are the differences between these lists? Today you will be focusing on both observations and interpretations."

Then, remind students of the question of the *Learning Set: How Do Biologists Study Animal Behavior?* and tie this in with the *Big Challenge,* to design a zoo enclosure that satisfies certain criteria and constraints. Let students know that they will also need to make good interpretations, and that they are also learning how biologists study behavior.

TEACHER TALK

"When you make observations and interpretations about animal behavior, you are being a biologist. Ethologists, biologists that study animal behavior, make many observations, and then try to interpret what these observations mean. You will need to know how biologists study animal behavior for designing your zoo enclosure. Remember that the zoo enclosure must allow the animal to live as closely as possible to its natural environment, and it must allow biologists to make observations to study the animal. To do this, you will need to learn a lot about the animal and its behaviors. You will need a good plan for observations, and you will also need to make good interpretations."

△ Guide

Introduce the activity by explaining that the students will be making observations of a picture of animal behavior. Let students know that they should use the observation methods they used in the previous section. Emphasize that they will need to look at all parts of the picture, consider whether the animal is alone or with others, consider the animals' environment, describe what the animal is doing, and consider how the animal's environment affects its behavior.

Observe

Pictures of Human Behavior

Begin this investigation by looking at one of the pictures of people below. Use the same techniques you learned in the last section. Pay attention to the details in the picture. What is happening around the people? What knowledge can you gain from the picture? On your own, write a description of the picture. Record at least two behaviors you observe in the picture.

Picture #1

Picture #2

Picture #3

Picture #4

Observe

5 min.

Have students observe animals in pictures.

Get Going

Begin by letting students know that each student should make their own observations of the pictures. They will share their observations later with their group.

Let students know how much time they have (no more than 5 minutes). They should independently write descriptions of their pictures, recording at least two behaviors.

Assign at least two groups to each of the pictures of human behavior in the student text. This will allow students to see that different groups are likely to have different ideas about what is depicted in the photographs.

☐ Assess

As students are working, become familiar with the types of interpretations and observations students are making. This will assist you in guiding students on the differences between observations and interpretations and about what makes accurate observations and interpretations in the following discussion.

NOTES

Conference

When you have completed your observations, discuss them with your group. Discuss what is happening in your picture and why you think the details you recorded are important. Could someone draw a picture based only on your description? If not, return to your picture and, with your group, write more descriptive observations. As you discuss your observations, work with your group to write descriptions on which you all agree. Make a list of these observations.

Communicate

Share your descriptions with the class. As you listen to one another's descriptions, notice where you agree and disagree. Think about how you might have described some of the pictures differently. Perhaps the class can identify some reasons why it is so hard to come to agreement about how to describe what you see.

interpretation: a description of the meaning of something.

Be a Scientist

Observation and Interpretation

When people say that animals are smiling, laughing, or talking to each other, they are giving animals human traits. Sometimes when people observe animal behavior, they describe the animal behavior as if the animals were people. Many people believe animals think and feel the way people do. It is natural for people to think animals have human feelings and emotions. They are interpreting the animals' behavior as though it is similar to human behavior. Thinking about animals as having human behaviors can become a problem when you study the behavior of animals scientifically. In a scientific study of animal behavior, it is important to observe what animals are actually doing. When ethologists record their observations, they work hard to separate observations and **interpretations**. They also make sure any interpretations they make are supported by many observations.

You make observations all the time. You notice that your best friend has a yellow shirt. You observe the teacher showing you how to do something. You look at a picture in a newspaper. You see the freckles on your arm. Descriptions of things you can see are observations. When you write accurate and detailed observations, someone else could almost draw a picture from your observations.

Conference

5 min.

Have groups discuss their observations and interpretations.

○ Get Going

Once students have recorded their observations, have them meet with their groups and discuss their observations with the goal of creating a list of observations and descriptions that the group agrees on. They should discuss which details are important and why, and they should determine if someone could draw a picture from the observations. If not, the group should refer back to the image and construct a more detailed observation. They should make a description with a list of observations all members agree on.

Assess

As students are discussing their observations, check to make sure that groups are discussing each group member's observations, and that they are discussing the reasons for any differences between observations.

Communicate

15 min.

Have groups share their observations with the class.

...ations.

Communicate

Share your descriptions with the class. As you listen to one another's descriptions, notice where you agree and disagree. Think about how you might have described some of the pictures differently. Perhaps the class can identify some reasons why it is so hard to come to agreement about how to describe what you see.

interpretation: a description of the meaning of something.

Be a Scientist

Observation and Interpretation

When people say that animals are smiling, laughing, or talking to each other, they are giving animals human traits. Sometimes when people observe animal behavior, they describe the animal behavior as if the animals were people. Many people believe animals think and feel the way people do. It is natural for people to think animals have human feelings and emotions. They are interpreting the animals' behavior as though it is similar to human behavior. Thinking about animals as having human behaviors can become a problem when you study the behavior of animals scientifically. In a scientific study of animal behavior, it is important to observe what animals are actually doing. When ethologists record their observations, they work hard to separate observations and **interpretations**. They also make sure any interpretations they make are supported by many observations.

You make observations all the time. You notice that your best friend has a yellow shirt. You observe the teacher showing you how to do something. You look at a picture in a newspaper. You see the freckles on your arm. Descriptions of things you can see are observations. When you write accurate and detailed observations, someone else could almost draw a picture from your observations.

META NOTES

It is important to let students experience the difference between observation and interpretation through discussions with each other of specific ideas about what an animal is doing. Let them realize that some of their observations do not really seem like observations before you introduce the concept of interpretations again.

Guide

Let students know that each group will be presenting their descriptions to the class.

Let the class know that, as each group presents, the audience should be focusing on how they might have described the pictures differently and why it is difficult to come to an agreement about how to describe what they see.

Students should have many disagreements about certain descriptions and should begin to notice that these are more from interpreted behaviors than observed behaviors. This should lead to a discussion of the difference between observation and interpretation.

When students have begun to discuss the fact that some of the descriptions are not really observations, then discuss the information in the *Observation and Interpretation* box with the class. You should emphasize the difference between observations (what you see, hear, taste, smell, and can measure) and interpretations (what you think your observations mean).

Ask students if they notice something about the interpretations they have made. Students should notice that it is very common to use human characteristics or qualities in our descriptions of animal behavior. Use the example of the image of the dog in the student text. Some students may have said that the dog is sad, guilty, lonely, etc. Point out that these are all interpretations and projections of human emotions.

Emphasize that, when ethologists record their observations, they work to separate observations and interpretations, and they make sure their interpretations are supported by many observations. Ask students what it means to support their interpretations with many observations and the importance of having other scientists have similar observations and interpretations. Students should realize this makes the interpretation more reliable.

META NOTES

One important part of interpretation is supporting your interpretation with reasons, but don't do this now with the class. Students will work with their groups, using the *Observing and Interpreting Animal Behavior* pages, to separate their interpretations from their observations and provide reasons for their interpretations.

NOTES

NOTES

Interpretations are different from observations. When you make an interpretation, you take what you see, add what you know from previous experience, and decide what is happening. Interpretations are essential in science. Interpretations can help scientists better understand animal behavior. They can then use this understanding to predict what similar animals might do or what an animal might do in a particular situation. Scientists share their interpretations with other scientists and check to see if other scientists agree with the way the interpretations have been made.

Analyze Your Data

Use *Observing and Interpreting Animal Behavior* pages as you analyze your data. It will help you separate observations from interpretations. The page will also provide you with space to record details about the animals you are observing and their environments. These other details may help you interpret and answer the question about how the animal's environment affects its behavior.

Using this new page, separate your original list of observations into two lists, one of observations and one of interpretations. Put observations in the first column and interpretations in the last. Remember that an interpretation provides the meaning of or reason for an action. In the picture of the dog at the beginning of this section, the interpretation might be that the dog is guilty of something because of how it looks. Look through your group's observations. See if any of the observations you wrote were actually statements that show meaning or describe reasons for behavior. If so, write these in the *Interpretations* column.

Observing and Interpreting Animal Behavior

Name: __________ Date: __________

Animal I am observing: __________

Observations	What about the environment and animal allows that behavior?	Interpretations

Analyze Your Data

10 min.

Have groups analyze their data, separating interpretations from observations.

△ Guide

Introduce the *Observing and Interpreting Animal Behavior* pages that help students make sense of their observations and interpretations. Describe how to use the pages.

- In the first column, students should write their observations — things they can see, smell, hear, measure, and they are confident everyone can agree on.

META NOTES

Students may not have much to list in the middle column right now. This column will become more important in the next two *Learning Sets*.

- In the middle column, students should list what they observe in the environment and observations about the animal that affect the behavior observed. These can be from common knowledge or personal experience.
- In the last column, students should record their interpretations. The interpretations should be supported by what students put in the middle column.

Have groups discuss observations and interpretations and separate their interpretations from their observations, using the *Observing and Interpreting Animal Behavior* pages.

Tell students to first record their observations and interpretations (in the first and third columns) and then record the factors that led to their interpretations. As the group discusses their ideas, they should revise the information in these columns.

△ Guide and Assess

As groups work on analyzing their observations, go around the room and get a sense of whether students are identifying interpretations. Emphasize that an interpretation will describe an emotion or a look or a reason for doing something, whereas an observation will simply describe something you can see in the picture.

NOTES

You can pick out interpretations because they do not describe the picture, but instead describe an emotion, a look, or a reason why a person or animal is doing what they are doing.

Use the middle column to record information about why you think your interpretations are good ones. If you have made observations about the environment in the picture, list those in this column. If you know something about parents, teams, children, or anything else that helped you interpret, that goes into the middle column, too.

You will use the *Observing and Interpreting Animal Behavior* page many times in this Unit. It will help you track what you are observing, interpreting, and learning about animal behavior. The *Big Challenge* of this Unit is to develop a new enclosure for an animal using what you learn about the animal's behavior. The *Observing and Interpreting Animal Behavior* pages you create throughout this Unit will be helpful in keeping your observations and interpretations organized. You will then use this information to support your design for the final challenge.

Stop and Think

1. How difficult was it for you to support your interpretations of people in the picture you looked at? What knowledge did you use for those interpretations?
2. When you discussed one another's observations, what was confusing? What were your disagreements about the observations? About the interpretations? Why do you think these happened?

These baby birds are waiting for food.

Stop and Think

10 min.

Have a class discussion on students' experience categorizing observations and interpretations and constructing reasons for their interpretations.

○ Get Going

Once groups have finished analyzing their data, have them answer the *Stop and Think* questions.

△ Guide and Assess

Lead a class discussion. Begin with groups sharing with the class their interpretations, along with the reasons behind their interpretations, and their responses to the two questions. Listen for the following:

1. Students should cite facts or ideas that everyone is familiar with or experiences they have had. They should say how these facts, ideas, or experiences led them to interpret their observations the way they did. For example, students might interpret the boy in *Picture #4* to have found a lost dog, but there is nothing to support this interpretation. One could interpret the boy to be happy because he is smiling.
2. Students should recognize that interpretations lead to more disagreements than observations, and they usually occur based on one's knowledge and experiences. Students should begin to understand that interpretations are more open to discussion because they rely on personal experiences.

META NOTES

Make sure the class recognizes when a group has misidentified an interpretation as an observation or vice versa. Students should now largely agree on observations of the pictures.

NOTES

△ Guide

Ask students why it was important to record their observations and their analysis. Review with students the *Keeping Records* box in the student text.

Be a Scientist

Keeping Records

It is very important that scientists carefully record their work. To record means to write, sketch, or diagram what is being done. This allows scientists to accurately report their findings to others and helps them answer scientific questions.

You have probably kept records of your work in science class before. Keeping records is also important for you as a student scientist. Recording your work helps you:

- share your work with others,
- remember what you did and decided along the way,
- remember what you saw and the environment surrounding your observations, and
- answer scientific questions.

Throughout this Unit, you will be collecting a lot of observational data. Because of the amount of data you will collect, you will need to keep it well organized. Scientists frequently use tables to record and organize their data. Tables help scientists keep track of what they are seeing. The information in the table may include the length of time a behavior happened, the surroundings of the animals, and perhaps even some interpretations.

□ Assess

Listen for some of the bulleted ideas about recording work from the *Keeping Records* box.

Observe

5 min.

Have groups use their observational plan to observe animals in pictures.

Be a Scientist

Keeping Records

It is very important that scientists record their work carefully. To record means to write, sketch, or diagram what is being done. This allows scientists to accurately report their findings to others and helps them answer scientific questions.

You have probably kept records of your work in science class before. Keeping records is also important for you as a student scientist. Recording your work helps you:

- share your work with others,
- remember what you did and decided along the way,
- remember what you saw and the environment surrounding your observations, and
- answer scientific questions.

Throughout this Unit, you will be collecting a lot of observational data. Because of the amount of data you will collect, you will need to keep it well organized. Scientists frequently use tables to record and organize their data. Tables help scientists keep track of what they are seeing. The information in the table may include length of time a behavior happened, the surroundings of the animals, and perhaps even some interpretations.

Observe

Pictures of Other Animal Behavior

Now you will use what you have learned about making good observations to observe and interpret what is happening in pictures of other animals. Again, you will look carefully at one picture.

Recording Your Data

Describe the picture using good descriptive words. Record as many details as you can. Record your descriptions of each behavior on the *Observing and Interpreting Animal Behavior* page. This time, be aware of the difference between observation and interpretation. Use the correct column on the *Observing and Interpreting Animal Behavior* page to show which of your comments about the picture are observations and which are interpretations.

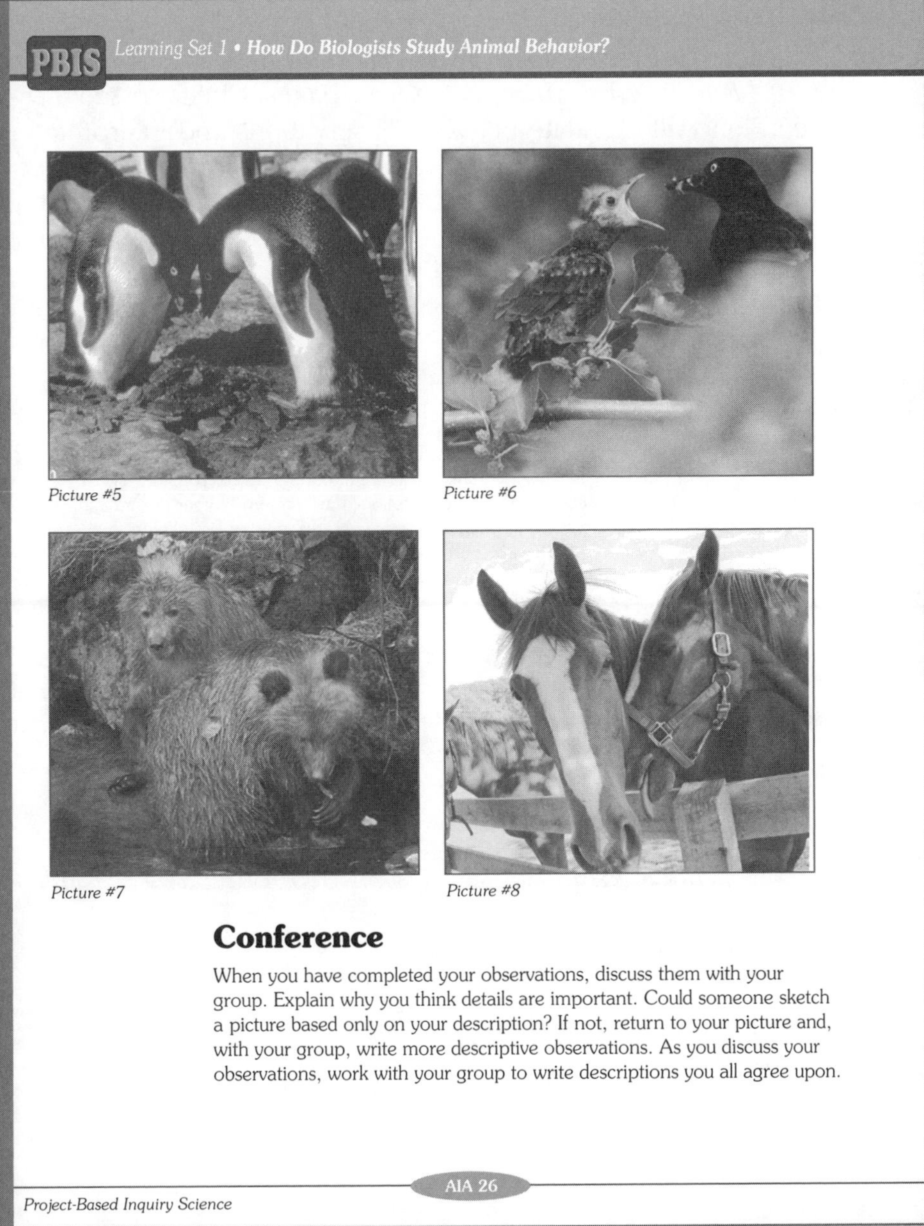

Picture #5

Picture #6

Picture #7

Picture #8

Conference

When you have completed your observations, discuss them with your group. Explain why you think details are important. Could someone sketch a picture based only on your description? If not, return to your picture and, with your group, write more descriptive observations. As you discuss your observations, work with your group to write descriptions you all agree upon.

⬡ Get Going

Next, assign two groups to each of the animal images in the student text. Have students use the *Observing and Interpreting Animal Behavior* pages to record their observations and interpretations. Let students know how much time they have. This should not take more than five minutes.

△ Guide and Assess

As students are making their observations, monitor their use of the *Observing and Interpreting Animal Behavior* pages. Check to see if students are able to differentiate between observations and interpretations and that they are using the correct column for each of these categories. Be aware of the observations students might be making that belong in the middle column. The middle column will provide evidence about the animal or the environment that supports their interpretation.

Conference

10 min.

Have groups discuss their observations, interpretations, and reasoning.

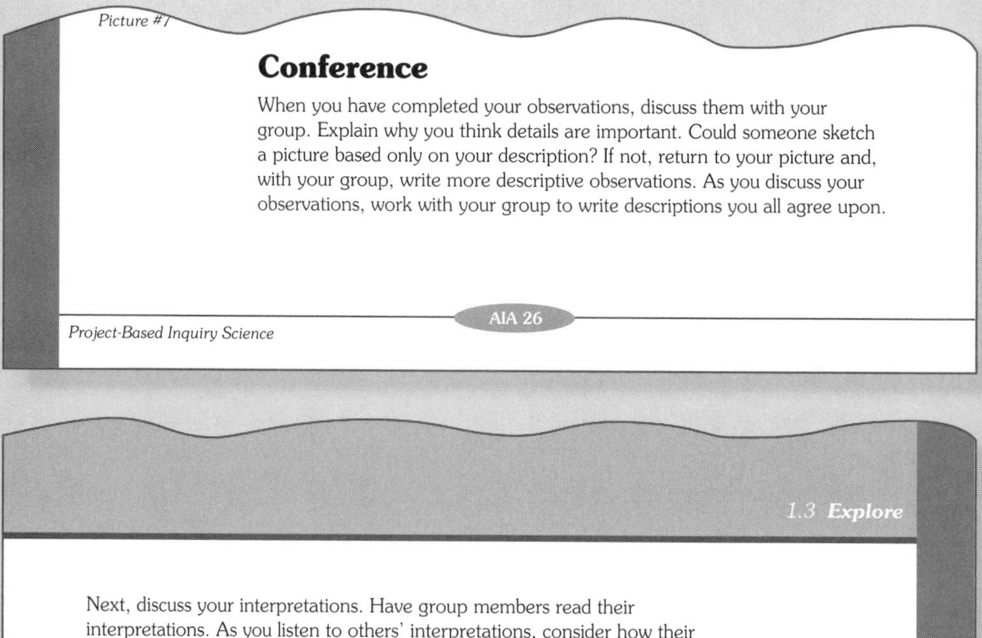
Picture #7

Conference

When you have completed your observations, discuss them with your group. Explain why you think details are important. Could someone sketch a picture based only on your description? If not, return to your picture and, with your group, write more descriptive observations. As you discuss your observations, work with your group to write descriptions you all agree upon.

Project-Based Inquiry Science AIA 26

1.3 Explore

Next, discuss your interpretations. Have group members read their interpretations. As you listen to others' interpretations, consider how their ideas might be different from yours. Together with your group, determine why the interpretations are similar or different.

△ Guide

Students will meet with their groups, will discuss their data details, and will identify why details are important. They should also discuss their interpretations and their reasons for their interpretations. As a group, they should create another *Observing and Interpreting Animal Behavior* page that everyone agrees on. Students should add the group's ideas and may clarify their ideas on their own *Observing and Interpreting Animal Behavior* page with colored pencil.

□ Assess

As students are discussing their interpretations, check to make sure groups are discussing each group member's interpretations and that they are discussing the reasons for any different interpretations group members had.

NOTES

Communicate

15 min.

Have a discussion on groups' presentations of observing and interpreting animal behavior.

Next, discuss your interpretations. Have group members read their interpretations. As you listen to others' interpretations, consider how their ideas might be different from yours. Together with your group, determine why the interpretations are similar or different.

Communicate

You will now share your observations and interpretations with the class. As you look at your classmates' pictures, think about whether you would have made the same interpretations. If you have a different interpretation or you do not understand a group's interpretation, you should ask questions to help you understand. Then discuss these questions.

1. You observed and interpreted two pictures, one of a human and one of another animal. Which of the two pictures did you find easier to observe? In which picture was it easier to interpret the behavior? Why do you think this is?
2. How did using the *Observing and Interpreting Animal Behavior* page make it easier to record your observations?
3. What is the difference between observation and interpretation? You might use a new example of both to show the difference.
4. Why is it important for scientists to think about the difference between observation and interpretation?
5. Why do you think it is sometimes easy to confuse observation and interpretation?

What's the Point?

Ethologists observe animal behavior. They watch and record what animals do. During their observations, they collect a lot of data and often use a table to keep track of the data. By keeping good records, ethologists can re-create what they have seen and use that information to determine why animals behave as they do.

Interpretation is different from observation. Interpretation includes the meaning or significance of an action. It is difficult, when making observations, to eliminate all interpretation. However, scientists try to separate their interpretations from their observations. Sharing ideas with others can also make separating observations and interpretations easier.

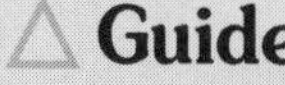

△ Guide

Let students know that each group will be presenting their group descriptions to the class. As each group shares their observations and interpretations, students should listen carefully and think about whether they agree or disagree. This time, when there are disagreements, students should use the factors that led to interpretations to support their arguments.

△ Guide and Assess

As the groups are presenting, lead students to understand the differences between observation and interpretation and what makes a good interpretation. Encourage students to describe differences they have in their observations and interpretations.

1. Students can select either picture, but they must support their answer with evidence from their experience. Pictures of humans may be more difficult to observe because it is more difficult not to interpret them; pictures of other animals may be more difficult to interpret because we know far less about their behaviors, and we may identify them with human behaviors.
2. Using an organizational tool helps make observations and interpretations easier to record. Students may find it is easier to know when something is an interpretation and when it's an observation if they think about the factors that led to their interpretations.
3. Observation should describe exactly what you see when you watch the animal. Interpretation should offer reasons why an animal might be doing something, including what the animal feels, what it's trying to do, and how it looks. Students should use examples from the image they observed. Examples may be the following:
   - Observations of *Picture #5* would be that the penguins are standing near each other and touching heads. Interpretations would be that they are greeting each other, fighting, or carrying out mating rituals. It would require more observations to decide which of these interpretations is correct.
   - Observations of *Picture #6* are that a young bird has its mouth open, the young bird is near the adult bird, and the adult bird has something in its closed beak. Interpretations would be that the adult bird is going to feed the young bird, and the adult bird is going to show the young bird how it eats. Because most people know adult birds feed their chicks, most students will agree that the feeding interpretation is correct.
   - In *Picture #7,* observations would be that two bears are at a river; one is leaning over looking into the water, and the other is sitting behind the first. Interpretations of this image would be that the bears are trying to catch fish, grooming themselves, or trying to cool off on a warm day. More observations are needed to determine if any of these interpretations are correct.

- Observations of *Picture #8* would be that three horses are near a wood fence. Two horses are wearing bridles. One horse has its tongue out and is touching the fence. Interpretations of this picture are that a horse is eating something on the fence, the horses are related, the horses are on a ranch, two of the horses have just been ridden or will be ridden soon. More observations are needed to determine which, if any, interpretations are correct. Some students may note that bridles are used for riding horses.

4. Scientists need to pay attention to the differences between observation and interpretation, because when they have moved beyond observation, they are interpreting what they see. When scientists make interpretations, they are using their background experiences and knowledge to draw conclusions about what is happening. Interpretations may change as new knowledge becomes available, or may be debatable or incorrect, but reliable observations can be used with confidence. Scientists need to know that interpretations may be different, depending on who is making the interpretation.

5. Students should recognize it is human nature to interpret behaviors, and we do it often without being aware of it.

What's the Point?

5 min.

Have groups discuss their observations, interpretations, and reasoning.

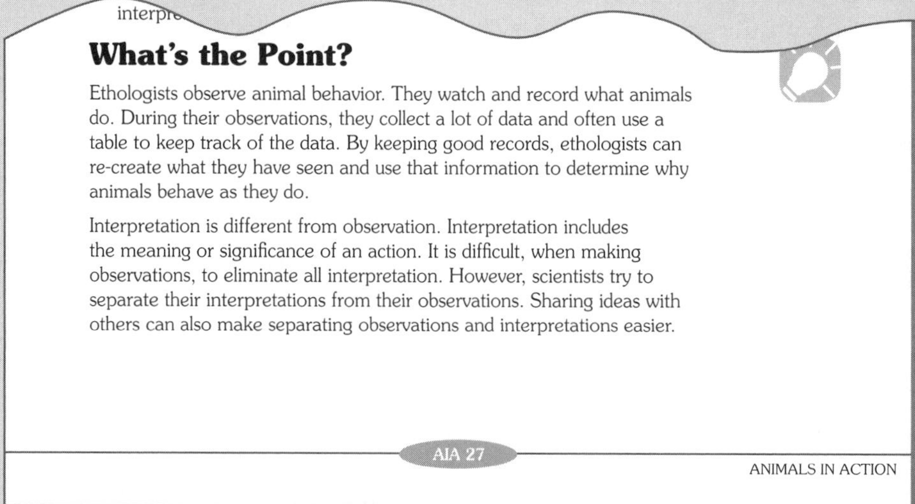

interpr
What's the Point?
Ethologists observe animal behavior. They watch and record what animals do. During their observations, they collect a lot of data and often use a table to keep track of the data. By keeping good records, ethologists can re-create what they have seen and use that information to determine why animals behave as they do.
Interpretation is different from observation. Interpretation includes the meaning or significance of an action. It is difficult, when making observations, to eliminate all interpretation. However, scientists try to separate their interpretations from their observations. Sharing ideas with others can also make separating observations and interpretations easier.
AIA 27
ANIMALS IN ACTION

◇ Evaluate

Make sure students understand the difference between observations and interpretations, the need for supporting interpretations, and the importance of good record-keeping.

Assessment Options

Targeted Concepts, Skills, and Nature of Science	How do I know if students got it?
Behavior is a type of response to internal or external stimulus. Behavior is determined by experience, physical characteristics, and environment.	**ASK:** What did you use to support your interpretations? What did you put in the middle column of your *Observing and Interpreting Animal Behavior* page? **LISTEN:** Students should have listed in the middle column of their *Observing and Interpreting Animal Behavior* page some environmental factors and some physical characteristics they observed or they are aware of. **NOTE:** It is unlikely for students to have many things listed at this time.
Scientists must keep clear, accurate, and descriptive records of what they do, so they can share their work with others, consider what they did and why they did it, and what they want to do next.	**ASK:** Why is it important to keep good records? **LISTEN:** Students should describe the importance of keeping accurate records, organizing their information, and using tools such as the *Observing and Interpreting Animal Behavior* pages for finding trends in their data and assisting them in discussing their observations and interpretations with others.
Scientists often work together and share their findings. Sharing findings makes new information available and helps scientists refine their ideas and build on others' ideas. When another person's or group's idea is used, credit needs to be given.	**ASK:** How has working in a group and with the class helped you make better observations and interpretations? **LISTEN:** Students should be able to describe how sharing their observations and interpretations with their group members and with the class has provided them with ideas on improving their own observations and interpretations.

Targeted Concepts, Skills, and Nature of Science	How do I know if students got it?
Tables are an effective way to communicate the results of a scientific investigation.	**ASK:** What was useful about the *Observing and Interpreting Animal Behavior* page and why? **LISTEN:** Students should be able to describe how the *Observing and Interpreting Animal Behavior* page helped them organize their observations, interpretations, and supporting reasons for their interpretations.
Scientists differentiate between observations and interpretations. They use their observations and interpretations to explain animal behavior.	**ASK:** What is the difference between observations and interpretations? Provide an example. **LISTEN:** Students should be able to describe the difference between observations and interpretations and give examples from their daily lives or from the images from which they made observations.

Teacher Reflection Questions

- How did you guide students when they had difficulty separating observations from interpretations? What would you do next time?
- How are students progressing with leading the class discussions rather than looking for the teacher to direct discussions?
- How were you able to keep discussions of disagreements constructive? What ideas do you have for next time?

SECTION 1.4 INTRODUCTION

1.4 Explain

Support Your Interpretation

◀ *2 class periods**

*A class period is considered to be one 40 to 50 minute class.

Overview

Students are introduced to the development of a scientific explanation. They construct explanations using what they have learned from their observations, interpretations, and factors that led to their interpretations. They make claims based on their interpretations and connect their claims to their observations (evidence) and science knowledge, using logical statements based on what they know. Students revise their claims if further evidence and science knowledge do not support them. They share their explanations with the class, and students consider whether all of the claims are valid and supported by the evidence and science knowledge included in the explanations.

Targeted Concepts, Skills, and Nature of Science	Performance Expectations
Scientists often work together and then share their findings. Sharing findings makes new information available and helps scientists refine their ideas and build on others' ideas. When another person's or group's idea is used, credit needs to be given.	Students should describe how their ideas are changing.
Tables are an effective way to communicate the results of a scientific investigation.	Students should describe how the *Create Your Explanation* page helped them organize their claims, evidence, and science knowledge, and construct explanations.
Scientific knowledge is developed through observations, recording, analysis of data, and development of explanations based on evidence.	Students should describe that science knowledge is based on repeated investigations that experts agree upon.
Scientists make claims (conclusions) based on evidence obtained (trends in data) from reliable investigations.	Students should make claims based on their observations.

Targeted Concepts, Skills, and Nature of Science	Performance Expectations
Explanations are claims supported by evidence. Evidence can be experimental results, observational data, and other accepted scientific knowledge.	Students should create an explanation for one of their interpretations of animal behavior, based on what they also observe about the environment, the structure of the animal, and scientific knowledge.
Behavior is a type of response to internal or external stimulus. Behavior is determined by experience, physical characteristics, and environment.	Students should use environmental (external) factors and characteristics of the animal (internal) in supporting their interpretations.

Materials	
1 per student	*Create Your Explanations* page

Homework Options

Reflection

- **Science Process:** What is a claim, and what is the supporting evidence in the following explanation? The orb spider weaves a web because the web allows it to capture prey, which it needs to survive. Orb spiders have repeatedly been observed capturing prey in their webs, and it has been shown that they are unable to eat anything but the small insects they capture in their webs. *(Claim: The orb spider weaves a web because the web allows it to capture prey. Evidence: Orb spiders have repeatedly been observed capturing prey in their webs, and they are unable to eat anything but small insects they capture in their webs.)*
- **Science Process:** Write an explanation for an animal behavior you observe outside the classroom. *(Look for claims supported by evidence and science knowledge. They should be logical and should contain no opinions.)*

Preparation for 1.5

- **Science Process:** What do you think animals need to survive? Provide examples by describing three different animals' needs. *(The purpose of this question is to prepare students for the next section and to provide you with students' initial ideas.)*

SECTION 1.4 IMPLEMENTATION

1.4 Explain

Support Your Interpretation

You have made accurate observations and learned how to separate observations from interpretations. When you were making observations of the animals in the pictures, you learned that it is difficult not to interpret behaviors as you are describing them. During your discussions with your team and the class, you were asked why you made your interpretations. They were asking you to support your interpretations with evidence.

explanation: a statement that connects a claim to evidence and science knowledge.

Now you will more formally connect your observations and interpretations to each other. You will write **explanations**. An explanation is a statement that connects a claim (a conclusion you have come to) with your evidence (the data you have collected) and science knowledge. The explanations you write will connect your observations and what you know about animals to your interpretations.

Be a Scientist

What Do Explanations Look Like?

Making claims and creating explanations are important parts of what scientists do. An explanation is made up of three parts:

Claim—a statement of what you understand or a conclusion that you reach from an investigation or set of investigations.

Evidence—data collected during investigations and observations.

Science knowledge—knowledge about how things work. You may have learned this through reading, talking to an expert, discussion, or other experiences.

An explanation is a statement that connects the claim to the evidence and science knowledge in a logical way. A good explanation can convince someone the claim is valid.

For example, suppose you have a pet hamster. You notice that when you remove the lid to the cage to feed the hamster, it runs toward the food cup. You learned in science class that some animals learn different behaviors. Dogs who are fed after a bell rings eventually respond to just

◀ *2 class periods**

1.4 Explain

Support Your Interpretation

30 min.

Introduce scientific explanations, which students will be constructing throughout the PBIS *curriculum.*

META NOTES

Constructing scientific explanations is an integral part of *PBIS*. It is important to lay down a good foundation during this section. Students are not expected to master constructing an explanation by the end of this section, but they should have a strong enough idea of its components to build on.

○ Engage

Begin by asking students what they think an explanation is, and have them record their ideas.

Then remind students that, in the previous section, they made observations and interpretations, and they supported their interpretations with evidence. Then tell them that now they will learn how to construct explanations for these interpretations. Emphasize how important it is to learn how to construct scientific explanations.

*A class period is considered to be one 40 to 50 minute class.

TEACHER TALK

"You've made observations of humans, particularly middle-school students, and other animals. You made interpretations that you supported with factors in the environment or physical characteristics. Today you will be writing explanations in which you will support your interpretations with your observations and any science knowledge you have."

△ Guide

Consider using as an example one of the images the students observed in the previous section. Then describe a scientific explanation by telling students that an explanation is a statement that connects claims (interpretations or conclusions from an investigation) to evidence (the data you have collected, or observations) and to science knowledge (accepted and verified ideas in the scientific community). The evidence and science knowledge must support the claim in a logical way.

TEACHER TALK

"Consider if you tried to explain the picture of the boy holding his clenched fists to his eyes. You would be able to use the observations that his fists are clenched and his mouth is open, but you don't know whether he's screaming. That might be a claim you could make and support based on other observations, but it isn't something you've observed. You might have other knowledge about children that could lead you to explain that the child is screaming. If so, you would need to provide that information to back up the statement that he is screaming.

When scientists make an explanation, they are answering a question about a situation or phenomenon. They make a claim, and they must support this claim, in a way that makes sense, with the evidence from their observations and with science knowledge. It is okay not to have complete or perfect explanations if you do not have complete or perfect understanding. Just like scientists, you can explain things based only on what you know. And, just as scientists, as you gather more information, you should revise your explanation."

Then, discuss claims, evidence, and science knowledge. Emphasize that a claim is a statement or conclusion reached from one or more investigations. Evidence is the data from the investigations, and science knowledge is knowledge about how things work, based on what experts have previously investigated and agreed upon.

Be a Scientist

What Do Explanations Look Like?

Making claims and creating explanations are important parts of what scientists do. An explanation is made up of three parts:

Claim—a statement of what you understand, or a conclusion that you reach, from an investigation or set of investigations.

Evidence—data collected during investigations and observations.

Science knowledge—knowledge about how things work. You may have learned this through reading, talking to an expert, discussion, or other experiences.

An explanation is a statement that connects the claim to the evidence and science knowledge in a logical way. A good explanation can convince someone the claim is valid.

For example, suppose you have a pet hamster. You notice that when you remove the lid to the cage to feed the hamster, it runs toward the food cup. You learned in science class that some animals learn different behaviors. Dogs who are fed after a bell rings eventually respond to just the bell. They begin to drool when they hear the bell, whether or not food is offered to them. You wonder if the hamster might have learned that the noise of the lid signals it is about to be fed. The next time you feed your hamster, you try to remove the lid very quietly and notice your hamster does not move toward the cup. You conclude that your hamster has learned that, when the lid makes noise, the food cup will be filled, and it is time to eat. You can now form an explanation.

Your **claim:** My hamster moves toward the food cup because of the noise of the lid.

Your **evidence:** I performed an experiment. My experiment showed that, when there is no noise, the hamster does not move. The hamster moves toward the food cup when I lift the lid normally (with noise). When I lift the lid quietly, the hamster does not move toward the food cup.

Your **science knowledge:** Some animals, like dogs, can learn different behaviors.

Your **explanation:** My hamster moves toward the food cup because of the noise of the lid. When it hears the noise of the lid, my hamster moves toward the food cup. This is because my hamster has learned that, when it hears me lifting the lid, I am going to feed it.

An explanation is what makes a claim different from an opinion. When you create an explanation, you use evidence and science knowledge to back up your claim. Then people know your claim is not simply something you think. It is something you have spent time investigating. You have found out things that show why your claim is likely to be correct.

Use the example explanation of why a hamster runs toward a food cup when you lift the lid to the cage to point out how each component (claim, evidence, and science knowledge) is connected in a logical way. Be sure to emphasize that explanations should contain the reason why, and they should not contain opinions.

NOTE: The explanation in the student text should contain the following two sentences at the end:

> This is also supported by another investigation in which a dog was trained to eat after hearing a bell. The dog began to salivate every time i heard the bell, even when food was not offered.

Then, provide students with the following example and have them pick out the claim, the evidence, and the science knowledge.

> Most middle-school students are happy interacting during lunchtime. I observed large numbers of middle-school students smiling, talking, and eating during lunch at school. Students smiling, talking, and eating are usually signs they are happy. The school is required to provide time for students to eat. The school has a large area where students gather durin the allotted time to eat.

Help students realize the claim that is the interpretation, the evidence that is backing it up, and the knowledge (in this case, it is not all science knowledge) that supports it.

> **Claim (interpretation):** Most middle-school students are happy interacting during lunch time.
>
> **Evidence (observations):** I observed large numbers of middle-school students smiling, talking, and eating during lunch at school. The school has a large area where students gather during the allotted time.
>
> **Knowledge:** Students smiling, talking, and eating are usually signs that they are happy (scientific knowledge). The school is required to provide time for students to eat (other knowledge).

Remind students that they can only construct explanations using science knowledge they are aware of. If they do not have enough information, then their explanation may be incomplete. Tell students that, at this point, a scientist would look for more information, and then have them revise their explanation based on the new information they obtain.

You may want to provide them with an example of a good and bad explanation.

> **BAD:** Most middle-school students have a binder because they need to organize their papers. I have observed students carrying binders in the hallways and bringing binders into the classrooms. Binders organize you papers. Most teachers at our school require a binder.

META NOTES

Our understanding of human behavior can be used to support claims about human behavior. It is important to be careful when interpreting animal behavior by using knowledge about human behavior. Other animals don't have the same goals and capabilities as humans, nor do they engage in the same activities as humans. Students may anthropomorphize (attribute human characteristics to animal behavior observed), as we all do. Sometimes describing animal behavior in human terms helps us understand animals better. When we do this, we realize how similar and dissimilar humans can be from other animals. If students begin to use their knowledge of humans to describe other animals, they should be asked to support their claims with additional knowledge. Usually, when pressed for more information, students realize that the anthropomorphized behavior is less human-like. Don't belabor the point about anthropomorphism, but help students recognize when they are making explanations with wrong assumptions.

What's wrong with this example? Binders do not organize your papers. This statement does not make sense, and it does not support the claim in a logical way.

GOOD: Most middle-school students have a binder because they need a place to keep their papers. I have observed students carrying binders in the hallways and bringing binders into the classrooms. Binders are designed for placing papers inside. Most teachers at our school require a binder.

What's wrong with this example? Nothing. The claim is supported by evidence (observations), knowledge of binders, and middle-school experience.

NOTES

Explain

10 min.

Have students construct explanations of the behaviors they observed.

the bell. They begin to drool when they hear the bell, whether or not food is offered to them. You wonder if the hamster might have learned that the noise of the lid signals it is about to be fed. The next time you feed your hamster, you try to remove the lid very quietly and notice your hamster does not move toward the cup. You conclude that your hamster has learned that, when the lid makes noise, the food cup will be filled, and it is time to eat. You can now form an explanation.

Your **claim:** My hamster moves toward the food cup because of the noise of the lid.

Your **evidence:** I performed an experiment. My experiment showed that when there is no noise, the hamster does not move. The hamster moves toward the food cup when I lift the lid normally (with noise). When I lift the lid quietly, the hamster does not move toward the food cup.

Your **science knowledge:** Some animals, like dogs, can learn different behaviors.

Your **explanation:** My hamster moves toward the food cup because of the noise of the lid. When it hears the noise of the lid, my hamster moves toward the food cup. This is because my hamster has learned that when it hears me lifting the lid, I am going to feed it.

An explanation is what makes a claim different from an opinion. When you create an explanation, you use evidence and science knowledge to back up your claim. Then people know your claim is not simply something you think. It is something you have spent time investigating. You have found out things that show why your claim is likely to be correct.

Explain

Now that you know more about what an explanation is, you are going to write an explanation of the animal behavior you just observed. Use a *Create Your Explanation* page to help you make sure your explanation takes into account your claim, evidence, and science knowledge. Your interpretation of the animal's behavior is your claim. Your observations of behavior and what you recorded about your animal's structure and environment are your evidence. You may have some science knowledge from your own experiences or from readings. Record all of these in the appropriate boxes.

○ Get Going

Next, have students work in groups to construct their own explanations for the behavior of the animal they observed, using their claims, data, and science knowledge. Students should use their *Create Your Explanation* pages.

Emphasize that an explanation should be convincing. The claim should be something they can back up with their observations. These observations should include what they see in the environment and characteristics of the animal. Students should also support their claims with science knowledge

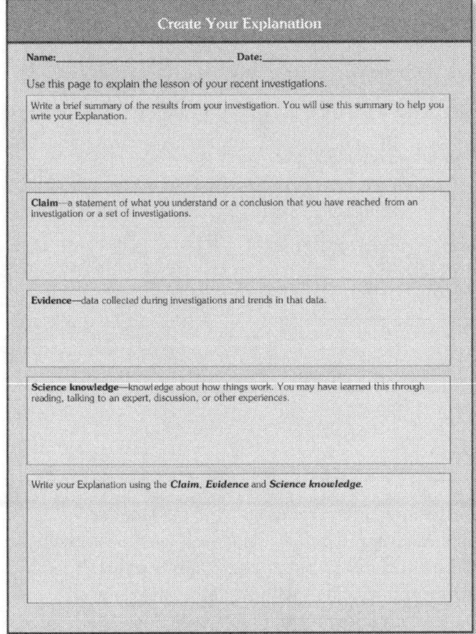

Create Your Explanation

Name:__________ Date:__________

Use this page to explain the lesson of your recent investigations.

Write a brief summary of the results from your investigation. You will use this summary to help you write your Explanation.

Claim—a statement of what you understand or a conclusion that you have reached from an investigation or a set of investigations.

Evidence—data collected during investigations and trends in that data.

Science knowledge—knowledge about how things work. You may have learned this through reading, talking to an expert, discussion, or other experiences.

Write your Explanation using the ***Claim***, ***Evidence*** and ***Science knowledge***.

Then write a statement using your evidence and science knowledge to support your claim. This is your explanation. A good explanation can convince someone else that your interpretation is good. If your statement doesn't seem convincing, revise your claim so your evidence and the science you know will support it. You can use the hamster example to know what to put in each part of your explanation.

Because your understanding of the picture you observed may not be complete, you may not be able to fully explain the animal's behavior. But use what you have read and what you know to develop your best explanation. Scientists finding out about new things do the same thing. When they only partly understand something, it is impossible for them to form a "perfect" explanation. They do the best they can based on what they understand. As they learn more, they make more accurate or clearer explanations.

This is what you will do now and what you will be doing throughout PBIS. You will explain your results the best you can based on what you know. Then, after you learn more, you will make your explanations more accurate.

Communicate

Share Your Explanation

Now you will share your observations, interpretations, and explanations with the class. Your classmates will listen carefully. They will want to know what your claim is. They will be checking to see if your claim is **valid**. You will have to make sure you support your claim with accurate and detailed observations and appropriate science knowledge. If you have done that, then your claim can be considered valid.

valid: well-grounded or justifiable.

of which they are aware. If students have trouble making a convincing explanation, they may need to revise their claims.

Let groups know they will be presenting their explanations to the class.

△ Guide and Assess

As groups are constructing their explanations, assist individual students and groups as needed. If you notice the majority of the class having difficulty in constructing a particular part of an explanation, stop the class and hold a class discussion.

Check if students have included opinions, and remind students that opinions are not to be included in an explanation. Remind students that claims (conclusions or interpretations) supported by evidence and science knowledge are allowed. Ask them if there is a way they could support their opinion by providing evidence or science knowledge.

Check that students are making logical connections by using phrases such as "this follows because..." Illogical factors used to support an interpretation or a claim should not be used to connect claims to evidence.

Encourage students to discuss their ideas, and require each one to support their claims with evidence. Also encourage students to discuss the validity and reliability of the evidence they use.

Communicate

30 min.

Have groups share their explanations with the class.

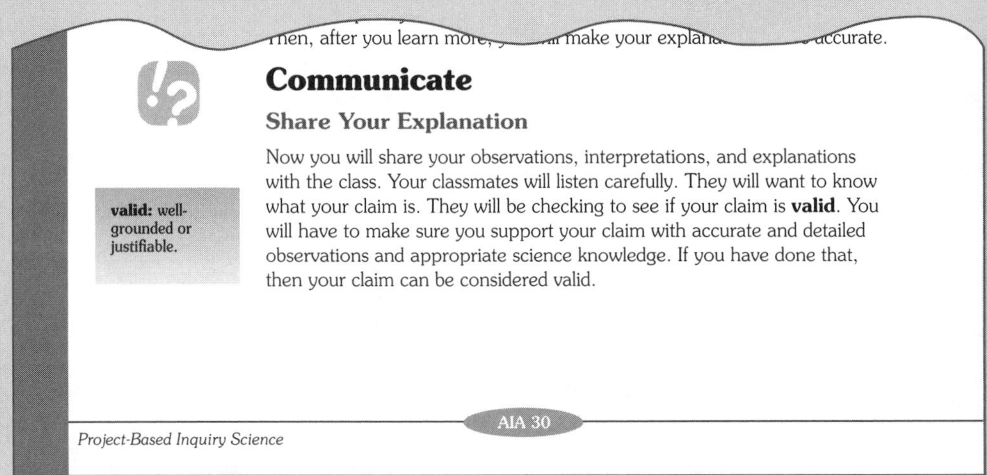

Then, after you learn more, you will make your explanation more accurate.

Communicate

Share Your Explanation

Now you will share your observations, interpretations, and explanations with the class. Your classmates will listen carefully. They will want to know what your claim is. They will be checking to see if your claim is **valid**. You will have to make sure you support your claim with accurate and detailed observations and appropriate science knowledge. If you have done that, then your claim can be considered valid.

valid: well-grounded or justifiable.

Project-Based Inquiry Science AIA 30

△ Guide

Let the class know that, while they are in the audience (not presenting), they should be looking for the parts of an explanation—valid claim(s), backed by observations and science knowledge in a logical way with no opinions. Highlight the term *valid claim*. Explain that scientists say a claim is valid if the following requirements are met:

- It is not an opinion.
- There is repeatable evidence supporting it.
- There is no evidence against it.
- It is logical.

Describe for students that, in this case, their claims are their interpretations.

TEACHER TALK

"Your interpretations are your claims. To be valid, they have to be backed up by multiple observations, including any you can find from the environment and the characteristics of the animal (such as physical features). There should not be any observations that go directly against your interpretation, and it should not be an opinion."

Then, model for the students how they should ask for clarification during the presentations.

TEACHER TALK

"I don't understand how the science knowledge backs up your claim. Could you walk me through it?"

△ Guide and Assess

Next, begin the presentations. After each presentation, have a discussion about the explanation presented. Ask students to pick out the claim, the evidence, and the science knowledge making up the explanation. Help groups clarify parts of their explanations. If the claim, evidence, and science knowledge is not clear, ask them if they can restate it in another way. Help students make their explanations clearer if needed.

During the discussion, assess students' understanding of claim, evidence, and science knowledge by observing how well they can determine each of these from the explanations presented.

Have the class discuss if the explanation is good by asking if the claim is valid, if the evidence supports it, if there is any evidence against it, if there are any opinions, and if the science knowledge supports it. If there are opinions in the explanation, point out that these should not be included, and let students know that everyone has opinions, but scientists work hard to keep their opinions out of their scientific explanations. Then, ask if anyone knows of any science knowledge that goes against the claim. If so, discuss this and the possibility of verifying this information by looking it up in accepted resource books, such as encyclopedias.

Check if the evidence and scientific knowledge support the claim in a way that makes sense, and check if students see this connection. Assist them with making these connections. Rewrite explanations as a group if the explanation is unclear, illogical, or unconvincing. Help students see how changing a few words can really make a difference in how their explanation reads.

Reflect

10 min.

Guide students to answer and discuss the Reflect *questions.*

Reflect

Explanations are critical for scientific understanding. Scientists in all science fields are trying to explain the way the world works. They develop explanations to help them inform other scientists of what they have learned from their observations.

1. After scientists make their observations and interpretations, they write explanations of what they have seen. Creating explanations can be difficult. What difficulties did you have developing an explanation of what you saw?
2. What questions do you still have about how explanations can be made?
3. One source of science knowledge might be your own experiments. List three more sources you might use for science knowledge.

What's the Point?

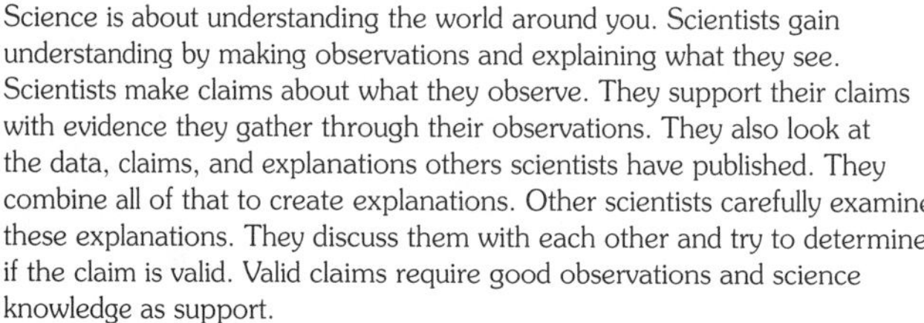

Science is about understanding the world around you. Scientists gain understanding by making observations and explaining what they see. Scientists make claims about what they observe. They support their claims with evidence they gather through their observations. They also look at the data, claims, and explanations others scientists have published. They combine all of that to create explanations. Other scientists carefully examine these explanations. They discuss them with each other and try to determine if the claim is valid. Valid claims require good observations and science knowledge as support.

Throughout PBIS, you will create explanations. Every explanation will include a claim, evidence, and science knowledge. Just like scientists, you will edit and improve your explanations.

To help others better understand what they learn, scientists must communicate their results effectively. When scientists share what they have learned, they allow others to question it and improve on the claims and explanations. By working together, scientists can develop a clearer understanding of the world.

⬡ Get Going

Students should first answer the questions on their own. Then they should discuss them with their group and develop their best group response. A class discussion will follow. Let the class know how much time they have. They should take no more than 5 minutes to answer the questions.

△ Guide

Lead a discussion of the *Reflect* questions. Explain to students that these questions are intended to help them think about the process of making explanations.

NOTE: Question number 3 is incorrect. A student's observations or experiments are called evidence in *PBIS,* not science knowledge. Science knowledge is knowledge about how things work. Students may have learned this through reading, talking to an expert, discussion, or other experiences. In general, science knowledge is knowledge accepted within the scientific community.

□ Assess

Listen for any remaining confusion about observations (evidence) and science knowledge (accepted ideas in the scientific community), interpretations (or claims), and making explanations. At this point, students should be able to determine claims, evidence, and science knowledge from a written explanation, but they may not yet be able to construct good explanations.

◇ Evaluate

At this point, students should understand the difference between observations and interpretations, and they should be able to support their claims with reliable observational data.

Students will have multiple opportunities to create explanations in *PBIS.* When categorizing their observations, interpretations, and explanations, students may still have difficulties with the exact language and may struggle with what to put where. It is important they see the need to support their explanation with evidence. Understanding of this categorization can be refined as the Unit and school year progress.

META NOTES

Be aware that students may consider something they heard an important adult figure saying as science knowledge. For example, someone may have told them the sky is blue because of reflections from the ocean, or that dinosaurs roamed Earth with humans. These are common misconceptions and are not scientifically accepted ideas. It is best to have students verify ideas they believe to be science knowledge by researching them.

What's the Point?

5 min.

Students will be writing explanations throughout PBIS, *because writing good explanations is a key part of doing science. It is important for them to understand the parts of an explanation and to be able to construct one. Writing explanations is not an easy task. They will have a chance to revise their explanations during the next section.*

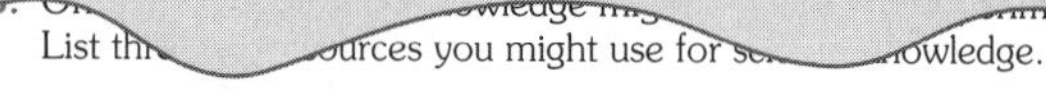

What's the Point?

Science is about understanding the world around you. Scientists gain understanding by making observations and explaining what they see. Scientists make claims about what they observe. They support their claims with evidence they gather through their observations. They also look at the data, claims, and explanations others scientists have published. They combine all of that to create explanations. Other scientists carefully examine these explanations. They discuss them with each other and try to determine if the claim is valid. Valid claims require good observations and science knowledge as support.

Throughout PBIS, you will create explanations. Every explanation will include a claim, evidence, and science knowledge. Just like scientists, you will edit and improve your explanations.

To help others better understand what they learn, scientists must communicate their results effectively. When scientists share what they have learned, they allow others to question it and improve on the claims and explanations. By working together, scientists can develop a clearer understanding of the world.

Assessment Options

Targeted Concepts, Skills, and Nature of Science	How do I know if students got it?
Scientists often work together and then share their findings. Sharing findings makes new information available, helps scientists refine their ideas, and helps them build on others' ideas. When another person or group's idea is used, credit needs to be given.	**ASK:** How did sharing your interpretations with the class help you make valid claims? **LISTEN:** Students should recognize that discussing your claims and explanations can be a helpful way of determining whether or not you can support a claim.
Scientists make claims (conclusions) based on evidence obtained (trends in data) from reliable investigations.	**ASK:** How did you make reliable claims based on your observations? **LISTEN:** Students should describe how they made a number of the same observations, or that a number of observations showed a trend in the animal's behavior.

Targeted Concepts, Skills, and Nature of Science	How do I know if students got it?
Explanations are claims supported by evidence and science knowledge in a logical way.	**ASK:** What could you use, other than observations, to support your claims? **LISTEN:** Claims can be supported by observations students have made about animal behavior, from nature programs, or from readings. Remind students that it is always good to verify information. Books and nature programs can make mistakes.
Behavior is a kind of response to internal or external stimulus. Behavior is determined by experience, physical characteristics, and environment.	**ASK:** What observations did you use, that come from the environment and from the animal, that support your interpretations? **LISTEN:** Students describe environmental (external) factors and characteristics of the animal (internal) in supporting their interpretations.

Teacher Reflection Questions

- What difficulties did students have understanding scientific explanation? How could you pinpoint the areas of difficulty and create opportunities for students to better understand?
- How was the group work prior to the presentations and class discussion helpful?
- How were you able to keep the class discussions student-centered and not a dialogue between you and individual students?

NOTES

SECTION 1.5 INTRODUCTION

1.5 Read

What Do Animals Need to Survive?

◀ ***1 class period****

*A class period is considered to be one 40 to 50 minute class.

Overview

Students read and discuss scientific writings about animal behavior. This will help students understand scientists' observations that an animal's need to survive leads the animal to behave in certain ways. Students then revise their explanations of the behaviors they observed in pictures of animals to incorporate this new information. Next, students present their revisions to the class, and they should realize their revised explanations are more convincing. From this, they learn that valid claims agree with and are supported by available science knowledge, as well as detailed observations.

Targeted Concepts, Skills, and Nature of Science	Performance Expectations
Scientists often work together and then share their findings. Sharing findings makes new information available, helps scientists refine their ideas, and helps them build on others' ideas.	Students should realize that sharing ideas allows them to refine their ideas. They can build on others' ideas by incorporating science knowledge into their new explanations.
Scientists must keep clear, accurate, and descriptive records of what they do so they can share their work with others, and consider what they did, why they did it, and what they want to do next.	Students should use their previous records and should be able to refer to the information and evidence from those records.
Scientific knowledge is developed through observations, recording and analysis of data, and development of explanations based on evidence.	Students should be able to add science knowledge from the section to support their previous claims or their revised claims.
Explanations are claims supported by evidence. Evidence can be experimental results, observational data, and other accepted scientific knowledge.	In order to support their claims, students should revise their explanations to include the accepted scientific knowledge introduced in this section.

Targeted Concepts, Skills, and Nature of Science	Performance Expectations
Behavior is a type of response to internal or external stimulus. Behavior is determined by experience, physical characteristics, and environment.	Students should use environmental (external) factors and characteristics of the animal (internal) in supporting their interpretations.
The structure and function of animals' bodies are complementary and affect animal behavior.	Students should specifically use observable physical characteristics in their evidence of animal behavior and in science knowledge where it pertains. They should then use these to support their claims.
Organisms need food to grow, reproduce, and maintain their bodies. They must be able to live in a changing environment.	Students should describe an animal's need for food and how this drives many of the animal's behaviors. This should be incorporated into their explanation.

Materials	
1 per student	*Create Your Explanation* pages

Homework Options

Reflection

- **Science Content:** What part of the environment are birds responding to when they roost? *(Birds are responding to the danger of enemies in the environment.)*
- **Science Content:** What are some reasons that some species may be threatened when their natural environments change? Think, for instance, of how a severe drought or the appearance of a new species might change an environment. *(Animals have adaptations that help them respond to their natural environment. They can gather or capture and digest certain foods found in their natural environment, and they can protect themselves from dangers in their natural environment. They may not be able to find the foods they need in an altered environment or be able to protect themselves from new dangers.)*

- **Science Process:** What did you learn that was most useful for revising your explanations? *(This question should get students thinking about what kinds of information are most useful and what makes that information useful.)*
- **Science Process:** What questions do you still have about the behavior of the animals you observed? Do you think you could still improve your explanations if you could get more information? *(This question should get students thinking about what they still do not know and about what would be most useful to know.)*

Preparation for 1.6

- **Science Process:** Now that you have had more practice with observing and explaining, how would you revise your explanation of the behaviors you observed in middle-school students? What have you learned about explanations that you could use? *(Students should have developed a better sense of what claims are valid, and they should be better able to support their claims.)*

NOTES

NOTES

SECTION 1.5 IMPLEMENTATION

1.5 Read

What Do Animals Need to Survive?

behavior: an animal's response to its environment.

adaptation: a special trait that allows an animal to survive in its environment.

species: a group of organisms that look alike and can breed with other members of the group and produce fertile offspring.

instinctive behavior: a behavior an animal is born with.

learned behavior: a behavior that comes from teaching or experience.

mammal: a warm-blooded animal with hair and in which the female has special glands to feed milk to its offspring

In this *Learning Set*, you have been exploring how scientists study animal behavior. One of the questions you are answering in this Unit is *Why do animals behave the way they do?* To answer that question, you need to understand what a **behavior** is from a scientific point of view. Scientists use the word behavior to describe an animal's response to its environment. One of the most important animal behaviors is survival.

Animals share certain needs for survival. These needs affect how they behave. They must obtain enough food for energy and to feed their young. They need protection from bad weather and enemies, and they need a safe place to raise a family. By using behaviors and other **adaptations**, or traits, to meet these basic needs, not only can an animal survive, but its family may survive, its group may survive, and its entire **species** may survive.

Lions live, travel, and hunt in groups called prides. This allows them to hunt cooperatively to catch bigger prey more often and protect each other.

Kinds of Behavior

A behavior can be **instinctive** or **learned**. Animals are born with instinctive behaviors. Learned behaviors come from teaching or experience. Most behaviors are a mixture of instinctive and learned. For example, **mammal** mothers have the instinct to care for their young, but some then learn from experience how to better care for their young. Whether a behavior is instinctive or learned, the most important thing an animal must do in its environment is survive.

◂ *1 class period**

1.5 Read

What Do Animals Need to Survive?

Guide students' reading through the science of animal behavior.

△ Guide

Begin by letting students know they will now learn about the basic needs of animals and about some of the behaviors that help them meet these needs. This will help them explain the behaviors they observe. Point out that scientists use information from the scientific community, as well as from their own experiments. Afterward, they will revise their explanations of the animal behaviors they observed in pictures.

*A class period is considered to be one 40 to 50 minute class.

TEACHER TALK

"You are going to learn more about animal behaviors from other scientists' observations and experiments that have been accepted in the science community. You should also think about how biologists study animal behavior. Biologists learn from observations and experiments, but they also use knowledge from the scientific community. There is a lot of information available from observations and experiments other scientists have done. You'll have a chance later to revise the explanations you made."

Before you discuss specific behaviors with the class, explain that scientists use the word *behavior* in a very specific way, to describe an animal's response to its environment. The most important thing for any animal to do in any environment is to survive. Emphasize that knowing what an animal needs for its survival helps explain the behaviors of the animal.

Also, explain that scientists talk about some behaviors as *instinctive* (behaviors an animal is born with) and some as *learned* (behaviors an animal gets from teaching or experience).

NOTES

Searching for Food

Animals need energy and materials to stay alive, grow, develop, and reproduce. They obtain the energy and the nutrients they need from the food they eat. The process by which animals break down their food to obtain energy and then use the energy is called **metabolism**. Finding enough of the right kind of food depends on many things: how and where animals live, how they find and gather or catch food, and how they digest that food.

An animal must live in a place that has the type of food it needs and it must have ways to gather or catch the amount of food it requires. Some animals have claws and big teeth; others are able to run fast. Animals might live in cooperative groups and help one another find food. Other animals live alone.

An animal must be able to digest the food it eats. Even if an animal finds a lot of food that has a lot of energy, it cannot use the energy if it cannot digest the food. Energy in food is measured in units called **calories**. A calorie is the amount of energy needed to raise the temperature of one gram of water by 1°C. Most foods contain thousands of calories of energy. For this reason, scientists use the **Calorie**, with a capital *C*, to measure the energy in foods. One Calorie is the same as 1 kilocalorie or 1000 calories.

Some animals eat only meat, and some eat only vegetable matter. Some eat both. All have different ways of getting their food and digesting it. You will learn more about feeding behaviors and adaptations in *Learning Set 2*.

metabolism: the combination of chemical reactions through which an organism builds up or breaks down materials converting energy to carry out its life processes.

calorie: the amount of energy needed to raise the temperature of one gram of water by 1°C.

Calorie: the amount of energy in foods. One Calorie is the same as 1 kilocalorie or 1000 calories.

roosts: communal resting places, mostly for a single species of birds.

Protection

In addition to finding food, animals must remain safe—from their enemies and from their external environment. Remaining safe from enemies is a primary concern for many animals. Organisms have developed different behaviors to keep them safe. Birds may form **roosts** that can number in the thousands. With many birds together in one place, many eyes are on the lookout, and each bird is safer than if it were sitting on a branch all alone. Some mammals that live in herds form a protective ring around their young if they are threatened by predators. Others like turtles retreat into their hard shells when they sense danger.

Some birds form roosts in trees at night for protection from enemies.

Searching for Food

5 min.

Have a discussion of how the need to find food affects animals' behavior.

META NOTES

The point of this section is not to learn the behaviors of a lot of different animals but to learn some general reasons animals have different behaviors and apply this knowledge to explaining the behaviors students have observed. You can use specific animal behaviors to illustrate the reasons for different behaviors, but you don't need to go into any depth.

△ Guide

Begin by emphasizing that all animals need food to produce energy in order to stay alive and that this is the most important need they share. You can use a few examples to illustrate the fact that the animals in any environment will be adapted to find food: in the sea, fish eat algae and plankton or other fish; in the plains, animals eat grass; and in the polar regions, penguins eat fish, and polar bears eat seals and other animals.

Next, discuss how behaviors or adaptations allow animals to gather or capture the type and amount of food they need in the environment where they live.

Let students know that animals may have hunting behaviors that allow them to capture the food they need, and they may have special adaptations that allow them to chew tough plant matter or to kill prey. Emphasize to students that behaviors are responses to an animal's environment, and adaptations are special traits that allow the animal to survive in its environment.

Ask students if they know of some animals that have special behaviors for getting food. Record their responses.

TEACHER TALK

"What are some different ways animals get food? *[Take time to discuss specific examples.]* Consider an animal's body. Tigers' bodies are good for hunting and eating raw food. Tigers can run fast, and they have claws and sharp teeth. Speed, claws, and teeth are all adaptations that tigers have.

Cows don't run fast, or have sharp claws or sharp teeth. They have teeth designed to chew grass and a digestive system that is specialized for grass. Cows have a four-chambered stomach to break down the grass and get the nutrients from it."

Let students know that these special physical characteristics, such as a tiger's claws and teeth and a cow's teeth and stomach, are also called adaptations that allow the animal to survive in its environment. Then, explain that animals must be adapted to digest the food they eat. They need to digest food to get the energy from it.

Next, describe the example of lions in the student text, and discuss the difference between learned behavior (from teaching or experience) and instinctive behavior (behaviors the animal is born with). For example, you can describe how mammal mothers have the instinct to care for their young.

learn more ... behaviors and adap... ...ning Set ...

Protection

In addition to finding food, animals must remain safe—from their enemies and from their external environment. Remaining safe from enemies is a primary concern for many animals. Organisms have developed different behaviors to keep them safe. Birds may form **roosts** that can number in the thousands. With many birds together in one place, many eyes are on the lookout, and each bird is safer than if it were sitting on a branch all alone. Some mammals that live in herds form a protective ring around their young if they are threatened by predators. Others like turtles retreat into their hard shells when they sense danger.

Some birds form roosts in trees at night for protection from enemies.

AIA 33 ANIMALS IN ACTION

PBIS ...w Do Biolog... ...havior?

Many animals living in cold climates, such as these polar bears, have fur, fat, and short limbs to conserve body heat.

Some animals, such as this golden retriever, pant to cool off their bodies.

But even if they are safe from enemies, animals might not be safe from their environment. Animals need a way to keep their internal temperature and fluids fairly constant. This process is called **homeostasis**. Some animals that live in hot areas have a special circulatory system that moves cooler blood across the brain, so they will not die from heat stroke while running. Or they may sweat or pant to cool off their bodies. Animals may shiver if they get too cold. The muscle action used to shiver produces heat that warms their bodies.

...ducti...

Protection

5 min.

Lead a discussion of how the need for protection affects animal behavior.

△ Guide

Discuss the fact that animals have adaptations to protect themselves and their young from enemies and from nature. Many animals have complex behaviors that help protect them and their young from enemies. The student text provides the example of roosting birds. These birds nest close to each other so that each has many neighbors on the lookout for danger. Ask students what examples of protective behaviors they know of, and record their responses. Some examples may include fleeing, being still, playing dead.

TEACHER TALK

"What are some basic ways animals protect themselves from other animals? What does a mouse do when it sees a cat or a person approaching?"

Then, introduce the need for animals to protect themselves from the climate and weather. Ask students for examples of animals that live in extremely cold environments. Ask what adaptations to the cold these animals have. *(Examples: penguins have round bodies and short limbs to avoid losing heat; seals and other sea mammals have thick layers of fat to keep in heat; polar bears have thick layers of fat and thick coats of fur to keep in heat. Some common heat adaptations are sweating, panting, or burrowing underground and sleeping through the heat of the day.)*

NOTES

Many animals living in cold climates, such as these polar bears, have fur, fat, and short limbs to conserve body heat.

Some animals, such as this golden retriever, pant to cool off their bodies.

But even if they are safe from enemies, animals might not be safe from their environment. Animals need a way to keep their internal temperature and fluids fairly constant. This process is called **homeostasis**. Some animals that live in hot areas have a special circulatory system that moves cooler blood across the brain, so they will not die from heat stroke while running. Or they may sweat or pant to cool off their bodies. Animals may shiver if they get too cold. The muscle action used to shiver produces heat that warms their bodies.

Reproduction

If animals have enough food and are safe, they can use their energy for reproducing and raising their young. Animals raise offspring in one of two ways. Either they feed and protect the young until they are old enough to go out on their own, or the mother leaves the young to manage for themselves.

One characteristic of mammals is that the mother raises the young on mother's milk. She produces the milk in her **mammary glands** and continues to feed her young until they are old enough to eat on their own. A mother must obtain enough energy from her food to meet her own needs and to produce milk for her young.

In many birds, both the mother and the father help feed and protect the young, which are called **hatchlings**. In some birds, the parents gather food for the young, process it in their own stomachs with **enzymes**, and then **regurgitate** it to feed to the young. These parents continue to feed their offspring until the young birds can gather food on their own.

homeostasis: the maintenance of stable internal conditions in an organism.

mammary glands: milk-producing glands found in female mammals that are used to feed the young.

Reproduction

5 min.

Lead a discussion of the need to reproduce.

△ Guide

Discuss how animals reproduce and rear offspring once they have met the need for food and the need for protection.

Other animals, like fish and turtles, lay many eggs in the water or on land, and then leave. When the eggs hatch, the young must take care of themselves.

Why do animals use different strategies for raising their young? In animals that rely on learned behaviors for survival, it may take a long time for the young to learn how to find food, how to find shelter, and how to protect themselves. In this case, the mother or both parents stay with and protect the young while they teach them what they need to know. In animals that rely mostly on instinctive behaviors, the young are born with the instincts they need to survive and do not need their parents.

hatchling: a very young baby bird.

enzymes: organic substances that cause chemical changes in other substances.

regurgitate: to bring partially digested food up from the stomach into the mouth.

A female turtle lays eggs, but she does not care for her eggs or the hatchlings. The hatchlings are born with the instintive behavior they need to survive. A female spider dies in the autumn, before the spiderlings (baby spiders) hatch the following spring. However, the spiderlings have the instinct to build a perfect web on their first attempt.

Many birds, such as these penguins, take care of their young and feed them regurgitated food.

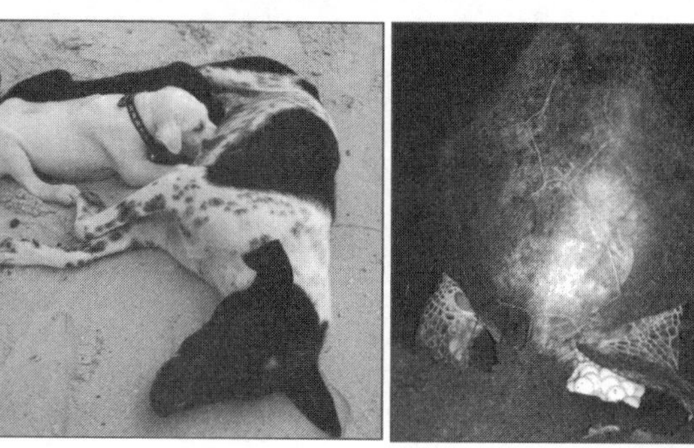

Mammals produce milk in mammary glands to feed their young (left).

Some animals, such as this sea turtle, lay many eggs but leave the young to manage for themselves (right).

Among those animals that care for their young, there are different ways to feed their young. Mammals have mammary glands, which produce milk for their young. Many birds process food with enzymes in their stomach before regurgitating it for their young.

Stop and Think

1. Some large animals, such as lions, live and hunt in groups. What are some of the advantages of hunting with others?
2. Living in a group can help animals with protection. Describe two ways that living in a group can protect individuals.
3. Some animals have only a few young at a time. Others have many offspring, even thousands. How do the numbers of offspring connect with how the animals care for their young?

Revise Your Explanation

You have just read about how an animal's behavior and adaptations help it survive in its environment and carry out its basic needs. With your group, look at the explanation of animal behavior you created on your *Create Your Explanation* pages after interpreting the pictures of animals. Now that you know more about the science of animal behavior, you can probably revise your explanation based on some of this science knowledge.

Review and rewrite your explanation based on your new science knowledge. First, check to make sure your claim is accurate. You may have just read information that shows that your claim was inaccurate. If your claim does not match the science you have read, revise it. Next, support your claim with the science knowledge you just learned.

Then, rewrite your explanation to make it more complete. Remember that an explanation is a statement that connects a claim to evidence and science knowledge in a logical way. Try to write your explanation so that it tells why your claim is true. Be sure that your explanation matches the science you just read.

Communicate

Share Your Explanation

Share your new explanation with the class. When you share your explanation, tell the class what makes this revised explanation more accurate than your earlier one. As each group shares their explanation, pay special attention to how the other groups have supported their claim with science knowledge. Ask questions or make suggestions if you think a group's claim is not as accurate as it could be, or if the group has not supported their claim well enough with observations and science knowledge.

Stop and Think

15 min.

Have a class discussion on students' responses.

○ Get Going

Let students know that they should first answer the questions on their own. Then, they should discuss their answers with their group and should record their best group answer. If group members cannot agree on an answer, then they should list all their answers. Tell students that groups will be sharing their answers during a class discussion. Let students know how much time they have to answer the questions.

△ Guide

Once groups have answered the questions, begin a class discussion of how some of the animal behaviors students read about and discussed help the animals to survive. Ask groups for their answers to the *Stop and Think* questions. Listen for the following responses:

1. Students' responses should include that animals, like lions, are able to catch large prey more often when they hunt cooperatively than when they hunt alone.
2. Students' responses may include the example of roosting birds. Roosting birds are protected by having a lot of neighbors looking out for danger. A large group of animals can more effectively protect the young than can an individual. For example, a herd often forms a protective circle around their young.
3. Students' responses should include that most animals with few young rely on learned skills for survival, while animals with many offspring usually rely on instinct. It is important for animals with few young to make sure that as many as possible survive to adulthood. These animals will usually spend a lot of time caring for their young and helping them learn survival behaviors. Those animals that have many offspring can rely on a few to survive and, in turn, those offspring will reproduce without their protection. These animals will usually leave their young to fend for themselves, using their instinctual behaviors.

Revise Your Explanation

10 min.

Guide students to revise their explanations and present their revisions to the class.

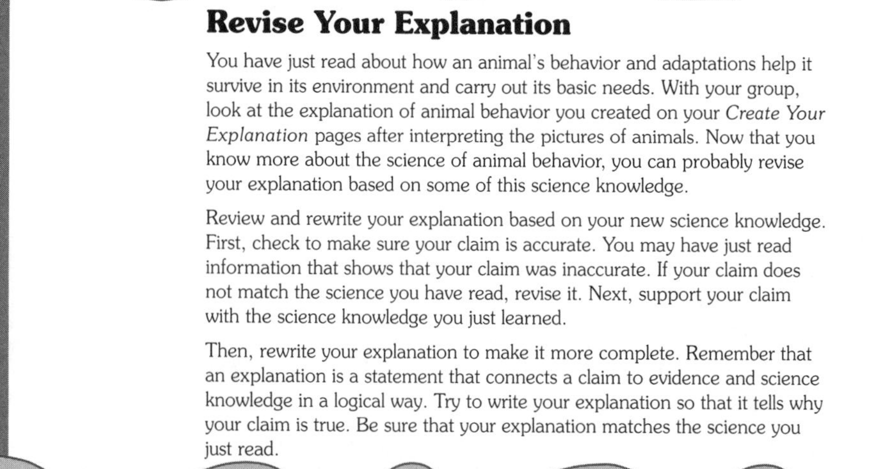

Revise Your Explanation

You have just read about how an animal's behavior and adaptations help it survive in its environment and carry out its basic needs. With your group, look at the explanation of animal behavior you created on your *Create Your Explanation* pages after interpreting the pictures of animals. Now that you know more about the science of animal behavior, you can probably revise your explanation based on some of this science knowledge.

Review and rewrite your explanation based on your new science knowledge. First, check to make sure your claim is accurate. You may have just read information that shows that your claim was inaccurate. If your claim does not match the science you have read, revise it. Next, support your claim with the science knowledge you just learned.

Then, rewrite your explanation to make it more complete. Remember that an explanation is a statement that connects a claim to evidence and science knowledge in a logical way. Try to write your explanation so that it tells why your claim is true. Be sure that your explanation matches the science you just read.

⬡ Get Going

Allow students time to review their explanation of animal behavior from the animal pictures in *Section 1.3*. Distribute new *Create Your Explanation* pages. Have students revise their explanations by providing more support for their claims from the new science content or by modifying them to make them more consistent with the new science content.

TEACHER TALK

"You have read about and discussed animal behaviors that you may not have known before you wrote your explanations. When you look at them now, you may see that information from your reading might be helpful in supporting your explanation.

First, check to make sure your claim fits with the new science knowledge. If your claim is inconsistent with what you learned, then you need to revise it.

Then, put any science knowledge you can use in the science knowledge box, and rewrite your explanation to make it more complete."

◇ Evaluate

As groups are editing their explanations, monitor students' progress. Check their explanations to make sure they are using the science knowledge they just learned to support their claims. If it looks like their claims are inconsistent with what they just learned, note this as something to discuss when the class discusses groups' presentations.

Students should now be able to construct explanations with valid claims, in which the evidence supports the claims. If groups are still using unfounded claims or using opinions for evidence, stop the class and review what they should know about constructing explanations.

Communicate

10 min.

Have groups present their explanations, and lead students in a discussion of the explanations.

...ead.

Communicate

Share Your Explanation

Share your new explanation with the class. When you share your explanation, tell the class what makes this revised explanation more accurate than your earlier one. As each group shares their explanation, pay special attention to how the other groups have supported their claim with science knowledge. Ask questions or make suggestions if you think a group's claim is not as accurate as it could be, or if the group has not supported their claim well enough with observations and science knowledge.

Project-Based Inquiry Science AIA 36

⬡ Get Going

When groups have finished revising their explanations, have each group present theirs to the class. While a group is presenting, the rest of the class should be taking notes on their own *Create Your Explanation* pages.

△ Guide

Encourage students to check that the explanations presented are composed of one or more claims supported by evidence and science knowledge in a logical way. Model for students how they should seek clarification or point out areas they may not understand.

META NOTES

If students ask you questions about other groups' explanations, redirect the question to the group presenting. In general, encourage students to discuss their explanations with each other and with the presenting groups.

TEACHER TALK

"I don't see how that fact backs up your claim. Could you clarify that for me?

I'm not sure I understand your claim. Could you describe it to me again?

Could you clarify the evidence and science knowledge that supports your claim?

Could you clarify how the evidence and science knowledge supports your claim?"

:ncourage students to point out where they think the group could have ısed some of the science knowledge they just learned to support their :laims.

◇ Evaluate

ʌ̃ake sure students incorporate information from the reading into their ·xplanations.

NOTES

Reflect

10 min.

Lead a class discussion of how students improved their explanations by including science knowledge.

Reflect

Explanation is an important scientific practice. Scientists use what they already know along with new evidence collected from investigations to explain how the world works. You will be doing a lot of scientific explanation in PBIS. It will get easier as the year goes on. For now, think about the differences between the first animal behavior explanations your class wrote and the new explanations. Identifying what makes the earlier explanations different from the newer ones will help you get better at explaining. Answer the following questions and be prepared to discuss them in class.

1. What are you able to explain now about the animal behaviors in the pictures that you were not able to explain well earlier?
2. What makes your revised explanations better than the earlier explanations?
3. In order to make a complete explanation, science claims need to be supported by evidence. What are some sources of evidence you might use to support your claims?

Be a Scientist

Good Explanations Tell Us How and Why

A good explanation uses what scientists know about how things work. The best scientific explanations use agreed-upon science knowledge in a logical way to support a claim. These kinds of explanations can usually convince others that a claim is valid. And with more science knowledge, you can write better explanations.

What's the Point?

You have learned that animals have many different behaviors and adaptations to help them survive in their environments. These behaviors can be instinctive or learned. Animals gather or catch food to obtain energy. They use other behaviors and adaptations to remain safe from enemies and to protect themselves against extreme weather. When energy gathering is efficient and animals are safe, they can reproduce. Animals care for their young in many ways, from total care and protection to laying eggs in water or on the ground and leaving the young to manage for themselves. Survival and success in animals depends on a wide variety of behaviors and adaptations.

⬡ Get Going

Let groups know how much time they have (about five minutes) to answer the questions, and be prepared for a class discussion.

☐ Assess

While groups are answering the questions, monitor their work and check their understanding of the questions. Decide how you will focus the discussion based on students' responses. If students seem to understand these questions, then lead a brief discussion.

△ Guide

Begin by discussing how groups improved their explanations using new science knowledge. Then, discuss how science knowledge is used in general.

Listen for the following in students' responses to the *Reflect* questions:

1. Students' responses should include at least some of the science knowledge about behaviors, adaptations, the search for food, protection, and reproduction presented in this section. The science knowledge they use should support their claims (interpretations) to improve their explanations.

2. Students should realize that the new explanations should support the claims better than the previous explanations, because there should be more science knowledge supporting their claim. Students may also have articulated their claims better than the previous time.

3. Students' responses should include sources of evidence, such as those from observations, and measurements from investigations. You might emphasize that science knowledge stems from many investigations on which scientists agree.

Emphasize what a good explanation is, using the *Be a Scientist* text box in the student text.

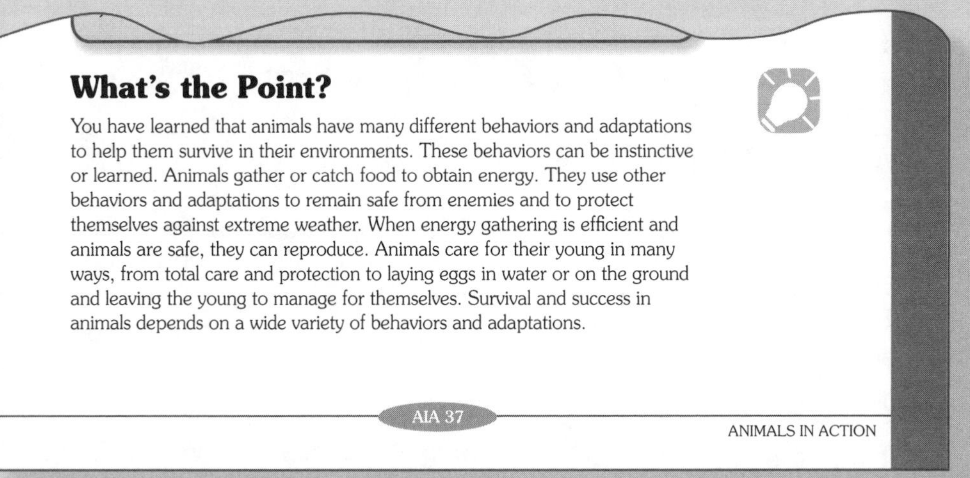

What's the Point?

You have learned that animals have many different behaviors and adaptations to help them survive in their environments. These behaviors can be instinctive or learned. Animals gather or catch food to obtain energy. They use other behaviors and adaptations to remain safe from enemies and to protect themselves against extreme weather. When energy gathering is efficient and animals are safe, they can reproduce. Animals care for their young in many ways, from total care and protection to laying eggs in water or on the ground and leaving the young to manage for themselves. Survival and success in animals depends on a wide variety of behaviors and adaptations.

AIA 37

ANIMALS IN ACTION

What's the Point?

5 min.

◇ Evaluate

Students should be able to describe how different behaviors and adaptations help animals survive in their environments.

Assessment Options

Targeted Concepts, Skills, and Nature of Science	How do I know if students got it?
Explanations are claims supported by evidence. Evidence can be experimental results, observational data, and other accepted scientific knowledge.	**ASK:** As it becomes available, how can you incorporate new scientific knowledge into explanations? **LISTEN:** Students should describe how new scientific knowledge can be used, as evidence either to support the claims in your explanations, or to modify them and support the revised claims.
Behavior is a kind of response to internal or external stimulus. Behavior is determined by experience, physical characteristics, and environment.	**ASK:** What did you use to support your interpretations that arose from the environment and those that arose from the animal's characteristics? **LISTEN:** Students should describe environmental (external) factors that lead to behaviors and characteristics of the animal (internal) as adaptations in supporting their interpretations.
The structure and function of animals' bodies are complementary and affect animal behavior.	**ASK:** How do the animal's physical characteristics affect its behaviors? **LISTEN:** Students should specifically describe observable physical characteristics and how these characteristics affect the animal's behavior. An example could be how tigers' claws and teeth help their feeding behavior.
Organisms need food to grow, reproduce, and maintain their bodies. They must be able to live in a changing environment.	**ASK:** How does an animal's need for food affect its behavior? **LISTEN:** Students should describe an animal's need for food, and how this drives many of the animal's behaviors. For example, lions hunt in packs.

Teacher Reflection Questions

- What concepts from this section did students have the most trouble understanding? As the class progresses, how can you help students with these concepts?
- Without correcting them, how were you able to help students recognize when their claims were not valid? Did the class recognize any invalid claims during the presentations? How can you help them do this in the future?
- How did you encourage collaboration between group members? What can you do to help foster a collaborative environment in future activities?

NOTES

NOTES

BACK TO THE BIG QUESTION INTRODUCTION

Learning Set 1

Back to the Big Question

◀ $1\frac{1}{2}$ *class periods**

*A class period is considered to be one 40 to 50 minute class.

Overview

Groups revise their plans for observing middle-school students by using what they have learned from the previous observations they made of students and other animals. Using these improved plans, they observe students from the class and record their data. They analyze their data and should recognize that they obtained accurate data using their improved plans. They categorize and interpret their observations and share them with the rest of the class, recognizing that some categories are the same, even when the behaviors are different.

Targeted Concepts, Skills, and Nature of Science	Performance Expectations
Scientists often work together and then share their findings. Sharing findings makes new information available and helps scientists refine their ideas and build on others' ideas. When another person's or group's idea is used, credit needs to be given.	Students should work with their groups to interpret their observations and should use the observations of other groups to see what they could do differently.
Scientists must keep clear, accurate, and descriptive records of what they do so they can share their work with others, consider what they did, why they did it, and what they want to do next.	Students should record their observations descriptively enough for someone to draw pictures from them and should keep records of the observation plans they've tried.
Scientists differentiate between observations and interpretations. They use their observations and interpretations to explain animal behavior.	Students should use the *Observing and Interpreting Animal Behavior* pages to separate their observations from their interpretations.

Materials	
1 per class	Snacks for a group of three students
2 per student; 1 additional per group	*Observing and Interpreting Animal Behavior* pages

Activity Setup and Preparation

Prepare snacks for one group of students. They will eat this snack while the rest of the class observes. Also, pick a topic of discussion for the snacking group, such as: What did you do over the weekend?

Select a different group of students to be observed. Every group will be creating an observation plan, but the group being observed will not be observing others in the class. Decide how this group will gain experience observing. You may want to have them observe students analyzing their data.

Homework Options

Reflection

- **Science Process:** Which of the changes that you made to your initial observation plans do you think made the biggest difference? How did they make a difference? *(Look for ideas that students can generalize from and can use to solve other problems in planning observations.)*
- **Science Process:** Without observing your classmates first, how easy do you think it would be to plan your observations? *(The point of this question is to get students to think about strategies for problem solving; in particular, it is often easier to solve a problem if we first try out some ideas.)*
- **Science Process:** How do you think what you have learned will help you design enclosures for animals? *(Students should recognize that they are now better able to find out what animals need from their environments.)*

BACK TO THE BIG QUESTION IMPLEMENTATION

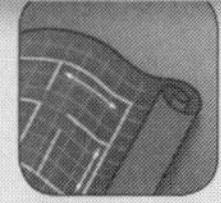

Learning Set 1

Back to the Big Question

How do scientists answer big questions and solve big problems?

The *Big Question* for this Unit is *How do scientists answer big questions and solve big problems?* You are answering the *Big Question* in the context of answering the science question: *Why do animals behave the way they do?* You will be applying your answers to those questions to a *Big Challenge*, creating an animal enclosure that allows an animal to behave as it would in its natural environment. To design an enclosure that meets all the criteria, it is important to know about how the animal behaves in the wild and what affects that behavior. You have learned that to better understand why animals behave as they do, scientists make observations. You have developed and used procedures for observing animal behavior. You will use those procedures as you learn more about animal behavior and address the *Big Challenge*.

iteration: a repetition that attempts to improve on a process or product.

Earlier in this *Learning Set*, you designed an observational plan to watch your classmates. Now that you've learned so much more about observing animal behavior, you will develop a new plan. It will probably make your observations a lot easier and allow you to collect more reliable data. Each time you develop a plan, you can use the successes and challenges from your previous experience to make a better plan. Scientists use their previous experiences to improve their procedures too. When they improve their procedures, the data they collect are usually more reliable. Scientists often perform almost the same investigation over and over again, each time using what they have learned earlier to make their next investigation and data more reliable. When someone redesigns a procedure or product based on what they have learned, it is called **iteration**. The word refers to the process of revising a plan or product. We also refer to each revision as an iteration.

You will now use all the things you have learned about how ethologists design their plans, make their observations and collect data, organize and analyze data, and describe their results, to develop a new observation plan. Then you will observe the students in your class one more time.

◀ $1\frac{1}{2}$ *class periods**

Learning Set 1

Back to the Big Question

Remind students of the Big Question *of the Unit.*

△ Guide

Ask students to describe what they have done so far. Record their answers. These should include: observing middle-school students snacking, making observations of animal behavior in pictures, and constructing and then revising explanations of animal behavior, using information they have read.

*A class period is considered to be one 40 to 50 minute class.

Then, remind students that ethologists (biologists who study animal behavior) work to answer four questions, and review those questions:

- How do different behaviors help animals survive?
- How do the animal's environment and learning affect different behaviors?
- How do behaviors change as the animal grows?
- How do animals that are similar to each other act in similar or different ways?

Ask students if, during all of this, they were acting like biologists. If students answer yes, say "Hooray!" and emphasize that they have been engaging in the practices of ethologists. If students say no, then guide them to understand that what they have been doing so far in class is the same thing ethologists do.

Next, remind students of the *Big Question: How do scientists answer big questions and solve big problems?* Remind students that they will not be able to fully answer this question until the end of the Unit. Then, ask them if they have any thoughts about the answer to this question. Create a list of students' ideas, and let them know they will revisit these at the end of the Unit, when they have all the information they need to formally answer the *Big Question*.

△ Guide

Point out that students have already engaged in practices of scientists, such as revising their ideas. Then, define iteration.

TEACHER TALK

"During this *Learning Set,* you have engaged in some of the practices of biologists. You've learned about how to make observations as ethologists do. You've developed plans for making observations, and each time you developed a plan, you were able to use its successes and failures to help you improve the next plan. This is a common practice in science. These revisions are called iterations."

Let students know they will be revising their observation plans based on all they now know, and then they will be observing their fellow students again.

Be a Scientist

Iteration

Scientists develop procedures for their investigations. They try to make sure procedures have the right steps that are followed in the right order. But sometimes procedures do not work out as they had planned. When that happens, scientists change their procedures based on where they might have had difficulty and run the investigation again to collect better data. They are continually learning from their investigations.

In your group, you wrote a procedure the first time you observed your classmates. There were things that worked well with your procedure and things that needed improvement. Iteration gives you a chance to revise the procedure and not make the same mistakes again.

Later you will use the skills and processes you are learning to study animal behavior. You will use the observations you make of different animals to understand why animals behave the way they do. Once you understand that, you will be able to apply your understanding to designing an animal enclosure.

Plan

You will continue your investigations of animal behavior by observing your classmates again as they are engaged in an activity. Your goal is to accurately observe and record the details of the students' behavior and to interpret their behavior, explaining why they are doing what they are doing.

With your group, revise your plan to make more accurate observations. Make sure all members of your group agree on the plan and know how to follow through on it.

- Will you watch an individual or all the students in the small group?
- How will you make sure you have observed all the members of the group?
- How will you make sure you have observed all the different behaviors?
- Will you each watch the whole scene or will it help to divide the observation task among all the members of your group?

Plan

15 min.

Have students design an observation plan to investigate the behavior of their classmates.

△ Guide

First, inform groups that they will be constructing an observation plan to observe their classmates. The goal is to have accurate observations, interpretations, and explanations for the behavior.

- How will you record what you see?
- Will you take notes on everything you see?
- How can you record quickly enough so you don't miss anything? Can you write key terms and not full sentences?
- Will you draw a picture?
- Will you keep track of the amount of time each behavior lasts? If so, how will you do this?

TEACHER TALK

"You will be observing one or more of your fellow classmates and will need to make accurate records of your observations. It is important to do this so you can make explanations of why, and interpretations of what, they are doing. To do this, you will need an observation plan based on what you now know about observing animal behavior."

Then, emphasize that, when constructing their observation plans, groups will need to consider all the points mentioned in the student text.

Let students know how much time they have to construct their plans (about ten minutes).

☐ Assess

As groups are constructing their plans, check to see if they are considering all the points in the student text.

- How will you record what you see?
- Will you take notes on everything you see?
- How can you record quickly enough so you don't miss anything? Can you write key terms and not full sentences?
- Will you draw a picture?
- Will you keep track of the amount of time each behavior lasts? If so, how will you do this?

Observe

As you watch your classmates, pay careful attention to the details of the situation. Follow the plan you made in your group as closely as possible. Record your observations as accurately as possible. Try to observe everything about the scene, your classmates, and the people around them. Watch what they are saying and what they are doing.

Analyze Your Data

After the observation time is over, discuss your observations with your group. Allow each group member time to describe the behaviors they saw. Decide if all of you saw the same things or if some of the observations were different. Use the same procedure you used before to analyze your group's data.

The first step is to copy each observation on to a separate sticky note. Put only one observation on each note. Your entire group will also copy their observations. If some are repeated, you can make just one sticky note for that observation.

Once each observation is on a separate sticky note, read through them and discuss how they might be grouped into categories by behavior. Combine observations of similar behaviors. If you disagree with the categories, ask others to be clear about why they think a behavior fits into that category. Sometimes it is necessary to reorganize categories as more observations are examined. Once you have each observation in a pile, use index cards to label the categories.

As a group, prepare a description of the behaviors in each category. The description needs to be specific enough so that other students in your class can recognize it. This will help you describe your groupings to the rest of your class.

Observe

5 min.

Have students observe classmates.

△ Guide

Emphasize that students should follow their group plan for making and recording observations, pay close attention to details, and record their observations accurately.

META NOTES

Select a different group of students to model snacking behavior so the previous group can practice making and recording observations. Provide the same directions, and ask them to perform similarly to the first group.

Get Going

Have a student group share a snack and talk for three minutes while the rest of the class observes them. If the snacking group has trouble talking, you should suggest a topic, such as movies, homework, sports, etc.

Distribute the *Observing and Interpreting Animal Behavior* pages to groups, and let them know they will be observing for three minutes.

Analyze Your Data

10 min.

Have groups analyze their data, classifying observations by behavior and writing descriptions of behaviors.

Analyze Your Data

After the observation time is over, discuss your observations with your group. Allow each group member time to describe the behaviors they saw. Decide if all of you saw the same things or if some of the observations were different. Use the same procedure you used before to analyze your group's data.

The first step is to copy each observation on to a separate sticky note. Put only one observation on each note. Your entire group will also copy their observations. If some are repeated, you can make just one sticky note for that observation.

Once each observation is on a separate sticky note, read through them and discuss how they might be grouped into categories by behavior. Combine observations of similar behaviors. If you disagree with the categories, ask others to be clear about why they think a behavior fits into that category. Sometimes it is necessary to reorganize categories as more observations are examined. Once you have each observation in a pile, use index cards to label the categories.

As a group, prepare a description of the behaviors in each category. The description needs to be specific enough so that other students in your class can recognize it. This will help you describe your groupings to the rest of your class.

AIA 40

Project-Based Inquiry Science

Back to the Big Question

Then, using an *Observing and Interpreting Animal Behavior* page, record your observations of the students' behavior, your interpretations of their behavior, and anything important about the environment and the students that helped you make each interpretation.

△ Guide

Tell students they will work with their groups to make sense of their data. As in previous investigations, they will make sure their methods are documented, will create categories for their data, and will identify trends in their data.

The first step will be for each group member to describe what they saw to the rest of the group. The group will discuss reasons for any differences in their observations and any difficulties they had with their observation procedures.

Then groups will categorize their observations. They should copy each observation onto a sticky note and group these sticky notes into categories by behavior. Then, using index cards, they should label the categories.

Next, groups will write descriptions of the behaviors in each category.

Finally, they will separate their interpretations from their observations, using the *Observing and Interpreting Animal Behavior* pages. Remind students to record their observations and interpretations (in the first and third columns) and then record the factors that led to their interpretations.

⬡ Get Going

Let groups know how much time they have to work. Analyzing the data should take about 10 minutes.

□ Assess

As groups work on analyzing their observations, monitor their work and their interpretations. Check to see that students are recording reasons for their interpretations and that they are not including opinions in their observations or interpretations.

META NOTES

To provide the group that was observed with an observing experience, consider having them observe their classmates analyzing the data for a minute or two. Also consider having them present their analysis after all the other groups have presented.

Communicate

15 min.

Have a discussion on groups' presentations, comparing their results.

Then, using an *Observing and Interpreting Animal Behavior* page, record your observations of the students' behavior, your interpretations of their behavior, and anything important about the environment and the students that helped you make each interpretation.

Communicate

Each group will now share the categories they created, the reasoning behind their categories, and the behaviors and interpretations they recorded. Listen carefully to see if others made the same observations you made. Also, listen as each group describes their category labels. Think about whether they used the same labels as your group. Consider whether their labels would be understandable by other biologists.

Think about how the categories are the same as or different from the categories you created earlier in the unit. The students you observed were behaving differently, so the categories may be very different. It is interesting that some of
the categories might also be the same, even though the overall behavior was very different.

Reflect

Iteration is an important part of scientific work. Repeating the same procedures again helps to check the validity of observations. Iteration also gives you a chance to revise your procedure when you learn it is difficult to do or when parts are missing.

1. What did you learn from revising and retrying your procedure?
2. In what ways were the observations and interpretations easier now than they were at the beginning of this *Learning Set?*
3. In what ways are your observations and interpretations more reliable? What did you do differently that made them more reliable?
4. What do you think you might do to make the observations and interpretations easier and more reliable next time?
5. How do you think iteration will help you when you design your animal enclosure?

△ Guide

Once groups have finished writing descriptions of the behaviors in each category and separating interpretations and observations, have each group share their categories with the class, explaining the reasoning behind their categories. As each group shares, the rest of the class should listen carefully to the labels groups used for their categories and the reasons for the categorization.

TEACHER TALK

"While other groups are sharing their categories, think about whether they were using the same categories you used. Did they divide things up the same way? Did they call their categories the same things? Also, think about what categorization would be most useful for communicating with other biologists and which labels are most understandable."

Point out that making observations again, when they know more, usually provides students with the ability to get better data. Let students know that, each time they revise their work, improve it, or test out another idea, it is called an iteration. Then, let them know that scientists go through many iterations in their work. Emphasize that an iteration is not just redoing the same thing. It is first modifying the plan in small ways and then trying the revised plan.

TEACHER TALK

"All of you have just observed middle-school student behavior again. You have tried to get better data, based on your experiences with previous observations. Scientists rework procedures and solutions to problems in order to improve them. That's what we mean by iteration. An iteration is a repetition with small changes. These observations and interpretations are your third iteration."

very di

Reflect

Iteration is an important part of scientific work. Repeating the same procedures again helps to check the validity of observations. Iteration also gives you a chance to revise your procedure when you learn it is difficult to do or when parts are missing.

1. What did you learn from revising and retrying your procedure?
2. In what ways were the observations and interpretations easier now than they were at the beginning of this *Learning Set?*
3. In what ways are your observations and interpretations more reliable? What did you do differently that made them more reliable?
4. What do you think you might do to make the observations and interpretations easier and more reliable next time?
5. How do you think iteration will help you when you design your animal enclosure?

AIA 41 ANIMALS IN ACTION

Reflect

15 min.

Have a class discussion of how students improved their results by revising their procedures several times.

⬡ Get Going

Let groups know how much time they have (no more than ten minutes) to answer the five questions, and be prepared for a class discussion.

☐ Assess

While groups are answering the questions, monitor their work and check their understanding of the questions. Decide how you will focus the discussion based on students' responses. If students seem to understand these questions, then lead a brief discussion of them.

△ Guide and Assess

Begin with a discussion of how groups' observation plans and observations changed from the first investigation to the last. Then, discuss how this challenge is representative of challenges in general and what students have learned about iterations, collaboration, and building on the ideas of others.

Listen for the following responses to the questions in the student text:

1. Students should realize that revising and retrying procedures gave them a chance to see what was effective and what was not. They may also have seen that one attempt, or iteration, was not enough to master the problems involved in making observations.
2. Students should realize they now know better what to look for, and since this is the third iteration, students should not have to think through how to do everything as they are making observations.
3. Students should identify the specific changes they made to their procedures to get more reliable observations. Students should also be able to define more reliable observations as those that are repeated over many trials or observed by many different people. You may have to refer students back to *Section 1.2,* where reliable data (observations) was introduced.
4. Students should respond to any problems they encountered in their latest observations, such as differences of opinion within the group.
5. Students should realize the iterative process will help them improve their zoo enclosure designs. During the process of designing their zoo enclosures, they will be gaining new information that will lead them to revise and improve their designs. In hearing others' designs, they will also come up with new ideas for their designs. You should also point out that, as they do this, they are behaving as biologists do when answering a big question and solving a big problem.

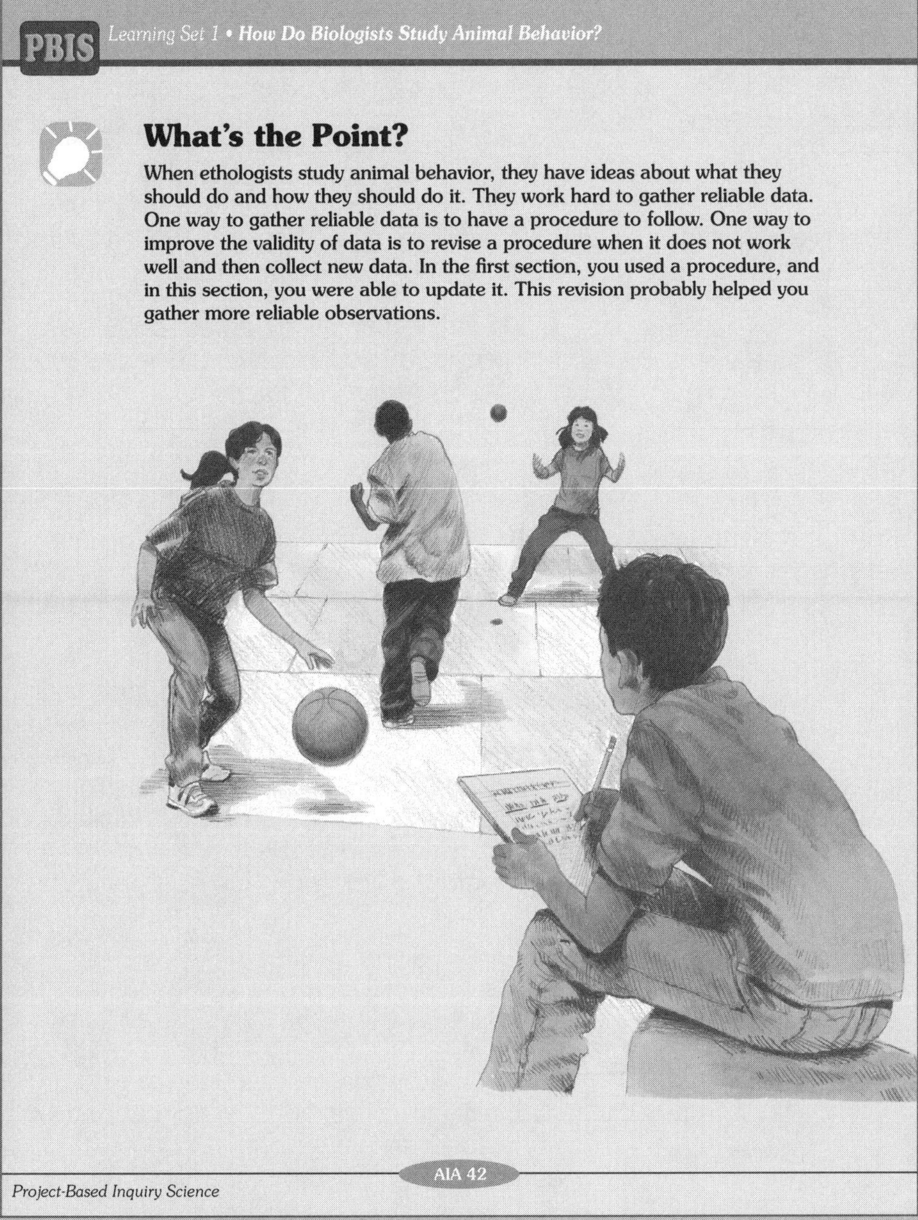

PBIS Learning Set 1 • How Do Biologists Study Animal Behavior?

What's the Point?

When ethologists study animal behavior, they have ideas about what they should do and how they should do it. They work hard to gather reliable data. One way to gather reliable data is to have a procedure to follow. One way to improve the validity of data is to revise a procedure when it does not work well and then collect new data. In the first section, you used a procedure, and in this section, you were able to update it. This revision probably helped you gather more reliable observations.

AIA 42

Project-Based Inquiry Science

What's the Point

5 min.

◇ Evaluate

Make sure students understand what reliable data is, why it is important to have a procedure to follow, and the purpose of the iterative process. If needed, refer back to *Section 1.2,* where reliable data was introduced.

Assessment Options

Targeted Concepts, Skills, and Nature of Science	How do I know if students got it?
Scientists must keep clear, accurate, and descriptive records of what they do, so they can share their work with others and consider what they did, why they did it, and what they want to do next.	**ASK:** When you made new observation plans, what revisions did you make to your original plan? On what were your revisions based? **LISTEN:** Students should describe how they revised their old plans. These revisions should be based on information they gathered during this *Learning Set.*
Scientists often work together and then share their findings. Sharing findings makes new information available, and helps scientists refine their ideas and build on others' ideas. When another person's or group's idea is used, credit needs to be given.	**ASK:** How did listening to other groups' observations help you plan your observations? How did discussing groups' observations help? **LISTEN:** Students should recognize that hearing other groups' observations allowed them to see ways they might not have thought of to make observations. The disagreements that came up in discussion should have helped them see where observations left room for interpretation.
Scientists differentiate between observations and interpretations. They use their observations and interpretations to explain animal behavior.	**ASK:** How do scientists use observations and interpretations? **LISTEN:** Scientists use observations to support interpretations of animal behavior. They use interpretations of observations to construct explanations of animal behavior.

Teacher Reflection Questions

- What difficulties did students have with the concept of iteration, and what ideas do you have to guide their understanding?
- What difficulties did students have in seeing the connection between the *Big Question,* the *Big Challenge,* and the *Learning Set Question?* What ideas do you have to assist them in making this connection?
- How did you manage allowing the students in the observed group to observe? What would you do next time?

LEARNING SET 2 INTRODUCTION

Learning Set 2

What Affects How Animals Feed?

◀ ***11 class periods****

*A class period is considered to be one 40 to 50 minute class.

Student ethologists make observations of animals feeding and focus on how their physical characteristics and environment affect their feeding behavior.

Overview

Student groups work as ethologists studying the feeding behaviors of animals and how animals' physical characteristics and environment affect their feeding. As they do this, they practice forming investigative questions as well as planning, collecting, analyzing and explaining observational data. They begin by considering how chimpanzees use tools during feeding, what physical attributes they have to allow them to use these tools, and how their environment affects their feeding. Students explore the foraging of bees and how a bee's special ability to see in the ultraviolet region of the electromagnetic spectrum helps their foraging. Student groups dissect a flower and learn about mutualistic relationships between bees and flowers as well as other mutualistic relationships. Student groups then explore a variety of feeding behaviors of other animals and share their results. Students also learn about how physical characteristics affect feeding, such as an animal's joints that acts as a simple machine.

LOOKING AHEAD

- **Students will observe animals using videos during this *Learning Set.* This requires appropriate equipment.**

Targeted Concepts, Skills, and Nature of Science	Section
Scientists often work together and then share their findings. Sharing findings makes new information available and helps scientists refine their ideas and build on others' ideas. When another person's or group's idea is used, credit needs to be given.	All Sections
Scientists must keep clear, accurate, and descriptive records of what they do so they can share their work with others, consider what they did, why they did it, and what they want to do next	All Sections
Tables are an effective way to communicate results of a scientific investigation.	2.2
Observations and measurements are considered reliable if the results are repeatable by other scientists using the same procedures.	2.2, 2.7

Targeted Concepts, Skills, and Nature of Science	Section
Scientists differentiate between observations and interpretations. They use their observations and interpretations to explain animal behavior.	2.3
Studying the work of different scientists provides understanding of scientific inquiry and reminds students that science is a human endeavor.	2.2, 2.3, 2.5, 2.6, 2.7
Scientists make claims (conclusions) based on evidence obtained (trends in data) from reliable investigations.	2.2, 2.3
Explanations are claims supported by evidence. Evidence can be experimental results, observational data, and other accepted scientific knowledge.	3.4, 3.5, 3.6, 3.7, BBC
Explanations are claims supported by evidence. Evidence can be experimental results, observational data, and other accepted scientific knowledge.	2.3, 2.5, 2.7, BBC
Scientists use models to simulate processes that happen too fast, too slow, on a scale that cannot be observed directly (either too small or too large), or that are too dangerous.	2.4, 2.5
Criteria and constraints are important in determining effective scientific procedures and answering scientific questions.	BBC
Behavior is a type of response to internal or external stimulus. Behavior is determined by experience, physical characteristics, and environment.	2.1, 2.2, 2.3, 2.4, 2.5, BBC
The structure and function of animals' bodies are complementary and affect animal behavior.	2.2, 2.3, 2.4, 2.5, 2.7, BBC
Organisms need food to grow, reproduce and maintain their bodies. They must be able to live in a changing environment.	2.1, 2.3, 2.4, 2.5
Animals' sense of sight is adapted to their environment. Some animals see things that humans cannot see.	2.5
To see an object, light reflected from the object must enter the eye. Eyes of various animal species work differently to make them effective in their environment.	2.5
White light is composed of all the colors of the rainbow. The sun emits electromagnetic radiation of which humans can only detect a small portion of with their eyes.	2.5

Targeted Concepts, Skills, and Nature of Science	Section
Flowers are the reproductive organs of plants, as well as food suppliers for many animals.	2.6
Animals' bodies have similarities to simple machines. These machines help the animal survive.	2.7

Students' Initial Conceptions and Capabilities

- Students may have difficulty understanding the difference between observations and inferences. (Allen, Statkiewitz, & Donovan, 1983; Kuhn 1991, 1992; Roseberry, Warrant, & Conant, 1992.)
- Students are familiar with many animals, but may be unaware of their behaviors and how to observe their behaviors.

LOOKING AHEAD

- In the *More to Learn* segment of *Section 2.6* there is a flower dissection lab for which you will need to acquire fresh flowers such as azaleas, lilies, gladioli, tulips, daffodils, geraniums, snapdragons, or sweet peas. Ideally they should be large, deep flowers with closed corollas, in which the parts are easily recognizable.
- The Big Challenge at the end of this Unit requires students to refer back to their *Observing and Interpreting Animal Behavior* pages. Students will benefit if they have some means of keeping these pages organized.

Understanding for Teachers

Animal behavior is affected by many different factors. In *Animals In Action,* students consider animals' body structures and their environment as influences on their feeding. Students use these two factors to develop explanations for feeding behaviors observed in this *Learning Set.*

All animals feed to survive. Some have complex feeding behaviors and others have very simple feeding behaviors. When animals feed they are constrained by their body structure and are forced to make adaptations to the environment. Animals use the resources of their bodies to obtain food. Bees, chimpanzees, and elephants are all foragers although they forage for different foods. They each use their bodies in specific ways to make sure they have enough to eat. Animals' adaptations make them efficient feeders

NOTES

LEARNING SET 2 IMPLEMENTATION

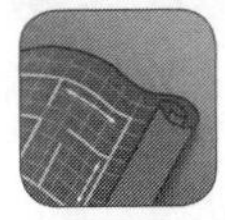

Learning Set 2

What Affects How Animals Feed?

The *Big Question* for this Unit is *How do scientists answer big questions and solve big problems?* In *Learning Set 1*, you observed some animal behavior. You also learned some skills ethologists use when they observe animal behavior.

What types of behaviors are ethologists interested in observing? They are most interested in looking at behaviors that are common to all living things. These behaviors include feeding, communicating, moving, playing, and taking care of their young. Each of these behaviors is very important to the life of the animal. It could take years to investigate each of these behaviors. Ethologists spend their lives observing animals. They use their collected data to make interpretations of what they see.

In this *Learning Set*, you are going to look at one animal behavior, feeding. You will observe several different animals feeding. You will answer the smaller question: *What affects how animals feed?* Through investigating, reading, and observing videos, you will begin to interpret feeding behaviors. You will need to consider the influence of the animal's body shape, size, and function on how it feeds. You will also observe each animal's environment to determine the effect of the environment on its feeding behavior.

Scientists can learn a lot by observing how animals feed.

Learning Set 2

What Affects How Animals Feed?

5 min.

Introduce students to the question of the Learning Set: What Affects How Animals Feed?

Engage

Begin by asking students what they observe in each of the pictures in the student text and if they can make a general statement about all the images.

TEACHER TALK

"How would you describe what's happening in each of these pictures? What do all these images have in common? What is different about them? Why do you think these animals eat the things they do? How do they eat? *(In each of the images animals are feeding, but their environments, food, and feeding behaviors are different.)*"

△ Guide

Inform students that ethologists study many behaviors and may spend their entire lives focusing on just one or two behaviors of one animal. Then, let the class know that they will just focus on two behaviors: feeding and communicating. Discuss how this *Learning Set* focuses on feeding, and that they will observe different animals feeding through videos, images, and reading. Then let them know that they will use what they learn to answer the question: *What affects how animals feed?*

NOTES

SECTION 2.1 INTRODUCTION

2.1 Understand the Question

Thinking about What Affects How Animals Feed

◀ ***1 class period****

*A class period is considered to be one 40 to 50 minute class.

Overview

Students observe chimpanzees feeding in a video and discuss their observations with their groups, identifying what they know and what they do not know. Based on these discussions, groups develop questions about animals' feeding behavior and select their two most interesting questions. Groups share their questions with the class and discuss how they will address the *Big Question* and *Big Challenge*. The class adds the questions to their *Project Board*, creating a record of what they intend to learn.

Targeted Concepts, Skills, and Nature of Science	Performance Expectations
Scientists often work together and then share their findings. Sharing findings makes new information available and helps scientists refine their ideas and build on others' ideas. When another person's or group's idea is used, credit needs to be given.	Students should collaborate with their groups to develop two investigative questions about animal behavior.
Scientists must keep clear, accurate, and descriptive records of what they do so they can share their work with others, consider what they did, why they did it, and what they want to do next.	Students should record questions about what they need to learn to meet the challenge on their *Project Board*.
Behavior is a type of response to internal or external stimulus. Behavior is determined by experience, physical characteristics, and environment.	Students should use chimpanzees' environment to help explain their behavior.
Organisms need food to grow, reproduce, and maintain their bodies. They must be able to live in a changing environment.	Students should use an animals need for food to explain chimpanzees' behavior.

Materials	
1 per class	*Animals In Action* DVD (*Chimpanzees* video) and a way to view the video
1 per student	*Observing and Interpreting Animal Behavior* page
1 per class	Class *Project Board*
1 per student	*Project Board* page

Activity Setup and Preparation

Have equipment set up to show the class the chimpanzee video. Consider how to arrange the class to allow groups to meet and watch the video.

Homework Options

Reflection

- **Science Process:** How could you get the information you need to answer your group's questions if you had unlimited resources? How could you get it in your local environment? *(Students should describe what they need to learn and how they can gather the information.)*
- **Science Process:** What other animals did you observe in the chimpanzee video? How would knowing the behaviors of those animals help you understand the chimpanzees' behavior? *(Students may identify the other animals as ants, termites, or insects. It is not clear from observing the video what is necessary to get the termites to cling to a stem of grass. If we knew that termites attack anything that enters the nest, then we could infer that the chimpanzee only has to stick the stem of grass in the nest.)*

Preparation for 2.2

- **Science Process:** If you were to watch the video of chimpanzees again, how would you change the way you make your observations of the chimpanzees? How would you share the work with your group members? *(Students should describe a way to capture what each of the chimpanzees is doing in as much detail as possible and a way to communicate their observations.)*

SECTION 2.1 IMPLEMENTATION

◀ *1 class period**

2.1 Understand the Question

Thinking about What Affects How Animals Feed

herbivore: an organism that eats only plants.

carnivore: an organism that obtains its food only from other animals.

omnivore: an organism that will feed on many different kinds of food, including both plants and animals.

forage: to search for something, especially food and supplies.

forager: an organism that searches for food.

predator: an organism that hunts and kills others organisms.

Feeding is one behavior common to all animals. From the smallest to the largest, from the least complex to the most complex, each animal needs food to provide it with energy.

Different animals eat different things. Some animals are **herbivores**. They only eat plant material. Some examples of herbivores are elephants, giraffes, squirrels, and bees. Other animals are **carnivores**. They only eat other animals. Lions, cheetahs, snakes, and spiders are carnivores. Still other animals are **omnivores**. Omnivores eat both plants and animals. Examples of omnivores are chimpanzees, bears, chickens, and flies.

Different animals also have different ways of feeding. Some animals **forage** or move from place to place looking for food under rocks or in the bark of trees. Many insects, including bees, are **foragers**. They move from plant to plant to find food. Many herbivorous animals (animals that are herbivores), such as giraffes and elephants, are also foragers.

A **predator** is an animal that eats other organisms. Some carnivores, like lions and tigers, are predators that hunt. Unlike foragers, these carnivores must move very quickly because their prey also moves quickly. Other predators hunt by waiting until the prey comes to them. For example, alligators are also carnivores, but they hide until their prey comes near, and then they pounce.

Scientists are interested in what animals eat and how they find food. They often observe animals while they are feeding to better understand how animals fit into their environment.

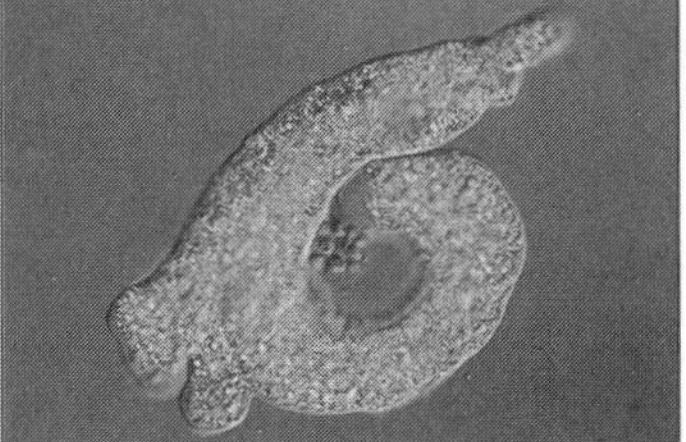

The single-celled amoeba engulfs its food.

The giraffe strips the leaves from trees with its powerful lips and long tongue.

AIA 44

Project-Based Inquiry Science

2.1 Understand the Question

Thinking about What Affects How Animals Feed

5 min.

Introduce the major types of feeding behaviors that students will study.

△ Guide

Emphasize that all animals feed to provide the animal with energy which it then uses to function. Also, emphasize that the way an animal feeds is specific to the needs of the animal.

Explain what a forager does and ask for examples of foragers that students' have observed. Many of these will be herbivores such as rabbits or squirrels.

*A class period is considered to be one 40 to 50 minute class.

META NOTES

Most domestic cats and dogs eat more than just the meat from other animals. Pet foods are made of meat as well as grains, such as corn. Domestic cats and dogs also eat other food items they may find in and around their home.

TEACHER TALK

"Foragers search for food wherever they can find it. They might look in the bark of trees, or in the ground, or under rocks. What foragers have you observed? Have you ever seen animals foraging in the park? (*squirrels, rabbits*)"

Next, introduce the term predator and ask students for examples of these. Predators are usually carnivores. Introduce the terms herbivore and carnivore, providing examples of each and asking students for examples.

TEACHER TALK

"Foragers don't hunt. Squirrels are an example of foragers. Squirrels are herbivores. They only eat plant matter. Animals that eat only meat are carnivores. What are some examples of carnivores? (*lions, tigers, dogs*)"

META NOTES

Students may ask, or you may want to tell students, what the difference between a predator and a scavenger is. Predators hunt for their food. Scavengers do not kill their food; they eat dead animals they find.

Ask students what humans are. (Omnivores that feed on both plants and other animals.)

TEACHER TALK

"Does anyone know what most humans are? We are omnivores, meaning that humans eat both plants and other animals."

You may want to mention that the amoeba shown in the student text is a predator.

NOTES

You will need to learn more about how animals feed for another reason too—so you can address the challenge of designing an enclosure for one of the animals you are studying. Remember that the habitat will have to allow your animal to feed as it does in the wild and will have to allow scientists and visitors at the zoo to observe your animal as it feeds.

Get Started

You will watch a short video of chimpanzees feeding. Before watching the video, discuss with other members of your group what you know about chimpanzees. Try to imagine what the life of a chimpanzee might be like. Use the following questions to guide your discussion.

- Where do chimpanzees live?
- What do chimpanzees eat?
- How do chimpanzees get their food?

Listen to others in your group as they share what they know about chimpanzees.

Watch the video. Pay attention to the feeding behavior of the chimpanzees. Try to figure out what the chimpanzees are doing. Observe what they are eating. Notice if all the chimpanzees are eating the same way.

Conference

Discuss with your group what you saw in the video. What surprised you? What did you not understand? What do you disagree about? Identify what you think you know about how animals feed. Remember your challenge—to design an enclosure for an animal. What more do you need to learn about chimpanzees to be able to design an enclosure for chimpanzees? What might you need to learn about other animals to be able to design enclosures for them? Watching the chimpanzees feeding might have reminded you of some of the things you know about how other animals feed and some of the things you have wondered about in the past.

Get Started

10 min.

Have students discuss what they know about chimpanzees, then show the Chimpanzee *video*

META NOTES

Students will be watching the video again in the next section.

○ Get Going

Inform students that they will watch a video of chimpanzees feeding, and give groups a few minutes to discuss what they know about chimpanzees' habitats, diets, and feeding behaviors.

After groups have discussed what they know, emphasize that they should carefully observe and interpret the behaviors of individual chimpanzees, including what they are eating.

Provide students with an *Observing and Interpreting Animal Behavior* page, then start the video.

You will wat... ...ideo of chimpanzees fe... ...ore watching the video, discuss w... other members of your group what you know about chimpanzees. Try to imagine what the life of a chimpanzee might be like. Use the following questions to guide your discussion.

- Where do chimpanzees live?
- What do chimpanzees eat?
- How do chimpanzees get their food?

Listen to others in your group as they share what they know about chimpanzees.

Watch the video. Pay attention to the feeding behavior of the chimpanzees. Try to figure out what the chimpanzees are doing. Observe what they are eating. Notice if all the chimpanzees are eating the same way.

Conference

Discuss with your group what you saw in the video. What surprised you? What did you not understand? What do you disagree about? Identify what you think you know about how animals feed. Remember your challenge—to design an enclosure for an animal. What more do you need to learn about chimpanzees to be able to design an enclosure for chimpanzees? What might you need to learn about other animals to be able to design enclosures for them? Watching the chimpanzees feeding might have reminded you of some of the things you know about how other animals feed and some of the things you have wondered about in the past.

AIA 45

ANIMALS IN ACTION

Conference

10 min.

Have groups discuss the Chimpanzee *video, and then construct questions to investigate*

⬡ Get Going

After showing the video, have groups discuss what they saw, what they learned, and what they still have to learn. Student groups should try to identify what they would need to learn to design enclosures for chimpanzees and for other animals.

☐ Assess

Visit groups as they discuss their observations, and assess how they are differentiating the behaviors of the different chimpanzees, and how they are interpreting the chimpanzees' use of twigs or grass stems. Later, they will learn that this use of tools by chimpanzees surprised biologists when it was discovered and changed the way biologists think about animal behavior.

Pay attention to those who state that the chimpanzees are eating termites or ants. There is no actual footage in the video of the chimpanzees eating insects off the grass, but there is enough evidence to interpret that they are eating something off the grass. Also, you might note that the insects in the video seem to be ants. Even though it is known that chimpanzees eat termites and ants, there is no direct evidence that the chimpanzees are eating either in the video, that would be an interpretation.

META NOTES

Keep a record of all students' questions to refer to during this *Learning Set.*

⬡ Get Going

Next, have each student individually develop two questions about what affects animal feeding. Emphasize that anything group members disagreed about or did not understand should be considered. They should keep in mind the criteria for their questions listed in the student text.

Have groups discuss members' questions and select the two that are most interesting. Let students know that they should keep a record of the rest of the questions as they may wish to include them on the *Project Board* or use them later in this *Learning Set.*

△ Guide and Assess

As groups are sharing their questions, take note of disagreements. Point out to the groups that these are areas that need to be investigated. Also, keep in mind that these items should be put on the class *Project Board.*

⬡ Get Going

Have each group share their questions with the class. Encourage students to think about and ask groups how their questions meet the criteria of good questions.

As you list questions, listen for and encourage students to share reasoning behind their ideas.

As groups share their questions and the class discusses them, record them on the *Project Board.*

◇ Evaluate

Make sure the questions listed focus on feeding and feeding behaviors of chimpanzees.

Update the *Project Board*

15 min.

Have a class discussion of groups' questions and update the Project Board.

META NOTES

Sharing one's reasoning is not easy but should be common practice after a few Units of *PBIS*.

Working by yourself, develop two questions that might help you better understand what affects how animals feed. During earlier discussions with your group, you might have found that you disagreed about some things. As you identify what else you need to learn about how animals feed, keep in mind what you disagreed about, what surprised you, and what you did not understand. When you write your questions, keep in mind that your questions should

- be interesting to you,
- require several resources to answer,
- relate to the *Big Question* and designing a new enclosure that will encourage the feeding and communication of an animal, and
- require collecting and using data.

Make sure your questions are not simply yes/no questions or ones you can answer with a single word or sentence.

When you have completed your two questions, meet with your group. Share all the questions with each other. Carefully consider each question and decide if it meets the criteria for a good question. With your group, refine the questions that do not meet the criteria. Choose the two most interesting questions to share with the class. Give your teacher the rest of the questions so they might be used later.

Update the *Project Board*

Recall that the *Project Board* helps you organize your ideas as you answer the *Big Question* and address the *Big Challenge*. You will now share with the class what you think you know about how animals feed and your group's two questions. Be prepared to justify why yours are good questions. Your teacher will add your questions to the *Project Board*. Throughout this *Learning Set*, you will work to answer some of these questions.

What's the Point?

All animals must eat to obtain the energy and nutrients they need to live. Therefore, feeding is a behavior common to all animals. What animals eat and what affects how they feed can be determined through careful observation. Some animals eat meat (carnivores), some eat plants (herbivores), and other animals eat both (omnivores). When animals feed, they do it in a variety of ways. Some animals that are predators move quickly to chase and capture their prey, while other predators wait until their prey comes near them, and then they pounce. Some animals, like chimpanzees, are foragers. When foragers eat, they move from place to place looking for food.

META NOTES

It is helpful to date the items recorded on the *Project Board*. That way, students can monitor changes in their ideas and their progress on the challenge. It is also important to draw lines connecting ideas across columns

two questions. Be prepa... why yours are ... Your teacher will add your questions to the *Project Board*. Throughout this *Learning Set*, you will work to answer some of these questions.

What's the Point?

All animals must eat to obtain the energy and nutrients they need to live. Therefore, feeding is a behavior common to all animals. What animals eat and what affects how they feed can be determined through careful observation. Some animals eat meat (carnivores), some eat plants (herbivores), and other animals eat both (omnivores). When animals feed, they do it in a variety of ways. Some animals that are predators move quickly to chase and capture their prey, while other predators wait until their prey comes near them, and then they pounce. Some animals, like chimpanzees, are foragers. When foragers eat, they move from place to place looking for food.

What's the point?

5 min.

The important aspects of this section are for students to begin observing the feeding behavior of chimpanzees and to think about how these ideas might affect their design for an enclosure. Students will view the video again in the next section.

Assessment Options

Targeted Concepts, Skills, and Nature of Science	How do I know if students got it?
Behavior is a type of response to internal or external stimulus. Behavior is determined by experience, physical characteristics, and environment.	**ASK**: What were some important things in the environment that affected chimpanzees' behavior in this video? **LISTEN**: Students should recognize that the termites and the stems of grass were necessary for the behavior they observed.
Organisms need food to grow, reproduce, and maintain their bodies. They must be able to live in a changing environment.	**ASK**: What need did the chimpanzees' behavior satisfy? **LISTEN**: The need for food to survive.
Scientists often work together and then share their findings. Sharing findings makes new information available and helps scientists refine their ideas and build on others' ideas. When another person's or group's idea is used, credit needs to be given	**ASK**: How did discussing what you saw with your group help you understand your observations? **LISTEN**: Students should have tested and refined their ideas through discussion with the class and their groups.

Targeted Concepts, Skills, and Nature of Science	How do I know if students got it?
Scientists must keep clear, accurate, and descriptive records of what they do so they can share their work with others, consider what they did, why they did it, and what they want to do next.	**ASK**: Why is it helpful to have our questions on the *Project Board*? **LISTEN**: The *Project Board* helps to organize students' ideas, questions, observations, and information, and helps them to see how their understanding has progressed.

Teacher Reflection Questions

- What difficulties did students have in developing investigative questions? What difficulties do you think they will face as they try to answer these questions?
- How are students progressing in engaging in the social behavior of scientists? Are they able to lead discussions rather than looking for the teacher to direct the discussions? How can you assist students to reach this goal?
- How were you able to keep students organized and focused during groups' discussions of the video?

NOTES

SECTION 2.2 INTRODUCTION

2.2 Explore

What Affects How Chimpanzees Feed?

◀ $1\frac{1}{2}$ class periods*

*A class period is considered to be one 40 to 50 minute class.

Overview

Students create plans for observing chimpanzees in the video and use the plans to make careful and detailed observations of the feeding behaviors of the chimpanzees. Then groups separate the observations they all agree on from those they do not agree on and categorize the unanimous observations. This helps them to organize their thoughts and create categories similar to scientists. Using posters, groups present their procedures, observations, and categories. The class evaluates the observations that were presented based on the procedures, and students think about how animals' body structures and environments affect their behavior

Targeted Concepts, Skills, and Nature of Science	Performance Expectations
Scientists often work together and then share their findings. Sharing findings makes new information available and helps scientists refine their ideas and build on others' ideas. When another person's or group's idea is used, credit needs to be given.	Students should share and categorize their observations with their groups, then they share their observations and categories with the class.
Scientists must keep clear, accurate, and descriptive records of what they do so they can share their work with others, consider what they did, why they did it, and what they want to do next	Students should keep clear and detailed records of their observations of chimpanzees' behavior and environment.
Tables are an effective way to communicate results of a scientific investigation.	Students use visuals such as tables, charts, and graphs in their posters to aid their *Investigation Expos.*
Observations and measurements are considered reliable if the results are repeatable by other scientists using the same procedures.	Students should point out reliable observations as those that other groups also observed.

Targeted Concepts, Skills, and Nature of Science	Performance Expectations
Scientists differentiate between observations and interpretations. They use their observations and interpretations to explain animal behavior.	Students should identify and separate their interpretations from their observations.
Scientific knowledge is developed through observations, recording and analysis of data, and development of explanations based on evidence.	Students record and analyze their observations of chimpanzees' feeding behaviors in their natural environment.
Scientists make claims (conclusions) based on evidence obtained (trends in data) from reliable investigations.	Students make claims about the feeding behavior of chimpanzees based on their observations.
Behavior is a type of response to internal or external stimuli. Behavior is determined by experience, physical characteristics, and environment.	Students should record observations of chimpanzees and consider how their physical characteristics and their environment affect their behavior.
The structure and function of animals' bodies are complementary and affect animal behavior.	Students should observe the structures of chimpanzees' hands and mouths and consider how these affect the chimpanzees' behavior.

Materials	
2 per group	*Animals In Action* DVD (*Chimpanzees* video) and a way to view the video. (This is the same video that students viewed in *Section 2.1.*)
1 per student	*Observing and Interpreting Animal Behavior* page
1 per group	Poster paper and markers

Activity Setup and Preparation

Read the *Teacher's Resource Guide* on *Investigation Expos.* Students are introduced to *Investigation Expos* during this activity. These will be used throughout the *PBIS* curriculum. Decide how you want to present these to the class.

Decide how you want to arrange the *Investigation Expo.* You should have groups display their posters around the room, and let the class circulate around room.

Homework Options

Reflection

- **Science Process:** What problems did you have following your observation plan? How could you change the plan to prevent these problems? Would your revised plan work for a different video? *(Students' responses should come from the discussions that groups held after making observations.)*
- **Science Process:** How were your categories different from other groups' categories? Were the differences because of different observations? *(Differences in the way observations were grouped should be noted. This will help students with refining their own observation plans and with developing questions to investigate.)*
- **Science Process:** What other animals did you observe in the chimpanzee video? How would knowing the behaviors of those animals help you understand the chimpanzees' behavior? *(Students may identify the other animals as ants, termites, or insects. It is not clear from observing the video what is necessary to get the termites to cling to a stem of grass. If we knew that termites attack anything that enters the nest, then we could infer that the chimpanzee only has to stick the stem of grass in the nest.)*

Preparation for 2.3

- **Science Process:** What is still unclear about the chimpanzee behaviors you observed in the video? What information would help you to explain the chimpanzee behaviors? *(This question should get students thinking about what they still have to learn and prepare them for the next section that introduces information about chimpanzees.)*
- **Science Content:** What more would you need to learn about the behaviors you observed to design a good enclosure for chimpanzees? *(This question should get students thinking about what they still have to learn and make connections for their designs for the* Big Challenge.*)*

NOTES

SECTION 2.2 IMPLEMENTATION

◀ $1\frac{1}{2}$ *class periods**

2.2 Explore

What Affects How Chimpanzees Feed?

You have seen chimpanzees feeding. Ethologists look for answers to why animals behave as they do. Now, it is time for you to begin looking for these kinds of answers. In this section, you will work with your group, watch the video again, and begin to look at the factors that might affect how chimpanzees feed.

Observe

You will now watch the video again. This time, you will make observations that will help you answer the question for this section. Start by making an observation plan with your group. How will you go about observing the chimpanzee group in this video? Remember that when you observed middle-school students' behavior earlier in this Unit, you developed a plan. You might use a similar plan this time. Create your plan, keeping in mind that you need to observe as much of the scene as possible, and that your observations are going to determine how well you understand the feeding of chimpanzees. Make sure that part of your plan includes using a page to record your notes.

Once you have your plan written, and all the members of your group understand the plan, your teacher will start the video again. Remember to take notes about the chimpanzees' behavior and habitat. Pay attention to the chimpanzees' bodies. Think about how their body structure helps them feed the way they do. Remember, too, to pay attention to how the chimpanzees' habitat affects how they eat. Remember to follow the plan your group decided on. Record your observations so you will be able to share them with your group and your class.

Analyze Your Data

You may have had difficulty making your observations. Each member of your group may have seen something different. By sharing your observations with others, you might have discovered things that you did not notice before. You may also have had a difficult time working with the plan you created.

AIA 47

ANIMALS IN ACTION

Learning Set 2

What Affects How Chimpanzees Feed?

5 min.

Introduce this section to the students

△ Guide

Let students know that they will be observing the video of chimpanzees again, looking for factors that might affect how chimpanzees feed. This time, they will develop a plan.

*A class period is considered to be one 40 to 50 minute class.

△ Guide

First, ask students what they learned about planning and making observations in *Learning Set 1* that they can use to improve their observations from the chimpanzee video. Then, ask students what they might be able to find out from improved observations from the video. Emphasize that they should observe chimpanzees' physical characteristics and environment as well as their behaviors.

Let students know that they will, again, record their observations on *Observing and Interpreting Animal Behavior* pages and that they should consider reviewing their previous observations when constructing their plan.

⬡ Get Going

Let groups know how much time they have to create plans for observing the chimpanzees in the video, emphasizing that they should record as much detail as possible in their observations.

Observe

10 min.

Have students plan their observations and show the video to the class again.

META NOTES

Students will use their observations to make explanations of chimpanzees' behavior based on the form and function of their bodies and on their habitat.

You have seen chimpanzees feeding. Ethologists look for answers to why animals behave as they do. Now, it is time for you to begin looking for these kinds of answers. In this section, you will work with your group, watch the video again, and begin to look at the factors that might affect how chimpanzees feed.

Observe

You will now watch the video again. This time, you will make observations that will help you answer the question for this section. Start by making an observation plan with your group. How will you go about observing the chimpanzee group in this video? Remember that when you observed middle-school students' behavior earlier in this Unit, you developed a plan. You might use a similar plan this time. Create your plan, keeping in mind that you need to observe as much of the scene as possible, and that your observations are going to determine how well you understand the feeding of chimpanzees. Make sure that part of your plan includes using a page to record your notes.

Once you have your plan written, and all the members of your group understand the plan, your teacher will start the video again. Remember to take notes about the chimpanzees' behavior and habitat. Pay attention to the chimpanzees' bodies. Think about how their body structure helps them feed the way they do. Remember, too, to pay attention to how the chimpanzees' habitat affects how they eat. Remember to follow the plan your group decided on. Record your observations so you will be able to share them with your group and your class.

Meet with your group to share your observations. Each member of the group should have a chance to read all of their observations to the rest of the group. Depending on your observation plan, you may have been watching different things than other group members were watching. Listen carefully for observations you might not have seen. Create a group list of observations that everyone agrees on. Make a second list of observations that were not agreed on. Make a third list that includes anything that was difficult about following your observation plan.

Using the same procedure you used in *Learning Set 1*, create a sticky note for each observation on which you agreed. Organize your observations into groups based on the type of behavior shown. Most observations will be about feeding. The important thing here is to provide as much detail as possible about the chimpanzees' feeding activity. For example, one group of observations might be about how the chimpanzee gets ready to eat. Another might be about how they use tools. If you think some behaviors might fall into two categories, you can list those behaviors more than once. Just make a copy of the sticky note so you have one for each category that a behavior fits into.

Communicate Your Results

Investigation Expo

You will share your observations and data analysis in an *Investigation Expo*. Each group will make a poster showing what they observed and the way they grouped behaviors. Then each group will present their poster to the class. Remember that each group created their own plan for making their observations and analyzing their data. Each group's observations may have been affected by their plan. Different groups' analyses may be different from each other.

For your presentation, create a poster that describes each of the following:

- questions you were trying to answer in your observations
- your observation procedure and how it helped you make observations
- what you observed
- the categories of behavior you identified, what behaviors fit into each category, and how confident you are about this analysis

One way you might present your questions, observations, and data analysis is in the form of a diagram. Place each group of observations in columns on your diagram. You can use the example on the next page to help you.

Analyze Your Data

15 min.

Have groups discuss their observations, categorizing the observations that they all agree upon.

☐ Assess

As groups work on their plans, go around the room and see how students are planning to observe chimpanzees' feeding behavior, as well as physical characteristics and factors in their environment that affect their feeding behavior. Make sure they also include using a page to record their notes in their plan.

Communicate Your Results: *Investigation Expo*

40 min.

Introduce the Investigation Expo.

Have students share their observations and categories on posters in their first Investigation Expo.

What affects how chimpanzees eat? Observations			
beginning to eat		eating	resting
chimps gather grass	chimps move to mound of insects	chimp puts grass in mouth	chimp waits 10 seconds—does nothing

As you present, be sure to be very detailed in presenting everything on your poster—your questions, your observation plan, your observations, your way of grouping those observations, how well you think your plan worked and why, and how much you trust your analysis.

As each group is presenting their poster, listen for the answers to the questions below. If a group does not answer all these questions, ask them questions to help you better understand what they found out and what their observation plan was. Remember to be respectful when you ask your questions.

- What was the group trying to find out?
- What procedure did they use to collect their data?
- How well were they able to make clear and detailed observations?
- How well did they group their observations? Does the way they grouped them make sense to you?
- What conclusions do their results suggest?
- Do you trust their observations? Why or why not?

⬡ Get Going

Once groups have finished creating their plans, distribute *Observing and Interpreting Animal Behavior* pages. Remind students to record all of their observations in detail. Emphasize that they should record any interpretations in the *Interpretations* column, and the supporting evidence for their interpretations in the middle column. Then show the video again.

△ Guide

Let groups know that they will be analyzing their data and then presenting it to the class. Explain that they will be constructing three lists. The first list should include the observations and interpretations they agree on. The second list should include all the observations and interpretations they do not agree on and their reasons. The third list should contain any difficulties they had with their observation plan.

Describe to the class how groups will categorize and label the agreed-upon observations, the way they did in *Learning Set 1* when they observed middle-school behavior. Remind them that they will be using sticky notes for the behaviors and index cards to label the categories.

Then provide students' with a time frame and have them begin their analysis.

△ Guide and Assess

Monitor groups as they discuss and organize their data. Identify inconsistencies in their results and/or problems with their procedures. Also, check to see if students are having trouble categorizing their data.

△ Guide

First, describe an *Investigation Expo*. It is similar to other presentations, but it is designed for sharing information about investigations. Explain that there are two parts: poster presentations and then discussions to share their procedure, observations, and categories. Emphasize that the student text has all this information in the *Be a Scientist* information box entitled *Introducing an Investigation Expo* and that they should refer to the student text while they are preparing for the *Investigation Expo*.

TEACHER TALK

"Throughout this class you will be using *Investigation Expos* to share ideas. During an *Investigation Expo* you work like scientists sharing observations and ideas. After we discuss everyone's ideas you will then have a chance to reflect on what you have heard and refine your ideas."

Emphasize that students will need to be able to inform the class of their investigative question, their predictions, their procedure, and what makes the procedure fair (to be discussed next), their results and how confident they are of them, and their interpretations and confidence in their interpretations. Also emphasize that they should be able to inform the class of their claims. Remind students that a claim is a statement of what a trend means or a conclusion of an experiment.

Investigation Expo: a presentation of the procedure, results, and interpretations of results of an investigation.

Be a Scientist

Introducing an Investigation Expo

An *Investigation Expo* is like other presentations you have done. However, it is specially designed to help you present results of an investigation. Presentations during *Investigation Expos* include your procedure, your results, and your interpretations of your results. *Investigation Expos* are similar to presentations scientists make to each other. Scientists present results of investigations to other scientists. This lets other scientists build on what they learned. You will do the same thing.

There are several things scientists usually want to know about investigations. These include the following:

- questions you were trying to answer in your investigation
- your predictions
- your procedure and what makes it fair
- your results and how confident you are about them
- your interpretation of the results and how confident you are of it

To prepare for an *Investigation Expo*, you will usually make a poster that includes all of the items listed above. You must present those things on your poster in a way that will make it easy for someone to follow. Others should be able to identify what you have done and what you found out. If you don't think your procedure was as fair or as accurate as you had planned, your poster should also have a report on how you would change your procedure if you had a chance to run the investigation again.

Sometimes, scientists make posters when they present their investigations and results. They set up their posters in a large room where other scientists have also set up their posters. Then other scientists walk around the room. They look at the posters and talk to the scientists who did the investigations. Another way scientists share results is by making presentations. For presentations, they stand in front of a room of scientists. They talk about their investigations and results. They often include visuals (usually slides) showing all the important parts of their procedures and results. They talk while they show the visuals. Then other scientists ask them questions.

Your *Investigation Expos* will combine these practices. Sometimes, each group will formally present their results to the class. Sometimes, each group will put their poster on the wall for everyone to walk around and read. In this *Investigation Expo*, because every group's observations were a little different, each group will present to the class.

Explain that an investigative procedure is considered fair if it is designed to answer the investigative question and if it compares things under the same conditions.

TEACHER TALK

“A fair investigative procedure is one where the procedures are designed to answer the investigative question and compares things under the same condition. For example, your investigative question is: What affects how chimpanzees feed? Your observation procedures should focus on answering this question. When you review your observations and try to make a statement from them, you should be comparing observations under similar conditions.”

Emphasize what groups will need to include on their poster for their presentation, using the bulleted lists in the student text. Groups will need to be able to inform the class of the questions they were trying to answer, their procedure, their observations and interpretations, and their categories.

Get Going

Show students an example of the poster materials they will be using. Let students know that their posters can include drawings, diagrams, and charts as well as written information. Show them the sample diagram in the student text.

Distribute poster materials and have groups get started.

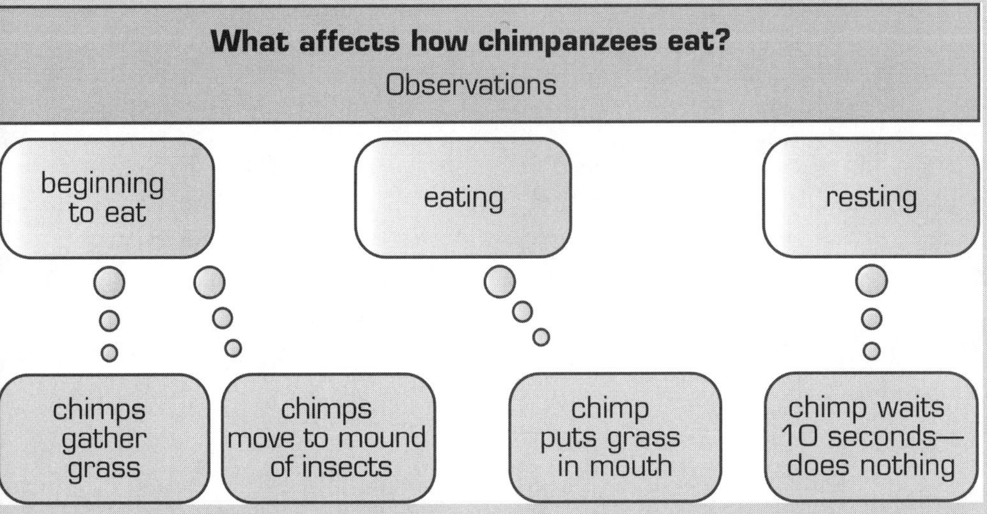

Also, point out the bulleted list of items the audience will be listening for during the presentation. Emphasize that they will have to answer these items during their presentations.

Guide and assess students as they prepare their posters and presentations.

△ Guide and Assess

Monitor and assist groups as needed. Some issues that may arise are listed below.

- Students might include opinions or interpretations in their observations. Ask them if anyone could disagree with their observations. If any of the observations are debatable, they probably include opinions or interpretations.

  An example of this is claiming that the chimpanzees are eating termites from the grass. Students cannot actually see termites on the blades of grass that the chimpanzees are eating, and the close up in the video of insects are of ants, not termites. It is known that chimpanzees eat ants and termites, but they have no direct evidence from the video that they are eating either. They may interpret this, but they should support this interpretation with evidence from the video. Their statements should not be definitive but more conditional, such as: "Perhaps the chimpanzees are eating the insects shown from the blade of grass." This is a good place to point out the importance of direct evidence supporting the interpretation. Students will obtain science knowledge based on other scientists' observations that allow them to make the claim that the chimpanzees are eating ants or termites off the blades of grass in the next section.

- Students may have trouble determining how confident they are of their analyses. They can base their confidence level on how careful they were to separate observations from interpretations and to keep opinions out of their analyses.
- Students may not give their observations and procedures in great enough detail. Ask them if the observations on their posters give a clear picture of what they observed and how they observed it.

Remind groups that everything they show on their poster or say in their presentation will have to be clear so that others can understand and follow their thinking.

Reflect

10 min.

Have each group present their observations and interpretations in an Investigation Expo.

There are two parts to an *Investigation Expo*: presentations and discussions. As you look at posters and listen to other groups present their work, look for answers to the following questions. Make sure you can answer this set of questions about each investigation:

- What was the group trying to find out?
- What procedure did they use?
- Was it a procedure that could help them answer their question?
- Was it a procedure that allowed them to collect accurate data?
- How consistent is their data?
- What did they learn?
- What conclusions do their results suggest?
- Do you agree with their results? Why or why not?

When looking at posters and listening to presentations, you should ask questions if you cannot identify a clear answer to any of the questions above. Ask questions that you need answered to understand results and to satisfy yourself that the results and conclusions others have drawn are trustworthy. Be sure that you trust the results that other groups report.

Reflect

Each group noticed some different things about chimpanzee behavior. By sharing your observations and interpretations with others, you probably discovered things you did not notice when you observed the chimpanzees. Answer the following questions. Be prepared to discuss your answers with your group and the class.

1. In what ways were the chimpanzees different from what you thought they would be?
2. The chimpanzees used different parts of their environment to help them feed. Describe how you saw the chimpanzees using the environment while they were feeding. What would be different if this part of the environment changed? How do you think the chimpanzees would react?
3. What parts of the chimpanzees' bodies were useful as they were feeding? How were these parts like human parts? How were they different? How would feeding be different if the chimpanzees' bodies were built differently?

META NOTES

Do not expect groups to have a lot of skill in presenting their observations, students' skills will improve with practice and time. However, it is important for the group and the class to notice differences in observations, conclusions, and procedures. The questions for the *Investigation Expo* will help to focus students on these points. It is important that the students lead the discussion as much as possible

◇ Evaluate

Before moving on, check to see if groups have completed the items required for their poster and ask them if they are ready for presentations. They should be prepared to say how their procedures helped them make their observations and whether they are confident of their analyses. Their posters should clearly show what their observations were and how they categorized them. Remind students that any interpretations (claims) made should be supported by their observations and no opinions should be included.

⬡ Get Going

Have each group display their posters around the room (see *Activity Setup and Preparation*). Allow everyone to visit each poster for a minute or so to become familiar with each groups' work.

△ Guide

Then, have each group present their results. Begin a short discussion after each presentation. Assist students with some of the language needed by modeling it for them or asking questions that guide them:

- Do you agree with what ... said? Why? or Why not?
- I agree with ... because ...
- I don't understand the reasoning behind your interpretation of your results. Could you start from showing the data and the trend?
- Why is this a fair procedure? Did your procedure answer the investigative question? Did you only compare similar events?

Encourage students to ask questions of the presenting group and the presenting group to respond to the student asking the question. Provide students with models of what you expect from them. Students should not make you the focal point during the discussion. For details on what is expected please refer to the *Teacher's Resource Guide.*

> **META NOTES**
>
> Questions should require more than one-word answers and should focus on a critical part of the investigation. Students will often want to question other students about color choice or layout. By modeling questions about content, you will also assist students in asking better questions.

☐ Assess Presentations

As you listen to groups' presentations, focus on the accuracy of the observations. Note any interpretations mixed in with the observations and look for interpretations that are supported by observations.

△ Guide

After all groups have presented, facilitate a class discussion of observations. Focus students' attention on the differences between different groups' observations and interpretations. Ask if these differences can be explained by differences in their procedures.

⬡ Get Going

After discussing the presentations with the class, let students know that they will now answer questions about what they saw and what their classmates saw. Give groups a time frame to answer the Reflect questions. Point out that for Question 1, they should compare their observations with what they initially thought prior to watching the video for the first time.

> **META NOTES**
>
> Keep the posters that groups created up for discussion in the next section.

4. The chimpanzee video was shot in the wild. Why do you think it is important for scientists to study animals in their natural environment?

What's the Point?

Scientists often observe animals in their natural habitat. When they observe animals, they take detailed notes. They pay attention to everything happening around the animal. But it is still difficult to make good observations in the wild, because things are constantly moving and the action is fast.

Chimpanzees are omnivores. They eat both plants and animals.

What's the Point?

5 min.

Emphasize the importance of detailed and accurate record keeping. Discuss how in the wild, things are constantly changing which makes record keeping difficult.

META NOTES

These questions are designed to help students think about what they saw, what others saw, and to begin to see how the animals' bodies and the habitat affect their behaviors.

META NOTES

You may want to mention how helpful modern technology, such as discreet video taping, has been in determining animal behaviors in the wild.

△ Guide and Assess

After groups have answered the *Reflect* questions, lead a class discussion of the answers. Listen for the following in students' responses:

1. Responses should compare students' initial ideas about chimpanzees discussed with their group before watching the video for the first time and the observations they have made since.

2. Students should have observed that the chimpanzees used grass stems, vines, and tree branches during feeding. Chimpanzees use a stem of grass to collect something (probably insects) on it, slid the stem of grass through their mouth, and then started chewing whatever was on the stem of grass during feeding. A chimpanzee also used vines or tree branches for support during feeding. The chimpanzees also sat during feeding. Students may suggest the following differences in the environment and their result.

- If there were no grass blades, chimpanzees may not be able to collect whatever they seemed to be eating from the blades.
- If there were no vines or tree branches, the chimpanzee would not use those for support while eating and may not support itself during feeding.

3. The chimpanzees used their hands and mouths. The chimpanzees' hands have long, jointed fingers and opposable thumbs (like humans' hands) that allow them to grip stems of grass. They also have lips (similar to humans' lips) that can hold things. Chimpanzees' mouths are larger than humans' in proportion to their faces and they protrude more. Their thumbs are separated from their fingers by a greater distance than humans' thumbs are. If chimpanzees didn't have opposable thumbs, they may not be able to hold or manipulate the grass to collect whatever they are collecting on the grass.

4. Students' responses should be based on their observations of how chimpanzees used their environment or were affected by their environment. For example, using grass to gather food is only possible because grass is available.

Assessment Options

Targeted Concepts, Skills, and Nature of Science	How do I know if students got it?
Behavior is a type of response to internal or external stimuli. Behavior is determined by heredity, experience, physical characteristics, and environment	**ASK:** What in the chimpanzees' environment or physical characteristics affected their behavior? **LISTEN:** Students should recognize that the stems of grass and chimpanzees' ability to hold them with their hands were important for the behaviors students observed.
The structure and function of animals' bodies are complementary and affect animal behavior.	**ASK:** How did the structure of chimpanzees' bodies affect their behavior? **LISTEN:** Students should describe a physical characteristic of the chimpanzee and how it helps, for example a chimpanzees' hands are shaped in a way that allows them to grip things. They use their hands to grip stems of grass.
Scientists often work together and then share their findings. Sharing findings makes new information available and helps scientists refine their ideas and build on others' ideas. When another person's or group's idea is used, credit needs to be given.	**ASK:** How did discussing your observations with your groups and then with the class help you understand what you observed? **LISTEN:** Any disagreements that arose should be mentioned here, as well as changes to groups' lists of observations and categories that were made during discussions.
Scientists differentiate between observations and interpretations. They use their observations and interpretations to explain animal behavior.	**ASK:** How can including interpretations in your observations lead to disagreements with other people's observations? **LISTEN:** Different people may interpret behaviors differently.

Targeted Concepts, Skills, and Nature of Science	How do I know if students got it?
Tables are an effective way to communicate results of a scientific investigation.	**ASK:** How did visual aids help you to communicate your ideas in your *Investigation Expo?* **LISTEN:** Students should discuss how they used tables, and charts in their *Investigation Expos*.
Scientific knowledge is developed through observations, recording and analysis of data, and development of explanations based on evidence.	**ASK:** How does analyzing your observations help you understand the behaviors you observe? **LISTEN:** Students should see that analyzing their observations helped them recognize where they had made assumptions and helped them to see how observations can be grouped by behavior.
Scientists make claims (conclusions) based on evidence obtained (trends in data) from reliable investigations.	**ASK:** What makes your interpretations (claims) good? **LISTEN:** Students should discuss how interpretations need to be supported by multiple observations and observations of the environment and physical characteristics of the animal that support the interpretation (claim).

- What behaviors or physical features of chimpanzees did students observe differently? How did their observation plans help them?
- What difficulties did students have in understanding the importance of making careful observations, keeping records, and sharing information? What ideas do you have to help students with these?
- What were some of the different ways students participated in analyzing data and constructing *Investigation Expo* posters? What ideas do you have for increased participation?

SECTION 2.3 INTRODUCTION

◂ *2 class periods**

2.3 Read

How Do Chimpanzees Feed and Why?

Overview

Students read an account of Jane Goodall's observations of chimpanzees in the wild. They also read some of the conclusions that the scientific community has drawn from Goodall's work. Using this information, students revise and update their interpretations based on their observations of the chimpanzee behaviors. Students then create explanations of how chimpanzees feed.

*A class period is considered to be one 40 to 50 minute class.

Targeted Concepts, Skills, and Nature of Science	Performance Expectations
Scientists often work together and then share their findings. Sharing findings makes new information available and helps scientists refine their ideas and build on others' ideas. When another person's or group's idea is used, credit needs to be given.	Students should share and discuss their observations with the class. They discuss all of the information they put on the *Project Board.*
Scientists must keep clear, accurate, and descriptive records of what they do so they can share their work with others, consider what they did, why they did it, and what they want to do next.	Students should record what they are learning and what they need to learn on the *Project Board.*
Scientists differentiate between observations and interpretations. They use their observations and interpretations to explain animal behavior.	Students should separate their interpretations from their observations on the *Observing and Interpreting Animal Behaviors* pages.

Targeted Concepts, Skills, and Nature of Science	Performance Expectations
Studying the work of different scientists provides understanding of scientific inquiry and reminds students that science is a human endeavor.	Students should consider the importance of Jane Goodall's work and think about how it is similar to their inquiry.
Scientists make claims (conclusions) based on evidence obtained (trends in data) from reliable investigations.	Students should construct claims based on observations and use these in their explanations.
Explanations are claims supported by evidence. Evidence can be experimental results, observational data, and other accepted scientific knowledge.	Students should create explanations, supporting their claims with evidence and scientific knowledge.
Behavior is a type of response to internal or external stimuli. Behavior is determined by experience, physical characteristics, and environment.	Students should include chimpanzees' environments and their physical characteristics in their explanations of chimpanzee feeding behavior.
The structure and function of animals' bodies are complementary and affect animal behavior.	Students should use the structure of chimpanzees' bodies to explain their behavior.
Organisms need food to grow, reproduce and maintain their bodies. They must be able to live in a changing environment.	Students should discuss how chimpanzees adapt to the behavior of termites and ants to get food.

Materials	
1-3 per student	*Observing and Interpreting Animal Behavior* page
1 per student	*Create Your Explanation* page
1 per class	Class *Project Board*
1 per student	*Project Board* page

Homework Options

Reflection

- **Science Process:** Describe the process you used to observe chimpanzee behavior and to develop explanations of how they feed. *(Students responses should include what they have done so far. Students began by watching chimpanzees feeding. They constructed an observation plan. Then they watched again, this time making observations using their observation plan. They categorized their observations according to behavior and discussed them with the class. Finally, they read about Jane Goodall's work and what she has learned, incorporated this into their interpretations, and created explanations of the behaviors.)*
- **Science Content:** What science knowledge did you use in your explanations and how did it help you to support your claims? *(Students may have used the fact that chimpanzees are omnivores, that the structure of chimpanzees' hands allows them to grasp and manipulate stems of grass, and that chimpanzees have been repeatedly observed using stems of grass as tools to get termites and ants out of their nests. Based solely on the video, students probably would not have had sufficient evidence to support their claims without additional science knowledge.)*

Preparation for 2.4

- **Science Content and Process:** What are some foragers that you are familiar with and how would you observe their feeding behavior? *(Students should describe the forager and consider how they could observe its feeding behavior.)*

NOTES

NOTES

SECTION 2.3 IMPLEMENTATION

◀ *2 class periods**

2.3 Read

How Do Chimpanzees Feed and Why?

One amazing scientist of our time is Jane Goodall. She is famous for studying chimpanzees in the Gombe, a wild, natural place in Africa. As a young woman, Jane Goodall became the student of another famous scientist, Louis Leakey. He suggested that there was important scientific work to be done observing the behavior of chimpanzees in Africa. Jane Goodall thought this sounded like quite an adventure. Over 35 years have passed, and Jane Goodall is still working with chimpanzees. She is now sharing her knowledge with her own students and the rest of the world.

Be a Scientist

Field Observations

Imagine you are a young scientist. Your name is Jane Goodall. You are working on your first scientific assignment on your own. That assignment is to observe the behavior of chimpanzees in their natural habitat in an area called Gombe in Africa.

AIA 53 ANIMALS IN ACTION

2.3 Read

How Do Chimpanzees Feed and Why?

10 min.

Introduce and discuss the work of Jane Goodall with the class.

META NOTES

Eliciting students' ideas can easily evolve into student discussions. Encourage this, guiding the discussions to be with the class.

○ Engage

Ask students if they know enough about chimpanzees' feeding behavior to know how to design a zoo enclosure for them that would allow them to feed as they do in their natural habitat. Record students' ideas of what they think they would still need to investigate.

*A class period is considered to be one 40 to 50 minute class.

META NOTES

Jane Goodall was chosen because she works with chimps and her work is interesting and current.

META NOTES

Case studies are often used as side notes or examples. This case study is used to provide scientific knowledge for students to use and build on.

META NOTES

This should help students connect their investigation to the investigations of scientists in the field and consider how analyzing and explaining the behaviors observed can lead to science knowledge.

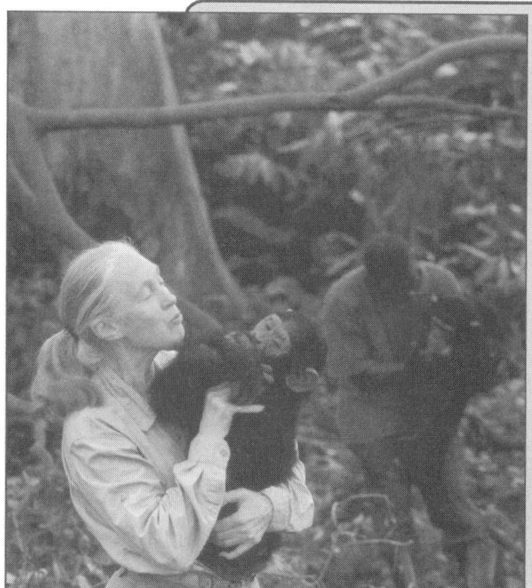

It took months of patient observations from a distance before Jane Goodall could get close to the chimpanzees.

You have prepared for months for this new position. You read about Africa so you would know what to expect there. You read as much as you could about chimpanzees to help you know what others before you had learned about their behavior. You have been interested in animals since you were a child. You are anxious to learn about chimpanzees in the wild. However, soon after you start your observations of chimpanzees, you discover that your job is not as easy as you may have thought. The chimpanzees do not let you get close enough to observe them.

You think that perhaps the chimpanzees are afraid because they do not know you. You decide to be patient. You start watching them from afar. You hope they will not be afraid of you. You follow them for days as they walk in the forest. You watch the chimpanzees as they play, eat, and socialize. After a long while, they allow you to get closer, and they actually approach you.

One rainy October day, you see something extraordinary. You are walking through tall, wet grass. Ahead of you is a chimpanzee you recognize. You have named him David Greybeard. He is hunched over a **termite** nest poking the entrance of the nest with a blade of grass and waiting patiently. When he pulls the blade out, it is covered with termites. David picks off the insects with his lips and eats them. You have observed David Greybeard using a tool to get his food!

termite: an insect similar to an ant that lives in large colonies. Termites feed on wood and can damage wooden structures.

Scientists such as Jane Goodall use questions to direct their investigations and plan what they need to observe. Jane Goodall wanted to know about chimpanzees. When she started her study, she watched everything the animals did. Because people did not know much about these animals,

Jane Goodall needed to watch them for a long time to develop good questions to investigate. When she saw David Greybeard use the twig to get termites, she realized that chimpanzees' feeding was very interesting. She was able to ask a more direct question, "How do chimpanzees feed?" All of Jane Goodall's careful observations allowed her to answer her question about how chimpanzees feed.

collaborate: to work together.

Be a Scientist

Collaboration

Scientists **collaborate** to better understand the world around them. Like all scientists, they build on the work of others. They share their ideas by writing papers, or articles, in scientific magazines called journals. They also attend meetings with other scientists and present their work. Presentations and articles are ways biologists tell others what they have discovered. When other scientists improve or add to an idea, they write a paper about it and publish it for others to read. They always acknowledge where they got the idea. Journal articles and presentations are good ways for scientists to share their ideas.

Although Jane Goodall does not make observations of chimpanzees in the wild anymore, she continues to share her understanding of chimpanzees with others. She is particularly concerned about the future of chimpanzees. In 1977, she founded the Jane Goodall Institute. One of the goals of the Institute is to expand research programs on chimpanzees. Many new scientists now work in the Gombe area. They continue the work Jane Goodall started and build on her observations. One of the several questions they are presently investigating is *How do female chimpanzees interact with each other?* Another goal of the institute is to preserve all of Jane Goodall's journals in digital form. They then can be available to other scientists, as well as to the public.

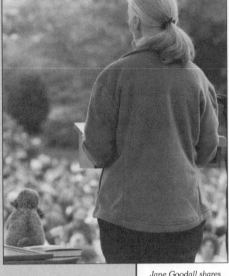

Jane Goodall shares her understanding of chimpanzee feeding in Gombe. Other scientists will then discuss her information and use it to develop their own knowledge.

META NOTES

This is a fictionalization of Jane Goodall seeing chimps eat with grasses or sticks for the very first time. She was prepared to see the behavior and was able to document and understand it. Hopefully, the whole story is intriguing and interesting to students.

△ Guide

Introduce what a case study is, how scientists use case studies, and how they will be using the case study of Jane Goodall.

Let students know that they will now learn about a scientist named Jane Goodall observing feeding behaviors in chimpanzees. Also let them know that after reading about Jane Goodall, they will read some of the science knowledge that Jane Goodall has contributed by studying chimpanzees' feeding behavior.

META NOTES

Students have been building on each other's work, sharing ideas and adding to them, just as biologists do.

TEACHER TALK

"You're going to read about the work of the scientist Jane Goodall. She is known for her work with chimpanzees. You should think about what you can learn from her observations. You should also think about what you can learn from her story about how scientists work. You'll have a chance later to use what you learn from studying Jane Goodall's work to explain your observations."

How Chimpanzees Feed

10 min.

Discuss the science knowledge about how chimpanzees feed from the reading.

Next, read the account of Jane Goodall's work with the class. Ask students how Jane Goodall's experience making observations is like theirs. Discuss with the class how Jane Goodall learned what aspects of chimpanzee behavior to study after she started observing chimpanzees in the wild. Emphasize that she began by watching chimpanzees and from her observations developed questions to investigate. She then was able to make careful observations to answer her question.

TEACHER TALK

"How did Jane Goodall begin her investigation of chimpanzees? What did she do after she saw David Greybeard using a tool?"

Then discuss how biologists collaborate to study animal behavior and how they share their work by writing articles and giving presentations. Ask students to point out similarities between how the biologists collaborate and how they have been collaborating in class.

○ Engage

Ask students why Jane Goodall was particularly interested to learn that chimpanzees use stems of grass to get termites. Then, ask them how the feeding behavior Jane Goodall observed is different from the feeding behaviors of most animals.

△ Guide

Next, discuss how chimpanzees' body structures and environment lead to this behavior. Chimpanzees have hands that allow them to grip stems of grass, and they can get termites out of their nest without damaging the nest by putting the stem of grass in an opening and then removing it, covered with termites.

Emphasize that this is an example of the chimpanzee using a tool to feed and that Jane Goodall's observations were the first of an animal, other than a human, using a tool to eat.

At one time, scientists thought that chimpanzees only ate fruit, leaves, flowers, and roots. However, scientists now know that chimpanzees are omnivores.

How Chimpanzees Feed

Jane Goodall and other ethologists have learned a lot about how chimpanzees feed. Chimpanzees are omnivores. This means they eat both plants and animals. Often, they eat fruits, leaves, and nuts from trees. At other times, they eat insects and small mammals. They choose what they eat based on what they can find. They move from place to place looking for food under rocks or in the bark of trees. Chimpanzees are foragers.

When Jane Goodall saw David Greybeard hunt termites, she was surprised to see that he used a tool to gather his food. Until that time, scientists thought humans were the only animals that used tools in that way. David Greybeard showed that other animals could learn to use tools.

David Greybeard and all the other chimpanzees could have just reached into the termite nests and grabbed the termites with their hands. Think about what might happen if they feed this way. The termite nest would get destroyed. By using a blade of grass or a twig to feed, the termite nest is not harmed. The termites can continue to live in it and make more termites.

If a chimpanzee ruins a nest in feeding, the supply of termites is lessened. Chimpanzees have found a way to adapt to their environment and protect it while they are feeding.

Chimpanzees are able to feed the way they do because of how their bodies work. They have hands similar to yours. They can pick up a single blade of grass or twig, hold it steadily and gently, guide it through a hole, and remove it carefully so the termites stay attached. Chimpanzees have been observed using other tools also. They crack nuts with rocks. Young chimpanzees learn to use tools by watching older chimpanzees.

Introduce the term omnivore. Chimpanzees are omnivores, and this affects their feeding behavior. They are foragers that will eat plants they find but will also eat termites, ants, and small mammals.

Stop and Think

15 min.

Have a discussion of Jane Goodall's observations and the chimpanzee behaviors they revealed.

○ Get Going

Let students know that they should first answer the questions on their own. Then they should discuss their answers with their group and record their best group answer. If group members cannot agree on an answer, then they should list all their answers. Tell students that groups will be sharing their answers during a class discussion. Let students know how much time they have to answer the questions.

△ Guide

Once groups have answered the questions, begin a class discussion of how some of the animal behaviors help the animals to survive and ask groups what their answers to the *Stop and Think* questions were. Listen for the following responses:

1. This question is meant to have students think about how Jane Goodall felt. Students' responses should show an understanding of how surprising Jane Goodall's discovery was.
2. Students should describe how chimpanzees' omnivorous diets allow them to eat many of the things they find foraging and how the structure of chimpanzees' bodies allows them to use tools. Students should also describe how termite nests present a problem (capturing termites without destroying the source of termites) and grass stems used as a tool presents a solution.
3. Chimpanzees' behaviors are responses to their environments; if chimpanzees' environment changes, their behaviors will likely change as well. The chimpanzees initially shied away from Goodall.
4. Students should use their experience of observing chimpanzees in the video and what they know from their reading about chimpanzee behavior to imagine Jane Goodall's observations.
5. Students should describe foraging behavior for herbivores and omnivores. A forager searches for food, and eats it wherever it can be found.

META NOTES

This discussion should illustrate and reinforce the difference between observations and interpretations. Make sure that examples of interpretations are discussed.

△ Guide

Next, ask students to review their old *Observing and Interpreting Animal Behavior* pages and consider their interpretations. Also, point out the interpretations in the students' posters from the previous section. Ask students what makes the interpretations different from the observations. Then, discuss some of the interpretations and what other observations about the environment or the animal allow the interpreted behavior.

Stop and Think

1. When Jane Goodall observed David Greybeard at the termite nest, she noticed something no one else had seen. Describe how you think Jane Goodall might have felt.
2. Describe the reasons why chimpanzees eat the way they do.
3. Jane Goodall watched the chimpanzees in the habitat where they lived. Why was it important to watch the animals in their own environment? What was difficult about watching the animals this way?
4. Jane Goodall made observations of the chimpanzees. Pretend you are Jane Goodall recording the observations. Record a descriptive observation of David Greybeard eating termites.
5. How does a forager obtain its food?

Reflect

Think about the observations and ideas presented in the previous section and what you read about in this section. Use an *Observing and Interpreting Animal Behavior* page, like the one shown, to record what you know about how chimpanzees eat. Include one behavior in each row of the chart. Then add what you know about what allows those behaviors and your interpretations of those behaviors in the other two columns. Collaborate with the members of your group to create one chart. Record all the ideas you have. Be prepared to share your observations with the class.

Observing and Interpreting Animal Behavior

Name: ______________________ Date: __________

Animal I am observing: ______________________

Observations	What about the environment and animal allows that behavior?	Interpretations

Reflect

15 min.

Have a class discussion on students' updated interpretations of the chimpanzee feeding behaviors they observed.

Get Going

Provide students with new *Observing and Interpreting Animal Behavior* pages and ask students to record what they know about the eating behavior of chimpanzees, what allows those behaviors, and the interpretations of those behaviors. Then have groups create a chart based on each student's ideas. Emphasize that students should include the science knowledge they learned from Jane Goodall's work.

Assess

As groups work with their *Observing and Interpreting Animal Behavior* pages, walk around the room and see if students are having trouble identifying interpretations. Also, note how different groups are doing things differently. Are some groups putting fewer observations on their sheets than other students? Are groups using different reasons?

TEACHER TALK

"Did you notice any interpretations in any groups' Investigation Expos? Can you see any now, looking at the posters? How do you know they're interpretations and not observations?"

Explain

15 min.

Have groups create explanations of their observations based on their interpretations and the science knowledge they learned.

Guide

When groups have finished, have a class discussion on each group's observations and interpretations. Ask groups who have the same interpretations, if their supporting reasons were the same. If not, they should describe how they were different. If disagreement arises concerning supporting reasons have groups describe why they chose their supporting reasons.

Now that groups have completed and discussed their interpretations, let them know that they will now create explanations of how chimpanzees feed using the *Create Your Explanation* pages. Remind students that an explanation is a statement that connects claims (interpretations or conclusions from an investigation) to evidence (the data you have collected, or observations) and science knowledge (accepted and verified ideas in the scientific community). Discuss the explanation template in the student text, which shows the structure of an explanation. Emphasize that students need to support their claim with reasons based on science knowledge and their observations. No opinions should be included. Students should use the example template to construct their explanation.

Emphasize to students that they should combine their interpretations to create a claim and use their observations as evidence and science knowledge to support their claim. Also, let students know that they will revise their explanations later in the *Learning Set.*

META NOTES

Students will revise their explanations in *Section 2.7* after learning more about what affects chimpanzee's feeding behavior.

Guide

Have a class discussion where at least two groups present their explanations for their interpretations. With the class, review each explanation and ask the class to pick out the parts of the explanation (claim, evidence, science knowledge). Encourage the class to discuss different ideas they have for the explanations presented. Ask groups with similar interpretations to share if they have anything different from the explanations presented.

TEACHER TALK

“The first time you listen to this explanation I want you to pick out all the parts of it — the claim (interpretations), the evidence (observations), and the science knowledge.

(After the explanation is presented:)

What are the parts of the explanation?

Let's listen to the explanation again. Can you think of any other observations (evidence) or science ideas that support or go against this interpretation (claim)? List these. Groups with similar interpretations, did you come up with the same explanations? Why or why not? List similarities and differences.”

After both explanations have been presented, ask the class if they have any different explanations and discuss these.

◇ Evaluate

Make sure that students are using evidence (their observations) and science knowledge to support their claims. Their evidence should not include any opinions or interpretations, and their claims should be valid (well supported by their evidence and science knowledge).

Update the Project Board

15 min.

Have a class discussion of what students are learning and update the Project Board.

○ Get Going

Draw students' attention to the *Project Board* and remind them that the *Project Board* is a way to organize their ideas and questions and to help them see how their ideas change as they gather more information.

Then, remind students what they have already recorded, and remind them of what the five columns are: *What do we think we know? What do we need to investigate? What are we learning? What is our evidence?* and *What does it mean for the challenge or question?* Also, remind students that while you update the class *Project Board*, they should update their *Project Board* pages

△ Guide

Model for students how to use columns three and four by emphasizing the direct link between the two columns. Remember to draw arrows directly linking the claim and evidence on the *Project Board*. You may want to use the example below.

Column three *(What are we learning?)* **Claim** (Interpretation): Chimpanzees use a stem of grass to get insects (such as ants and termites) to eat.

META NOTES

Students may disagree about what to put on the *Project Board*. Encourage discussions of anything students disagree about. Anything that students can't agree on should be added to the *What do we need to investigate?* column.

Create Your Explanation

Name:________________ **Date:**____________

Use this page to explain the lesson of your recent investigations.

Write a brief summary of the results from your investigation. You will use this summary to help you write your Explanation.

Claim—a statement of what you understand or a conclusion that you have reached from an investigation or a set of investigations.

Evidence—data collected during investigations and trends in that data.

Science knowledge—knowledge about how things work. You may have learned this through reading, talking to an expert, discussion, or other experiences.

Write your Explanation using the ***Claim***, ***Evidence*** and ***Science knowledge***.

Explain

You now have learned a lot about chimpanzees and how they feed. You know that they are foragers and that they eat all different types of food. You also know that they can use tools to help them find food. Now you will write an explanation of how chimpanzees feed and what helps determine how they feed. Using a *Create Your Explanation* page, write an explanation of how chimpanzees feed and why they feed that way.

Your claim will combine the interpretations you recorded on your chart. It will state how chimpanzees feed. You have evidence from your observations showing that they feed this way, and you have learned a lot from your reading about chimpanzee feeding. The science knowledge you record should state the reasons chimpanzees feed the way they do.

Include in the science knowledge for your explanation what you now know about how a chimpanzee's body and habitat affect the way it feeds. Use all this information to help you develop your explanation of why chimpanzees feed the way they do. Your explanation will be a logical statement that connects your claim to your evidence and science knowledge. It will look something like this.

> Chimpanzees feed by [*tell how*] because [*give reasons from your science knowledge*]. We can tell they do it this way because we can see [*tell what you observed*].

Your explanation should match your claim, your science knowledge, and your evidence. If they don't match, revise your claim and your explanation until everything makes sense together.

Develop the best explanation you can using the evidence and science knowledge you have available. You will get a chance to write another one later in the Unit when you know more.

Column four *(What is our evidence?)* **Evidence:** We saw a chimpanzee holding stems of grass down and then putting the grass in their mouth and sliding it out, and chewing. We also read that Jane Goodall observed a chimpanzee using a blade of grass or sticks to collect termites from their mound to eat. She observed the chimpanzee placing the grass or stick into an opening in the mound and then eating the insects that climbed onto it.

Ask students what they have learned about how chimpanzees behave and how biologists answer questions and solve problems. Students can review the readings and their explanations for ideas. The *Project Board* should

How do scientists answer big questions and solve big problems?				
Why do animals behave the way they do?				
What do we think we know?	What do we need to investigate?	What are we learning?	What is our evidence?	What does it mean for the challenge or question?

Update the *Project Board*

Earlier you began a *Project Board* focusing on the idea of creating an enclosure for animals. Now you have done some investigations and reading, and you know about how chimpanzees feed. You are now ready to record your knowledge about chimpanzee feeding on the *Project Board*.

You will focus on the third and fourth columns: *What are we learning?* and *What is our evidence?* When you record what you are learning in the third column, you will be answering some questions in the *What do we need to investigate?* column. But you cannot just write what you learned without providing evidence for your conclusions. Put your evidence in the fourth column. Some evidence will come from your data, and some will come from your reading.

You may use the text in this book to help you write about the science you have learned. However, make sure to put it into your own words. As your class works at the large *Project Board*, record the same information on your own *Project Board* page.

The *Project Board* is a great place to start discussions. You may find that you disagree with your classmates about what you have learned and the evidence for it. This is a part of what scientists do. These discussions help participants identify what they still don't understand well and what else they need to learn or investigate. Put any new questions your class develops in the *What do we need to investigate?* column.

·eflect everything that they know about chimpanzees. Update the *Project Board* with ideas from the class, draw arrows indicating the flow of ideas ınd remember to date the entries.

◇ Evaluate

Make sure students' entries pertain to claims and evidence about :himpanzees feeding behavior, especially the use of tools and why they night use those tools. Explanations should say that it is to not harm the ermite mound in order to protect their food supply.

What's the Point?

5 min.

What's the Point?

Feeding is a critical behavior for all animals. When animals feed, they eat food that gives them the nutrition they need. Different animals eat different foods.

Animals find food in different ways. Some animals forage. Chimpanzees are foragers. They gather plants and look for food in termite hills. Chimpanzees use tools to help them get termites out of their nests. Most other animals do not have hands that allow them to use tools the way chimpanzees do. Those animals have to get their food in different ways. The ability to use tools is one way chimpanzees are different from many other animals.

Because interpreting observations can be difficult, it is important for scientists to share their observations and ideas. This type of collaboration makes it possible for scientists to make sure their data is reliable. It also helps them build on each other's work. Building on the work of others allows scientists to answer questions they have developed but that they did not find answers to themselves. It also allows them to ask more detailed questions over time. For example, after Jane Goodall discovered that chimpanzees use tools, other scientists began to ask questions about whether other animals also use tools.

◇ Evaluate

Make sure students understand that it is important to share observations and ideas to improve interpretations.

Assessment Options

Targeted Concepts, Skills, and Nature of Science	How do I know if students got it?
Behavior is a type of response to internal or external stimuli. Behavior is determined by experience, physical characteristics, and environment.	**ASK:** What in the environment affects chimpanzees' behavior? What physical features of chimpanzees affect their behavior? **LISTEN:** Students should describe that the behavior of chimpanzees feeding on termites depends on the availability of grass stems or sticks and on the structure of chimpanzees' hands, which allows them use of grass stems or sticks to gather the termites.
The structure and function of animals' bodies are complementary and affect animal behavior.	**ASK:** How does the structure of chimpanzees' hands affect their behavior? **LISTEN:** Students should recognize that the structure of chimpanzees' hands allows them to grasp and manipulate stems of grass.
Organisms need food to grow, reproduce and maintain their bodies. They must be able to live in a changing environment.	**ASK:** How are the behaviors you observed adaptations to chimpanzees' environment? **LISTEN:** Students should see that the observed feeding behavior of chimpanzees is an adaptation to the nesting behavior of termites and ants and the availability of grass stems. It depends on the structure of chimpanzees' hands, which allows them to grasp and manipulate stems of grass.
Scientists must keep clear, accurate, and descriptive records of what they do so they can share their work with others, consider what they did, why they did it, and what they want to do next.	**ASK:** How does using the *Project Board* help your investigation? **LISTEN:** It allows students to keep records of what they have learned and what they still need to learn. It also allows them to consider what they did, why they did it, and what they want to do next.

Targeted Concepts, Skills, and Nature of Science	How do I know if students got it?
Scientists often work and then share their findings. Sharing findings makes new information available and helps scientists refine their ideas and build on others' ideas. When another person's or group's idea is used, credit needs to be given.	**ASK:** How did discussing observations with your group and then with the class help you understand what you observed? **LISTEN:** Any disagreements that arose should be mentioned here, as well as changes to groups' lists of observations and categories that were made during discussions.
Studying the work of different scientists provides understanding of scientific inquiry and reminds students that science is a human endeavor.	**ASK:** How did Jane Goodall begin her investigation? **LISTEN:** Jane Goodall began by watching chimpanzees without looking for anything in particular. By watching them, she developed a question. Then she made careful observations to answer her question.
Scientific knowledge is developed through observations, recording and analysis of data, and development of explanations based on evidence.	**ASK:** How does analyzing your observations help you understand the behaviors you observe? **LISTEN:** Students should see that analyzing their observations helped them recognize where they had made assumptions and helped them to see how observations can be grouped by behavior.
Explanations are claims supported by evidence. Evidence can be experimental results, observational data, and other accepted scientific knowledge.	**ASK:** How can you incorporate new scientific knowledge into explanations as it becomes available? **LISTEN:** New scientific knowledge can be used as evidence either to support the claims in your explanations or to modify them and support the revised claims.

Targeted Concepts, Skills, and Nature of Science	How do I know if students got it?
Scientists differentiate between observations and interpretations. They use their observations and interpretations to explain animal behavior.	**ASK:** How can including interpretations in your observations lead to disagreements with other people's observations? **LISTEN:** Different people may interpret behaviors differently.
Scientists make claims (conclusions) based on evidence obtained (trends in data) from	**ASK:** Why do we call interpretations claims? **LISTEN:** Students should describe how interpretations are based on many observations pointing to a subjective idea.
Explanations are claims supported by evidence. Evidence can be experimental results, observational data, and other accepted scientific knowledge.	**ASK:** How is an interpretation different from an explanation? **LISTEN:** In an explanation, claims are supported by the observations (evidence) and science knowledge. An interpretation is a claim based on many observations pointing to an idea that is subjective.

Teacher Reflection Questions

- What signs did students give of understanding the significance of Jane Goodall's work? What ideas do you have to guide their understanding?
- What ideas do you have that may guide students to construct explanations by supporting ideas with evidence?
- How were you able to keep discussion constructive while updating the *Project Board*? What would you do next time?

NOTES

SECTION 2.4 INTRODUCTION

2.4 Investigate

How Do Bees Forage?

◀ *1 class period**

*A class period is considered to be one 40 to 50 minute class.

Overview

Students read about how bees forage and then groups simulate the foraging habits of bees by using a nectar-collecting model. During the simulation, groups decide on a strategy to most efficiently collect nectar. Groups then share their strategies and results with the class. The class discusses the best strategy for bee foraging. The class also discusses that models generally have strengths and limitations and the limitations of their own model.

Targeted Concepts, Skills, and Nature of Science	Performance Expectations
Scientists often work together and then share their findings. Sharing findings makes new information available and helps scientists refine their ideas and build on others' ideas. When another person's or group's idea is used, credit must be given.	Students should work in groups to create strategies for collecting the most nectar from flowers and present their strategies to the class.
Scientists must keep clear, accurate, and descriptive records of what they do so they can share their work with others, consider what they did, why they did it, and what they want to do next.	Students should keep descriptive and accurate records of their strategies for the nectar-foraging simulation. Students should keep descriptive and accurate records of their strategies for the nectar-foraging simulation.
Scientists use models to simulate processes that happen too fast, too slow, on a scale that cannot be observed directly (either too small or too large), or that are too dangerous.	Students should describe what a model is and the limitations of the nectar-gathering model they used. They should also describe the usefulness of the model.
Behavior is a type of response to internal or external stimuli. Behavior is determined by experience, physical characteristics, and environment.	Students should describe how the environment affects the feeding behavior of bees

Targeted Concepts, Skills, and Nature of Science	Performance Expectations
The structure and function of animals' bodies are complementary and affect animal behavior.	Students should describe their thoughts on how the bee's body affects its feeding behavior. They should also be able to describe what types of food is needed in different development stages of the bee
Organisms need food to grow, reproduce and maintain their bodies. They must be able to live in a changing environment.	Students should describe how necessary it is for bees to be efficient feeders in order to survive.

Materials	
1 per group	Set of flower cards

Activity Setup and Preparation

The spy pens have two sides. One side has ink in it that is sensitive to a particular part of the spectrum. Marks made with this pen cannot be easily seen unless an ultra violet (UV) light is shined on them. The other side of the spy pen has a flashlight in it that allows you to see the markings made with the pen.

Use a spy pen to mark the cards numbered: 1, 2, 3, 7, 12, 13, 14, 15, 20 and 23. (You can mark them however you like, with a swirl, a pattern, or a simple line.) The marks should not be noticeable. Divide the cards into twelve-card sets. One set should contain the cards numbered from 1-12 and the other should contain cards 13-24. Students will not be using the spy pens in this section. They will be viewing the cards with the spy pens in the next section.

Review the rules for the nectar gathering simulation.

Homework Options

Reflection

- **Science Process:** How does the nectar-gathering model simulate what bees see to help you understand bee behavior? What are some limitations of using the flower cards? *(The flower cards allow you to explore how different bee behaviors could help them forage successfully or could prevent them from foraging successfully. Limitations occur because it doesn't account for* *bees' anatomy and it doesn't account for the differing distances between flowers bees might have to travel or the accidents they might encounter on the way.)*

- **Science Content:** How does the bee's environment affect its feeding behavior? *(Answers will vary but should contain features such as the bee's foraging depends on which flowers are available and that some flowers contain more nectar for the bee than others – this can depend on the color of the flower.)*

Preparation for 2.5

- **Science Content**: How do you think bees' eyesight affects their behavior? *(Students should recognize that if bees are unable to see certain colors, they will be unable to differentiate flowers of those colors. If they couldn't see blue and red, for instance, they would be unable to learn that blue flowers have nectar and red flowers do not.)*

NOTES

NOTES

SECTION 2.4 IMPLEMENTATION

◀ *1 class period**

2.4 Investigate

How Do Bees Forage?

You have learned about the feeding behavior of chimpanzees. How do you think the feeding behavior of bees is similar to and different from the feeding behavior of chimpanzees? Interestingly, like chimpanzees, bees are foragers. Unlike chimpanzees, which are omnivores, bees are herbivores, which means they eat only plants. During the spring and summer, bees fly many kilometers away from their hive, searching for food. But how do bees know which flowers have the **nectar** and **pollen** they need? How do they choose which flowers to land on? Making these decisions will determine how successful a bee is at foraging.

Bees use nectar to make honey.

In this section, you are going to use model flowers, called flower cards, to simulate the foraging behavior of bees and the decisions that bees make when they are foraging. You will learn more about how bees forage and the special ways that bees are able to use their bodies to make foraging efficient.

Be a Scientist

Using Models and Simulations

A **model** is a representation of something in the world. A globe is one model that you know. The parts of the globe represent parts of Earth. Scientists use models to investigate things that are too difficult, too dangerous, too large, or too small to examine in real life. Models are also used to investigate events that would take too long or occur too quickly to observe in real life. The best models are designed so people can easily examine them to better understand something in the real world.

To learn from using a model, the model needs to be similar to the real world in ways that are important for what the scientist is investigating. Sometimes what you want to model is a situation or event. To do this, you create a model that includes the things that are part of an event and then use that to act out the event. This is called a **simulation**.

nectar: a sugary liquid produced by plants.

pollen: small, powdery grains that contain the male sex cells in seed plants.

model: a way of representing something in the world in order to learn more about it.

simulation: use of a model to imitate, or act out, real-life situations.

2.4 Investigate

How Do Bees Forage?

10 min.

Introduce students the use of models and to some behaviors of bees.

META NOTES

This discussion helps students to make connections between the previous section and this one.

○ Engage

Begin by eliciting students' ideas of the similarities between how bees feed and how chimpanzees feed. Guide the class to discuss how both animals are foragers, but bees are herbivores and chimpanzees are omnivores.

Next, describe the foraging behavior of bees, emphasizing that bees fly many kilometers from their hives looking for flowers with nectar. Not all of these flowers have the nectar and pollen they need, and they have to make good choices to be successful at foraging.

*A class period is considered to be one 40 to 50 minute class.

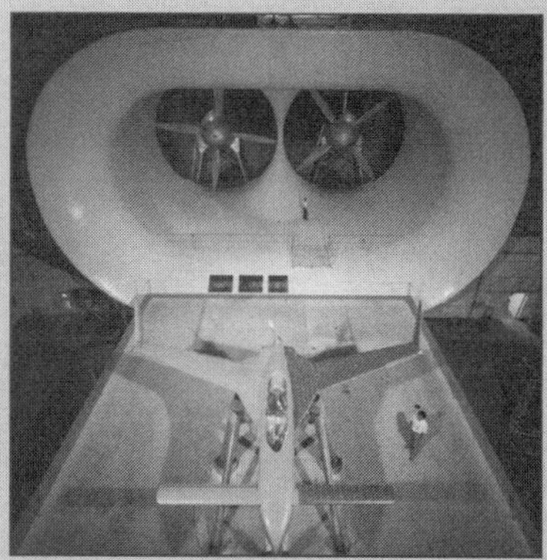

A wind tunnel is used to simulate the effects of air moving over or around objects, such as airplanes or cars.

Simulations use a model to imitate, or act out, real-life situations in a way that is similar to real life but that allows you to examine what is happening without causing any harm or danger. Scientists use simulations when what they want to study is too big or too small, too fast or too slow, or too dangerous to investigate directly.

Models and simulations show the natural world as best they can, but there are always limits to how well a model can show what happens in nature. Nature is very complex, and sometimes models simplify nature too much. But to understand something complex, scientists design models that are simpler than the natural world. When you are working with models, it is always important to note how the model you are using is similar to and how it is different than the real world.

Now you will model how bees forage. Before you can understand the model and make comparisons between the model and the life in the hive, you need more information about bee life. Read the following carefully to better understand what bees do in the hive and how they are responsible for foraging for food.

colony (plural, colonies): a group of similar organisms living or growing together.

larvae (singular, larva): the newly hatched form of an insect (in this case, the bee).

Life in the Hive

Honeybees live in large groups, called **colonies**, inside a hive. There are three types of bees in the hive: the queen, the drones, and the workers. The queen bee lays all the eggs. The drones, the male bees, fertilize the eggs. Worker bees are all females. They build and guard the hive, take care of the **larvae**, produce the honey, clean the hive, feed the queen and drones, and collect nectar and pollen from flowers. They use the nectar to make honey to feed the colony. They use the pollen to feed the larvae.

△ Guide

Let students know they will be using a model to simulate how bees forage and discuss models. The *Using Models and Simulations* box in the student text will help to guide the discussion. Describe to students how and when scientists use a model.

Discuss how scientists use models to simulate things that they are studying, provide examples such as the wind tunnel shown in the student text. Then have a class discussion on why they will use a model to simulate the choices bees have to make. Record students' ideas.

Bees gather pollen from flowers. They are attracted to colorful patterned flowers. Bees recognize the color of the flowers they visit and whether or not that color or pattern flower has nectar or pollen. When a bee sees a color or pattern of flower that has nectar or pollen, it lands on that flower and collects its nectar or pollen. When it sees a color or pattern of flower with no nectar or pollen, it avoids that flower.

When a bee sees an unknown flower on a foraging trip, it has to make a decision about whether or not to land on it. Flying and collecting nectar and pollen take a lot of energy. If the unknown flower has nectar and pollen in it, the bee can collect its nectar and learn about another kind of flower it can collect nectar from. If the unknown flower does not have any nectar or pollen, landing on it uses up energy that could have been used for finding other food. But if a worker bee knows exactly where to find a food source, it saves a lot of energy and time. So sometimes it is useful to check a new kind of flower to see if it has nectar or pollen.

Each time they go out to collect nectar and pollen, they have to bring back enough nectar to feed themselves and help feed the queen and drones. The bees in a hive will starve unless the worker bees are good at finding nectar and pollen.

This means that bees have to find flowers with nectar. Landing on flowers without nectar is dangerous, because it can use up a bee's energy before it has a chance to collect enough nectar. But learning which flowers have nectar and which don't will help a bee become good at nectar collection.

META NOTES

Do not distribute the flower cards to students yet.

TEACHER TALK

"Scientists use a model when they are interested in studying something that is too difficult, too big, too small, too fast, too slow, or too dangerous to study. By using models scientists can learn about a situation of event. The model must be similar to the real world."

Emphasize that before they can understand the model they will use, they will first need to know more about the life of bees. Using the information in the student text entitled *Life in the Hive*, discuss beehives and bee colonies.

Stop and Think

5 min.

Have a discussion of students' ideas about bees' feeding behaviors.

Stop and Think

Before you begin the investigation, use these questions to help you think about bees and how they forage.

1. Describe the environment of the bee and how you think that environment might affect how a bee will find food.
2. One reason chimpanzees are successful foragers is that they use their body in ways that help them forage. How do you think a bee's body can help it forage?

The Model

You are going to imagine you are a foraging bee. You have found a patch of flowers and need to collect as much nectar as possible while conserving energy. To do so, you must select the flowers with nectar and avoid the empty ones.

You and your partner will be given a set of 12 flower cards. Each card represents a different kind of flower. You will simulate the work of a foraging worker bee by selecting which of your flower cards the bee will visit.

You will want to select flowers with nectar while avoiding flowers without nectar. When you encounter an unknown flower, you will have to decide whether to check it for nectar. To do this, you will have to be able to identify which flowers have nectar. Read the box on the next page to know what information bees have about different flowers. This information will help you make good foraging decisions.

As you begin your foraging, pay careful attention to the differences between the flowers. Use what you know to land on flowers that contain nectar and avoid those that do not contain nectar. Remember that every visit to a flower with no nectar uses up energy without collecting more food.

Be a Scientist

Simulations Have Limitations

Simulations come close to how things occur in nature but also always have some limitations. In this simulation, you do not know if you have gathered nectar until your teacher tells you. A bee would know right away and might be able to visit another flower to gather more nectar. Also, as you will read later, the flower cards you use in this investigation are simpler than flowers are in nature.

META NOTES

Students will learn about the flowers bees feed from and the structure of bees' eyes later in the Unit. Right now, they should just begin to think about what they know about bees' feeding behavior and begin to wonder about what they don't understand.

Point out that every bee colony has a queen, drones (males), and workers (females). The workers care for the larvae, feed the queen and the drones, and build and maintain the hive. They use nectar to make honey to feed the colony and they feed the larvae pollen.

Provide students with some background knowledge of how bees collect pollen and nectar. Describe how bees use the colors and patterns of flowers to help them learn which flowers have nectar and pollen. Emphasize that if a bee sees an unknown flower, it has to make a decision about whether or not to land on it. Since it costs the bee energy to land on a flower, the bee needs to make smart choices about where to land.

⬡ Get Going

Let students know that they should answer the two questions and a class discussion will follow. Then let students know how much time they have.

△ Guide

Once groups have answered the questions, begin a class discussion of how bees' environment and bodies affect how they forage. Record students' ideas.

NOTES

The Simulation

15 min.

Explain the flower-card simulation and get groups going.

Simulation Rules

What you (a bee) know about the flowers:

- Blue flowers have nectar.
- Violet flowers do not have nectar.
- Purple and pink flowers are new. As a bee, you do not know if they have nectar until you land on them. You have to take your chances and choose which of these flowers to visit.

The Simulation

Remember that on each foraging trip the bee makes, it needs to bring back enough nectar to replenish its own energy and enough extra nectar and pollen to help feed the queen, the drones, and the larvae. In this simulation, you will forage for nectar by choosing cards from your deck to visit. You will not forage for pollen.

You will get 1 point for each flower you select that has nectar. You will lose 1/2 a point for each flower you choose that has no nectar. Your goal is to get at least 5 points—3 to represent the amount of food you (a bee) need to eat to get back the energy used in foraging, and 2 to contribute to the hive.

Procedure

1. Working with your partner, decide which of the 12 flower cards in your deck you will visit. Select between 5 and 9 cards. You may think you have a better chance of getting a full load of nectar if you choose more flowers. But remember that selecting the wrong flower results in subtracting points because of the extra energy a bee uses up.
2. Record your reasons for selecting each card. You and your partner may have some interesting discussions about which flowers to select. Perhaps you will even disagree about it. If you record what you talk about in these discussions, you will be able to participate in a discussion later about how bees might make their decisions.

△ Guide

Show students the flower cards and explain that they will simulate the way a bee makes choices using the flower cards as a model of the flowers bees might feed from. Ask students what the differences between the simulation and the feeding behaviors of real bees are and record their responses. Then, discuss the limitations of the simulation using the *Be a Scientist* box in the student text entitled *Simulations Have Limitations.*

3. After you have chosen your cards, look at the card numbers at the bottom right corners of the cards. Arrange your cards from lowest number to highest number. Put aside the cards you did not choose.
4. When everyone is finished, your teacher will tell you which of the flowers contain nectar and which ones do not. You will determine your foraging score based on this information.

Explain that some of the flowers have nectar and some do not. During the simulation, students will have to try to pick as many flower cards with nectar as possible while avoiding the ones without nectar.

△ Guide

Show students several examples of the flower cards. And, introduce the rules of the game.

- Blue flowers have nectar.
- Violet flowers do not have nectar.
- Purple and pink flowers are new. As a bee, you do not know if they have nectar until you land on them. You have to take your chances and choose which of these flowers to visit.

Discuss that the goal is to select flowers that will provide the bee with nectar and pollen for energy. Emphasize that it costs a bee energy to land on a flower with no nectar. You can demonstrate this by putting a card on one side of the room and another card on the opposite side of the room and asking if anyone would feel like going back and forth between the two cards. Then, ask students when it would be worth the effort. Emphasize that there would have to be a good reason to motivate anyone to go from card to card. For the bee, the reason would be nectar.

Next, review the steps in the procedure with the class.

META NOTES

Students should not use the numbers in the corners of the cards to choose, they should use the colors or patterns on the cards.

TEACHER TALK

"Your goal will be to get at least 5 nectar points. You will need 3 points for foraging and 2 should go back to the hive. To do this, you will work with your partner and decide on which of the cards you will visit. The maximum number of cards you can pick is nine. You will need to record your reasoning behind the cards you chose."

⬡ Get Going

Have groups pick their cards and record their reasons.

△ Guide

When groups have finished selecting their cards, have them put aside the cards they did not select and then explain the scoring guidelines. Have groups determine their foraging scores.

Guidelines for Scoring

NOTE: The maximum efficiency score is 8. It pertains to the bee that has selected five cards in this way:

a) Set of cards #1-12: 3 blue cards, 0 violet cards; cards #7 and #12

b) Set of cards #13-24: 3 blue cards, 0 violet cards, cards #20 and #23

Condition	Points
If you have selected #1	Add 1 point
If you have selected #2	Add 1 point
If you have selected #3	Add 1 point
If you have #4, but did NOT select it	Add 1 point
If you have #5, but did NOT select it	Add 1 point
If you have #6, but did NOT select it	Add 1 point
If you have selected #7	Add 1 point
If you have selected #8	Subtract ½ point
If you have selected #9	Subtract ½ point
If you have selected #10	Subtract ½ point
If you have selected #11	Subtract ½ point
If you have selected #12	Add 1 point
If you have selected #13	Add 1 point
If you have selected #14	Add 1 point
If you have selected #15	Add 1 point
If you have #16, but did NOT select it	Add 1 point
If you have #17, but did NOT select it	Add 1 point
If you have #18, but did NOT select it	Add 1 point
If you have selected #19	Subtract ½ point
If you have selected #20	Add 1 point
If you have selected #21	Subtract ½ point
If you have selected #22	Subtract ½ point
If you have selected #23	Add 1 point
If you have selected #24	Subtract ½ point

Stop and Think
15 min.

Have a class discussion on students' responses and results.

Stop and Think

You and your partner made decisions about picking flowers based on a strategy you chose. Different students used different strategies and may have had different foraging scores. The higher the score, the more efficient your foraging was. Answer the following questions to prepare for a classroom discussion.

1. How did you decide which flower cards to choose?
2. How did you decide how many flower cards to choose?
3. What was your final foraging score? What could have helped you improve your foraging score?
4. Is there a way for a bee to successfully bring back enough nectar if it lands only on flowers that have nectar? Record how you think a bee could, or could not, do that.
5. How did the color of flowers help you as you looked for food? Record how you think the color helps the bees.

What's the Point?

Models and simulations help scientists better understand the world. Scientists use models and simulations to determine how things work. Models and simulations come close to how things occur in nature but also always have some limitations. In this simulation, you made decisions about what flowers to land on. Bees also need to make these decisions. Bees use the information they learn to help them make decisions. One way this simulation was different from nature is that the flowers you had were much simpler than flowers are in nature.

In this simulation, you had to make decisions with your group. The decision discussions may have been difficult, and you may not have agreed with your group. Scientists also have disagreements. Scientists sometimes disagree about scientific understandings. Usually, this is because they are working from different, or incomplete, evidence. In the next section, you will find more evidence to support your ideas.

META NOTES

Question 4 is intended to get students to think about the decision making process and prepare them to find out that when they can see the marks, they will be able to make better, more efficient choices

⬡ Get Going

Let the class know that there will be a class discussion on the groups' responses. Also, let them know how much time they have to complete the questions.

△ Guide and Assess

While groups are working, check on their answers to see what you might want to focus on during the class discussion.

)nce groups have answered the questions, have a class discussion on their esponses. Encourage discussion of the groups' answers. Model for students)olite discourse as needed.

.isten for the following in students' answers:

1. Groups should have used some kind of criterion, such as the pattern or the color.
2. Students should have based this choice on the likelihood that the unknown cards contained nectar.
3. Groups should explore strategies for maximizing their score.
4. Since there are only three flowers that students know have nectar and they need to get nectar from five flowers, the bee has to take a chance on the mystery flowers.
5. Students should describe how they used the information of the color of the flowers in their decision-making process and then describe how they think it helps the bees to make their decision of which flower to visit. Encourage students to discuss and debate their decisions. Some students may suggest that bees can't see color the way humans do. Let them know that they will be learning about bees' vision in the next section.

5. How d... ...or of flowers help you as yo... ...a for food? Record how you think the color helps the bees.

What's the Point?

Models and simulations help scientists better understand the world. Scientists use models and simulations to determine how things work. Models and simulations come close to how things occur in nature but also always have some limitations. In this simulation, you made decisions about what flowers to land on. Bees also need to make these decisions. Bees use the information they learn to help them make decisions. One way this simulation was different from nature is that the flowers you had were much simpler than flowers are in nature.

In this simulation, you had to make decisions with your group. The decision discussions may have been difficult, and you may not have agreed with your group. Scientists also have disagreements. Scientists sometimes disagree about scientific understandings. Usually, this is because they are working from different, or incomplete, evidence. In the next section, you will find more evidence to support your ideas.

AIA 67 ANIMALS IN ACTION

What's the Point?

5 min.

Have a discussion of the usefulness and limitations of simulations.

△ Guide and Assess

Have a class discussion on how scientists use models and simulations. Simulations help scientists see how things work in nature, but they have limitations. Ask students what some of the limitations of the simulation they used were.

One limitation students should discuss is that you had to tell them after their cards were chosen whether they had nectar from them or not. Real bees get immediate feedback when they go foraging. They either get nectar or they do not. In our model, bees had to wait until someone else told them if they got nectar.

Discuss with the class that scientists often disagree and this is an indicator of scientists having different or incomplete evidence and the need for more investigations.

Assessment Options

Targeted Concepts, Skills, and Nature of Science	How do I know if students got it?
Behavior is a type of response to internal or external stimuli. Behavior is determined by experience, physical characteristics, and environment.	**ASK:** How would bees' behavior change if all of the flowers with nectar changed color? **LISTEN:** Bees would have to adapt using strategies like those discussed in class. They would have to start looking for the new colors associated with nectar.
Scientists use models to simulate processes that happen too fast, too slow, on a scale that cannot be observed directly (either too small or too large), or that are too dangerous.	**ASK:** What is a model, and how did the model you used compare with real bee-foraging behavior? **LISTEN:** Students should describe what a model is and compare the limitations of the model with what they currently know about bee foraging-behavior.

Teacher Reflection Questions

- What difficulties did students have making connections between animal behavior and the animal's environment? What ideas do you have to help students make that connection more easily?
- How could you guide students in understanding the importance of models?
- What would you do to improve participation in class discussions?

SECTION 2.5 INTRODUCTION

2.5 Investigate

What Do Bees See that Helps Them Forage?

◀ *2 class periods**

*A class period is considered to be one 40 to 50 minute class.

Overview

Students simulate bee foraging again, this time using ultraviolet (UV) flashlights to see markings sensitive to UV light on the flower cards. Students calculate their efficiency scores for their foraging simulation and are shown how much better everyone does when the model takes into account what bees see. The class then learns that bees are capable of seeing in the UV region of the spectrum and that many flowers have UV patterns visible to bees. Students then create explanations of bees' foraging behavior using their data from the simulations for evidence and using the science knowledge presented about what bees see.

Targeted Concepts, Skills, and Nature of Science	Performance Expectations
Scientists often work together and then share their findings. Sharing findings makes new information available and helps scientists refine their ideas and build on others' ideas. When another person's or group's idea is used, credit needs to be given.	Students should work in groups to run simulations on how bees forage and discuss bees' foraging behaviors and then discuss this as a class. The class should update their Project Board with claims about what they have learned and the evidence that supports these claims.
Scientists must keep clear, accurate, and descriptive records of what they do so they can share their work with others and consider what they did, why they did it, and what they want to do next.	Students should keep descriptive and accurate records of their strategies for the nectar-foraging simulation, and record their selection of flowers to visit for collecting nectar. Students should also keep records of their explanations, which will help them design a zoo enclosure.
Scientists differentiate between observations and interpretations. They use their observations and interpretations to explain animal behavior.	Students should separate their interpretations from their observations.

Targeted Concepts, Skills, and Nature of Science	Performance Expectations
Scientific knowledge is developed through observations, recording and analysis of data, and development of explanations based on evidence.	Students should be able to describe how scientific knowledge is developed and recognize that this is the process they have been using.
Scientists use models to simulate processes that happen too fast, too slow, on a scale that cannot be observed directly (either too small or too large), or that are too dangerous.	Students should use the nectar-gathering model to simulate bee-foraging behavior, adding to the model a UV light to simulate how bees see flowers. The model contains cards with UV sensitive ink.
Explanations are claims supported by evidence. Evidence can be experimental results, observational data, and other accepted scientific knowledge.	Students include observations and scientific knowledge in their explanations, linking their claims to their evidence.
Behavior is a type of response to internal or external stimulus. Behavior is determined by experience, physical characteristics, and environment.	Students should consider flower and bee respond and how the bee's experience in its environment affects its feeding behaviors.
The structure and function of animals' bodies are complementary and affect animal behavior.	Students should explain how the structure and function of bees' bodies, particularly their eyes, affects their feeding behavior.
Organisms need food to grow, reproduce and maintain their bodies. They must be able to live in a changing environment.	Students should include on their *Project Board* the reason for bees' foraging is for the survival for the entire hive.
Animals' sense of sight is adapted to their environment. Some animals see things that humans cannot see.	Students should recognize that bees' ability to see ultraviolet light allows them to distinguish ultraviolet patterns on flowers.
To see an object, light reflected from the object must enter the eye. Eyes of various animal species work differently to make them effective in their environment.	Students should recognize that a requirement of seeing an object is for light from the object to enter the eye. They should also be able to describe some of the features of the eye that allow animals to see and how these differ between bees and humans.
White light is composed of all the colors of the rainbow. The sun emits electromagnetic radiation of which humans can only detect a small portion with their eyes.	Students should recognize that the colors they see are a small portion of the electromagnetic spectrum, how white light is made up of all the colors of the rainbow, and why objects appear as different colors.

Materials	
1 per group	Set of flower cards
1 per group	Spy pen
1 per student	*Create Your Explanation* page
1 per student	*Observing and Interpreting Animal Behavior* page
1-2 per student	*Project Board* page
1 per class	Class *Project Board*

Activity Setup and Preparation

If you have not done this yet, use a spy pen to mark all the cards numbered: 1, 2, 3, 7, 12, 13, 14, 15, 20, and 23. (You can mark them however you like, with a swirl, a pattern, or a simple line.) Divide the cards into twelve-card sets. One set should contain the cards numbered from 1-12 and the other should contain cards 13-24.

Students should not open the pen side of the spy pens during class. Consider taping these shut. Keep in mind that a regular flashlight will not work, students will need to use the light on the spy pen.

Homework Options

Reflection

- **Science Content:** Why does a red apple appear the way it does? *(Responses should include that the color an object appears comes from the color of light reflected from the object (predominantly red for a red apple – the other colors are absorbed) and how the reflected light must enter an eye and stimulate a cone that is sensitive to that light.)*
- **Science Process:** What evidence did you use in your explanation? *(Students should describe observations from their simulations.)*
- **Science Content:** How could you use light bulbs of all the colors of the rainbow to get white light? *(Students should recognize that they should be able to combine the light from these bulbs to get white light.)*

- **Science Content**: How does light affect your behavior? How do you think light affects the behavior of bees or other animals? *(This question should get students to think about how their behaviors might change if they perceived light differently. Students should also describe how they think light affects the behavior of bees in particular animals.)*

Preparation for 2.5

- **Science Content**: Why do you think flowers might have evolved to have patterns that are visible to insects such as bees? *(This question should engage students in thinking about the relationship of insects to flowers.)*

NOTES

SECTION 2.5 IMPLEMENTATION

◀2 class periods*

2.5 Investigate

What Do Bees See that Helps Them Forage?

When you were simulating bee foraging, you probably wished you had more information about each of the flowers. Maybe you wished you knew what circles inside circles represented. Maybe there were other things you wanted to be able to see. In the next simulation, you are going to be able to see the flowers more like bees do.

The Model and Simulation

Just like last time, you are going to be a foraging bee. You and your partner will have 12 flower cards. Your goal will be to collect as much nectar as you can while avoiding landing on flowers that have no nectar.

One thing will be different this time. You will have a small flashlight that you can think of as a pair of "bee's eyes." In the last simulation, you used your human eyes to look for flowers. This time you will use the "bee's eyes" to look for flowers.

Procedure

1. Using the same flower cards as before, select between 5 and 9 cards to bring a full load of nectar back to the hive. Remember that selecting the wrong flower card results in subtracting points because of the extra energy the bee uses.
2. This time use the bee's eyes to help you make your foraging decisions. Select your flower cards and record the reasons for making your selections. Put the cards you did not select in a separate pile.
3. Listen as your teacher reads the scoring again. Keep track of your foraging efficiency.

Stop and Think

This time, you and your partner made decisions while using the bee's eyes. Think about how this was different from the first time you did the simulation. Answer these questions.

2.5 Investigate

What Do Bees See that Helps Them Forage?

5 min.

Introduce the revised model to students and show them how to use spy pens in the simulation.

Engage

Remind students of the nectar-gathering simulation they did in the previous section and have a discussion on how they think they might be able to improve the model so that they can better simulate bee foraging. Record students' ideas

*A class period is considered to be one 40 to 50 minute class.

△ Guide

After collecting students' ideas, let students know that one way they can improve the foraging simulation is to try to model how bees see flowers. Introduce students to the flashlight (spy pen) they will use to simulate how bees see and demonstrate how it should be used.

TEACHER TALK

"This time you will have a special flashlight that will allow you to simulate how a bee sees. This will allow you to see the flower cards using a model of bees' eyes to simulate how a bee sees flowers."

Procedure

5 min.

Review the procedures and have students simulate bee-foraging behavior using a model for a "bee's eyes."

META NOTES

Students should not have their notes out from the previous section

...thing will be differe... You will have a ... that you can think of as a pair of "bee's eyes." In the last simulation, you used your human eyes to look for flowers. This time you will use the "bee's eyes" to look for flowers.

Procedure

1. Using the same flower cards as before, select between 5 and 9 cards to bring a full load of nectar back to the hive. Remember that selecting the wrong flower card results in subtracting points because of the extra energy the bee uses.
2. This time use the bee's eyes to help you make your foraging decisions. Select your flower cards and record the reasons for making your selections. Put the cards you did not select in a separate pile.
3. Listen as your teacher reads the scoring again. Keep track of your foraging efficiency.

Stop and Think

This time, you and your partner made decisions while using the bee's eyes. Think about how this was different from the first time you did the simulation. Answer these questions.

Project-Based Inquiry Science AIA 68

△ Guide

Describe to the class that during this nectar-gathering simulation groups should choose flowers to visit using their "bee's eyes." Then, they should record their reasoning for choosing the cards they did. Let the class know that you will give them the scoring information after they finish selecting their cards. Remind students that they must record the reasoning behind their decisions.

Remind the class of the simulation rules that are located in the student text in *Section 2.4*.

TEACHER TALK

"While you are foraging for nectar you should remember the following:

- You get 1 energy point for each flower that has nectar.
- You lose ½ energy point for each flower that has no nectar.
- You need at least 5 energy points (3 energy points for your foraging and 2 energy points to bring back to the hive.)
- You are allowed to select between five and nine flowers to visit out of the twelve flowers your group will be given to choose from.

And, as a bee you know the following:

- Blue flowers have nectar.
- Violet flowers do not have nectar.
- Purple and pink flowers are new. As a bee you do not know if they have nectar until you land on them. You have to take your chances and choose which of these flowers to visit.

These rules are located in the student text in the Section 2.4."

Get Going

Next, distribute sets of twelve cards and flashlights (spy pens) to groups. As before, blue flowers have nectar; violet flowers do not have nectar; and purple and pink flowers may or may not have nectar. Groups should pick between five and nine cards, using the flashlight to help them decide which cards to pick.

Let the class know they have a couple of minutes to select their cards and let them begin.

After a few minutes, let groups know the scoring and have them figure out their score.

Guidelines for Scoring

NOTE: The maximum efficiency score is eight. It pertains to the bee that has selected five cards in this way:

a) Set of cards #1-12: 3 blue cards, 0 violet cards; cards #7 and #12

b) Set of cards #13-24: 3 blue cards, 0 violet cards, cards #20 and #23

Condition	Points
If you have selected #1	Add 1 point
If you have selected #2	Add 1 point
If you have selected #3	Add 1 point
If you have #4, but did NOT select it	Add 1 point
If you have #5, but did NOT select it	Add 1 point
If you have #6, but did NOT select it	Add 1 point
If you have selected #7	Add 1 point
If you have selected #8	Subtract ½ point
If you have selected #9	Subtract ½ point
If you have selected #10	Subtract ½ point
If you have selected #11	Subtract ½ point
If you have selected #12	Add 1 point
If you have selected #13	Add 1 point
If you have selected #14	Add 1 point
If you have selected #15	Add 1 point
If you have #16, but did NOT select it	Add 1 point
If you have #17, but did NOT select it	Add 1 point
If you have #18, but did NOT select it	Add 1 point
If you have selected #19	Subtract ½ point
If you have selected #20	Add 1 point
If you have selected #21	Subtract ½ point
If you have selected #22	Subtract ½ point
If you have selected #23	Add 1 point
If you have selected #24	Subtract ½ point

NOTES

1. What information did the bee's eyes give you that you did not have before?
2. How did that information help you make decisions?
3. Look at the cards that you now know have nectar and the ones that do not. Try to find patterns on them that might help you identify next time which flowers have nectar. If you are having difficulty doing that, describe why.
4. What was your final foraging score? Why was your score different from the first time you did this simulation?

receptor cells: the cells that receive information from the world and send it to the brain.

ultraviolet (UV) light: a kind of light not visible to the human eye.

How Bees See

Bees and other insects have eyes adapted for the tasks they need to do. A bee's vision is important for foraging. When a bee lands on a flower that has food, it will remember the size, shape, and smell of that flower. It will then find other flowers with the same characteristics. Special patterns on flowers direct the bee to where the nectar is in the flower.

However, bee vision is very different from human vision. Bees see different things than humans do because the **receptor cells** in bees' eyes are different than those in human eyes. Humans see the range of light from red to violet. A bee's receptor cells are sensitive to green, blue, and **ultraviolet (UV) light**. UV light is not visible to the human eye, but it is visible to the eye of the bee.

Bees' eyes allow them to see UV light.

Look at the pictures of geraniums in visible light and UV light. UV light reveals a secret. Some flowers have a dark spot in the center and lines leading from the petals to the center. These are visible only under UV light.

Geranium flower in visible light	The same flower in UV light

The same thing happened in your simulation. The flashlight you used is a UV flashlight. It allowed you to see markings on the flower cards not normally visible to you. The UV flashlight "changed" the UV markings into visible marks that you could see. A bee can see those markings without any special tools.

Stop and Think

15 min.

Have a class discussion of how using the bee's eyes changed students' results.

△ Guide and Assess

Have groups share their answers with the class. Encourage discussion between class members. During the discussion, listen for the following in students' answers:

1. The new model included the flashlight to simulate how bees see and allowed students to see patterns on the flowers that they could not see before. This simulates the fact that bees can see in the ultraviolet region, which students will learn more about later on.

META NOTES

Most flowers have patterns visible in UV light that are called nectar guides that show pollinators, such as bees, where to find the nectar. This is not the only way flowers attract bees for pollination.

2. Students should describe trends in their observations of flowers that showed patterns when the UV light was on and which flowers did not have patterns. They may have determined that flowers with patterns visible under the UV light had nectar.

3. The cards with patterns made visible by the special flashlight have nectar. The model is supposed to assist students in understanding that bees use their ability to see in the UV light to help them gather nectar. Remind students that most flowers in nature have a UV pattern visible to bees. These patterns are called nectar guides and bees use this in addition to the other colors and patterns they see to determine if the flower has nectar.

4. Students should look for reasons in their procedures for choosing cards. Students should have higher scores if they realized that the cards with the markings are the ones that they could observe under the flashlight. In general, students should have higher scores than last time for two reasons: 1) they have a better model and 2) they have done this once before and might remember which cards have nectar.

How Bees See

5 min.

Have a class discussion on how bees see.

the first ... did this simulation?

How Bees See

Bees and other insects have eyes adapted for the tasks they need to do. A bee's vision is important for foraging. When a bee lands on a flower that has food, it will remember the size, shape, and smell of that flower. It will then find other flowers with the same characteristics. Special patterns on flowers direct the bee to where the nectar is in the flower.

However, bee vision is very different from human vision. Bees see different things than humans do because the **receptor cells** in bees' eyes are different than those in human eyes. Humans see the range of light from red to violet. A bee's receptor cells are sensitive to green, blue, and **ultraviolet (UV) light**. UV light is not visible to the human eye, but it is visible to the eye of the bee.

Bees' eyes allow them to see UV light.

Look at the pictures of geraniums in visible light and UV light. UV light reveals a secret. Some flowers have a dark spot in the center and lines leading from the petals to the center. These are visible only under UV light.

The same thing happened in your simulation. The flashlight you used is a UV flashlight. It allowed you to see markings on the flower cards not normally visible to you. The UV flashlight "changed" the UV markings into visible marks that you could see. A bee can see those markings without any special tools.

Geranium flower in visible light

The same flower in UV light

Reflect

Use an *Observing and Interpreting Animal Behavior* page, like the one shown, to record what you know about how bees forage. Use your observations from the previous section and what you read in this section. Include one behavior in each row of the chart. Then add what you know about what allows those behaviors and your interpretations of those behaviors in the other two columns. Collaborate with the members of your group to create one chart. Record all the ideas you have. Be prepared to share your observations.

Observing and Interpreting Animal Behavior

Name: ______________________ Date: ____________

Animal I am observing: ______________________

Observations	What about the environment and animal allows that behavior?	Interpretations

Explain

You have written several explanations. You are probably getting more used to developing explanations and supporting your claim with evidence and science knowledge. Develop an explanation about the foraging behavior of bees. Be sure to include information about bees' bodies and about the environment that might affect how they forage. Use information on your *Observing and Interpreting Animal Behavior* page to support your claim. Remember that your evidence and science knowledge can come from your experiences with the bee investigation and the science reading you have done. Reread the hamster example from *Learning Set 1* if you need more help with the different parts of the explanation.

Keep in mind that your explanation should be based on what you know now. You will have a chance to develop a more detailed explanation later as you learn more about bees and about what affects animal feeding.

Reflect

10 min.

Have a discussion of groups' observations and interpretations.

META NOTES

Humans are incapable of seeing ultraviolet light. Ultraviolet flashlights do not allow us to view UV light. We only see in the visible region of the spectrum. However, instruments have been made that allow us to detect the UV light reflected off objects (such as flowers). We can view this in visible light (with false coloration) to get a sense of the amount of UV light being reflected and where it is reflected from. Again, the information from the instrument (such as a digital camera with filters and hot mirrors) is imaged for us to "see" in visible light.

The way the UV pen/light works is through fluorescence. The ink absorbs UV light emitted by the flashlight and emits visible light. We see the visible light, not the UV light.

△ Guide

Discuss the differences between bee vision and human vision using the information in the student text. Then summarize the discussion by reviewing the main points.

Let students know that few animals—birds, butterflies, and some other insects—have evolved to be able to see ultraviolet light.

Emphasize that by using the special flashlights and special markings the nectar gathering model was improved.

Explain

5 min.

Have students create and discuss explanations of bees' foraging behavior.

META NOTES

Constructing a scientific explanation is not something most students have done prior to this curriculum. It is important to scaffold for them how to construct an explanation by reminding them of the components of an explanation, discussing what these components are, and providing them with examples.

TEACHER TALK

"Bees' eyes have receptor cells that are sensitive to ultraviolet light, as well as green and blue light. Like humans, they have only three receptor cells. Humans' eyes are sensitive to the range of colors from red to violet. Humans have blue, green, and red receptor cells. Most flowers have patterns that are only visible in UV light, and bees use these patterns to help them recognize flowers that have nectar and to locate where the nectar is. The bees' vision has adapted to help them forage more efficiently."

△ Guide

Emphasize how the structure and function of bees' eyes affect bees' feeding behavior.

TEACHER TALK

"If bees could not see UV light, they would not be able to differentiate the patterns on most flowers. These patterns help bees to forage for nectar more efficiently."

Let students know that they will be working with their partners and recording their observations from the previous section and this section in the *Observations* column on an *Observing and Interpreting Animal Behavior* page. Remind students that these are not observations of actual bee behavior, but observations from the simulation about what foraging strategies are effective. Let students know that in the second column they should record what they know about bees' environment that allows those behaviors. In the third column, they should record their interpretations.

⬡ Get Going

Distribute the *Observing and Interpreting Animal Behavior* pages and let students know how much time they have.

□ Assess

As groups are working, check that students are recording their observations and science knowledge in the correct columns. Determine whether or not students have appropriate supporting information about the bee's environment or physical characteristics that support their interpretations.

△ Guide and Assess

Have a class discussion on groups' interpretations and their supporting evidence. Encourage groups to discuss, challenge, and clarify their interpretations based on the class discussion.

△ Guide

Remind students what an explanation is—a claim or statement that is supported by evidence and science knowledge in a logical way. They should use their interpretations to develop a claim and then use their observations and science knowledge from their reading to support the claim. Emphasize that every explanation is comprised of a claim (a conclusion or interpretation), evidence (observations), and science knowledge (ideas accepted in the scientific community).

Then provide students with an example of an explanation.

TEACHER TALK

"Let's first discuss an example of an explanation before you begin writing your own. Here is an explanation:

You notice that your iguana sits in the windowsill in the afternoons. You have read that some animals, especially cold-blooded reptiles like your iguana, warm themselves by lying on warm rocks in the sun. You wonder if the iguana is staying warm in the sunlight that comes through the window in the afternoons. To test this, you close the blinds in the afternoon. The iguana does not go to the windowsill. Then, with the blinds still closed, you put a warm electric blanket in the windowsill. The iguana returns and lies on the electric blanket. You conclude that the iguana lies in the windowsill in the afternoons because it is warm.

Your claim: My iguana lies in the windowsill because it is warm.

Your evidence: I performed an experiment. When the blinds were closed in the afternoon, the iguana did not lie in the windowsill. When the blinds were open and sunlight was entering or when the blinds were closed and an electric blanket was placed on the windowsill, it did.

Your science knowledge: Some animals, especially cold-blooded animals like my iguana, lie in the sun on warm rocks to gain heat.

Your explanation: My iguana lies in the windowsill when the sun is shining through because my iguana is warming itself in the sunlight. I tested this. When I closed the blinds my iguana did not sit in the windowsill in the afternoon. When the blinds were open and sunlight entered or when the blinds were closed and an electric blanket was placed in the windowsill my iguana would sit there. It is known that cold-blooded animals such as my iguana warm their bodies by sitting on warm rocks or in the sun."

⬡ Get Going

Have groups create explanations of bees' foraging behaviors using *Create Your Explanation* pages.

△ Guide and Assess

As groups are working on their explanations, go around the room and ask students what their claims are and what evidence they are using to support their claims. Their claims should be valid—they should be based on the evidence and their evidence should not include any opinions.

Below is an example of explanations about bee foraging.

Your claim: Bees use the colors and patterns they see to determine which flower they will visit to gather nectar.

Your evidence: I performed a simulation using a nectar-gathering model that consisted of flower cards. The only cards with nectar were cards with patterns in the ultraviolet spectrum visible to "bee's eyes". This was a simulation to show that bees use their ability to see in the UV spectrum to gather nectar. (Most flowers have UV patterns called nectar guides that help bees forage for nectar.)

Your science knowledge: Bees' eyes can see light from green to ultraviolet. Most flowers have patterns in the ultraviolet region that lead from the petals to the center of the flower where the nectar is. Bees use the visible color and patterns, as well as the UV patterns to help them discern if they should visit a flower to try to gather nectar and pollen. Bees must efficiently gather food for themselves and the hive.

Your explanation: Bees determine which flowers to visit based on the colors and patterns they see. The bees' eyes can see from green to ultraviolet. When the bee recognizes a flower's color or pattern it lands on it to gather pollen and nectar. In this way, bees can efficiently gather food for the hive.

Hold a class discussion on group's explanations. Allow a few groups to share their explanations and describe their evidence and science knowledge.

Update the *Project Board*

You have now learned a lot about how bees forage. Add what you have learned to the third column of the *Project Board*. Make sure you add evidence from your investigations and reading in the fourth column. Then add questions to the second column. For example, you might wonder why bees see differently than people. As your class records their learning on the class *Project Board*, add the information to your own *Project Board* page.

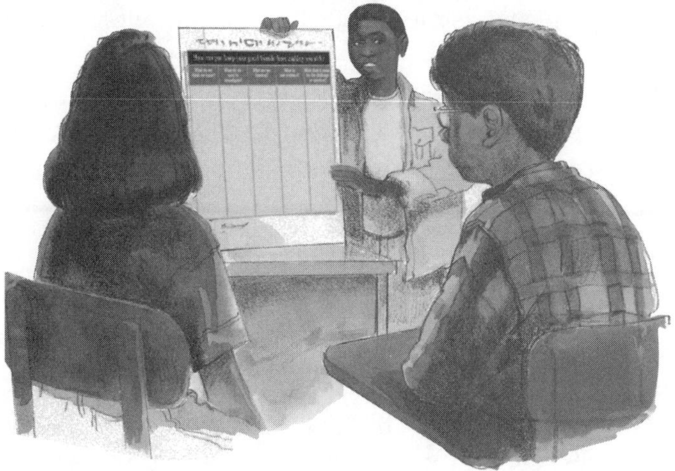

What's the Point?

Foraging is one way animals feed. Foraging animals move from place to place to find their food. Some foragers, including bees and other insects, are herbivores. Some omnivores, like chimpanzees, are also foragers.

Bee foraging behavior is efficient because bees' bodies have adapted to finding pollen and nectar in flowers. Bees' eyes help make bees very efficient foragers. When bees look at flowers, they see them differently from the way other animals see them. Bees' eyes are tools that help them to be efficient in their environment.

Update the *Project Board*

10 min.

Record students' ideas on the class Project Board.

△ Guide

Have a class discussion on what students have learned and what they think they should add to the *Project Board.* Focus on the third and fourth columns—*What are we learning?* (our claims) and *What is our evidence?* (our observations and science knowledge).

Record ideas on the class *Project Board* and have them record their ideas on their own *Project Boards.*

Students should also add information from *Section 2.4.*

TEACHER TALK

"We've been investigating bees and how they forage. What do we know about bees' feeding behaviors now and what is our evidence? Think about all of the explanations you heard and our class discussion on them to help you decide what to put up on our Project Board."

META NOTES

Consider writing the information on the board first, so that you can edit it before writing it on the *Project Board.*

Example:

Claim: Bees visit flowers with patterns in the ultraviolet spectrum.

Evidence: During our simulation, the only flowers with nectar were ones that had patterns visible to bees in the ultraviolet spectrum. (Most flowers have patterns in the UV light, these patterns help bees forage for nectar.) Bees can see in the ultraviolet light. Flowers have patterns visible in ultraviolet light called nectar guides. These patterns consist of lines from the petals to a dark center where the nectar is.

META NOTES

Project Board entries should be dated and arrows drawn to link the evidence (column 4) to the claims (column 3).

Claim: Bees live in large groups called colonies and need to collect nectar and pollen for the bee colony to survive.

Evidence: Bees work together for the survival of the colony. The worker bees are in charge of collecting food (nectar and pollen) for the entire colony. To do this efficiently, they must use their eyes' adaptations. Bees' eyes can see from green to ultraviolet light in the spectrum. Bees can recognize colors and patterns that help them to determine which flowers to gather nectar from.

Evidence: Bees work together for the survival of the colony. The worker bees are in charge of collecting food (nectar and pollen) for the entire colony. To do this efficiently, they must use their eyes' adaptations. Bees' eyes can see from green to ultraviolet light in the spectrum. Bees can recognize colors and patterns that help them to determine which flowers to gather nectar from.

What's the Point?

5 min.

Emphasize that animals adapt to their environment to find food. Bees' eyes, for example, have adapted to help them efficiently forage for nectar.

◇ Evaluate

Make sure the class *Project Board* contains information about bees' bodies, in particular their eyes, and how they are suited for foraging nectar.

What's the Point?

Foraging is one way animals feed. Foraging animals move from place to place to find their food. Some foragers, including bees and other insects, are herbivores. Some omnivores, like chimpanzees, are also foragers.

Bee foraging behavior is efficient because bees' bodies have adapted to finding pollen and nectar in flowers. Bees' eyes help make bees very efficient foragers. When bees look at flowers, they see them differently from the way other animals see them. Bees' eyes are tools that help them to be efficient in their environment.

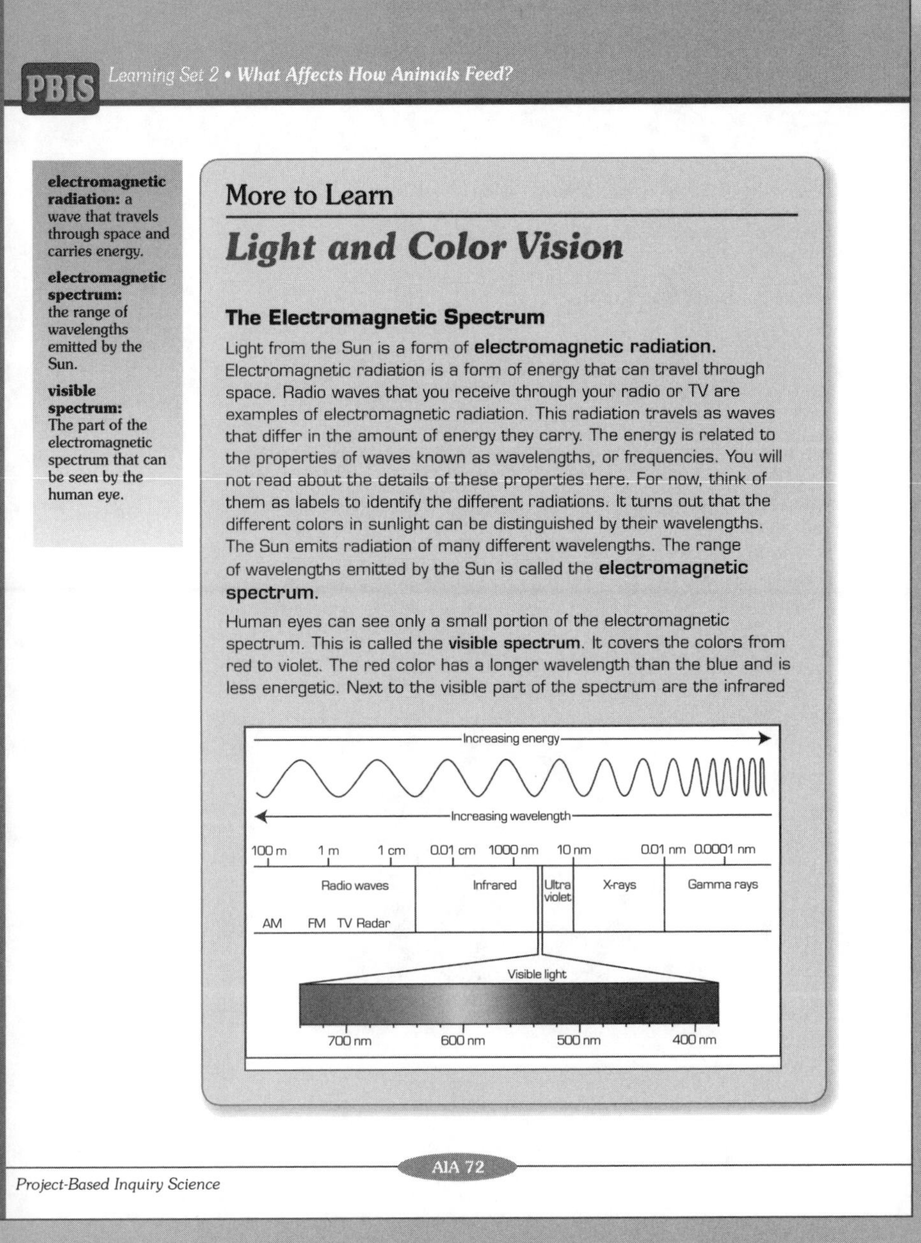

electromagnetic radiation: a wave that travels through space and carries energy.

electromagnetic spectrum: the range of wavelengths emitted by the Sun.

visible spectrum: The part of the electromagnetic spectrum that can be seen by the human eye.

More to Learn

Light and Color Vision

The Electromagnetic Spectrum

Light from the Sun is a form of **electromagnetic radiation.** Electromagnetic radiation is a form of energy that can travel through space. Radio waves that you receive through your radio or TV are examples of electromagnetic radiation. This radiation travels as waves that differ in the amount of energy they carry. The energy is related to the properties of waves known as wavelengths, or frequencies. You will not read about the details of these properties here. For now, think of them as labels to identify the different radiations. It turns out that the different colors in sunlight can be distinguished by their wavelengths. The Sun emits radiation of many different wavelengths. The range of wavelengths emitted by the Sun is called the **electromagnetic spectrum.**

Human eyes can see only a small portion of the electromagnetic spectrum. This is called the **visible spectrum**. It covers the colors from red to violet. The red color has a longer wavelength than the blue and is less energetic. Next to the visible part of the spectrum are the infrared

More to Learn: Light and Color Vision

5 min.

Discuss light and how humans see colors.

○ Engage

Begin by asking students where the ultraviolet light that bees see come from. Ask them what they know about white light and how we see colors. Record their ideas.

△ Guide

Then, explain that the only way an animal can see an object with its eyes, is if light from the object enters the animal's eye.

More to Learn

Humans cannot see IR and UV wavelengths, but some other animals can. For example, birds and some insects, like bees and some butterflies, can see UV light. Some animals, like snakes, can sense infrared light.

How Humans See Colors

Although small, the human eye is a remarkable organ. Basically, it works like a small camera. The light enters through a single transparent lens. The lens focuses the image on a layer that is sensitive to light called the **retina**. The retina lines the back portion of the inside of the eye. It contains two types of cells, **rods** and **cones**. The rods are the sensors that allow you to see in low light. The cones allow you to see colors.

The cones in the human retina are sensitive to red, green, and blue wavelengths of light. When the light strikes the retina, it generates an impulse that is transmitted to the brain by a nerve, called the optic nerve. By combining impulses received from the red, green, and blue cones, the human eye can see almost any color.

Objects appear to be a particular color because they reflect some of the wavelengths of the light they receive more than others. A red apple is red because it reflects light from the red end of the spectrum, and absorbs light from the blue end. When light reflected from an object strikes the retina, it stimulates the cones in such a way that you see the color.

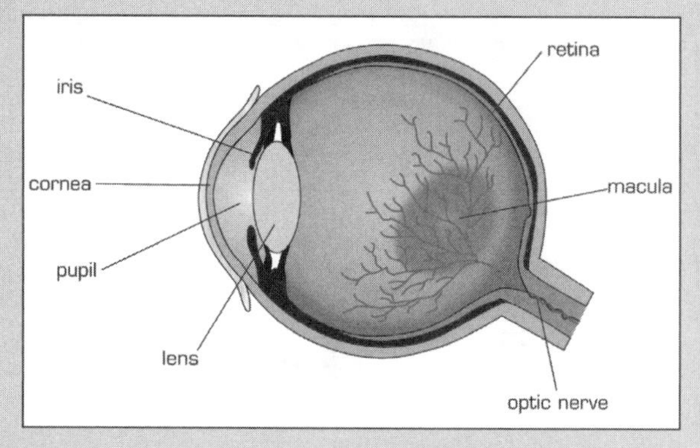

retina: a membrane sensitive to light that lines the back portion of the inside of the eye.

rod: a type of cell found in the retina. Each rod is sensitive to low levels of light, but cannot see colors.

cone: a type of cell found in the retina. Each cone is sensitive to one color: red, green, or blue.

AIA 73 ANIMALS IN ACTION

The light from the object either has to be created by the object (e.g., the sun creates its own light), or reflected by the object (the moon reflects sunlight). Let students also know that light travels in straight lines.

Next, use the information in the student text and discuss the electromagnetic spectrum. Let students know that the light humans can see is called visible light and it is electromagnetic radiation. Describe how visible light makes up only a small portion of the electromagnetic spectrum, using the image in the student edition to emphasize this point. Then, describe all the other types of radiation that compose the electromagnetic spectrum: radio waves, infrared light, ultraviolet light, and X-rays. Point out that radiation with high frequencies and short wavelengths (like gamma rays) have more energy than radiation with low frequencies and long wavelengths (like radio waves).

Let students know that white light is composed of all the colors of the rainbow.

TEACHER TALK

"When all the colors of the rainbow are combined, you get white light. This is not like paint. White light can be separated into all the colors of the rainbow by using a prism or water droplets when a rainbow appears."

Remind students that eyes need light to enter in order to see. Ask students what else they know about how the human eye works. Describe the human eye to the class using the information in the student text *How Humans See Colors*. Describe how human eyes have a lens, much like a camera lens, and that this lens focuses light on the retina, a layer of light-sensitive tissue at the back of the eye. Then describe how the retina contains rods and cones; rods allow us to see in low light, and cones allow us to see colors.

Provide an example of how we see a particular color by describing how some frequencies (or colors) of the white light get absorbed by an object and others get reflected. What gets reflected enters our eye and we see it as that color. An example is someone's red shirt. Light at the red end of the spectrum gets reflected by the shirt and goes into our eyes so we see red. The other light (the green and blue part of the spectrum) gets absorbed by the shirt. A white shirt appears white because most of the light is reflected off of it and enters our eyes. A black shirt appears black because most of the light is absorbed by the shirt and none or very little of it enters our eyes.

Describe how the light entering our eye strikes the red, green, and blue cones that send impulses to the brain and our brain/eye system then perceives the color.

Assessment Options

Targeted Concepts, Skills, and Nature of Science	How do I know if students got it?
Scientists must keep clear, accurate, and descriptive records of what they do so they can share their work with others, consider what they did, why they did it, and what they want to do next.	**ASK:** How can you use what you've learned in future explanations? **LISTEN:** Students should recognize that keeping records of their work allows them to use it for future work.
The structure and function of animals' bodies are complementary and affect animal behavior.	**ASK:** How does the ability to see ultraviolet light affect bees' behavior? **LISTEN:** Because bees are able to see ultraviolet light, they can differentiate flowers that have ultraviolet patterns and use this information to find nectar.
Animals' sense of sight is adapted to their environment. Some animals see things that humans cannot see.	**ASK:** What do bees see that humans do not and how does it help them in their environment? **LISTEN:** Bees can see ultraviolet light that helps them forage for nectar. Flowers have patterns on them that are visible in the ultraviolet region of the spectrum.
Scientific knowledge is developed through observations, recording and analysis of data, and development of explanations based on evidence.	**ASK:** What steps did you have to use to develop your explanation? **LISTEN:** Students should describe how they had to run a simulation and record and analyze their observations, and formulate a claim.
Explanations are claims supported by evidence. Evidence can be experimental results, observational data, and other accepted scientific knowledge.	**ASK:** How did you support your claims? **LISTEN:** Students should describe how they used observations and scientific knowledge to support their claims.

Teacher Reflection Questions

- What concepts in this section confused students, and what ideas do you have to enhance their understanding of these concepts?
- What evidence do you have that students made connections between the models they created and how bees forage? What ideas do you have for guiding students to understanding the importance of models
- How did you manage the reading segments? How would you change this next time?

SECTION 2.6 INTRODUCTION

2.6 Read

What Adaptations Do Bees Have that Affect Their Feeding Behavior?

◀ $1\frac{1}{2}$ *class periods**

*A class period is considered to be one 40 to 50 minute class.

Overview

Students consider and discuss the mutualistic relationship between bees and flowers. Students read about how flowers have adapted to attract bees and other small animals that will spread their pollen, and how bees have adapted to discern which flowers have nectar. The class discusses mutualistic relationships, learning that this is one way living things adapt to other living things in their environments. The class updates the *Project Board* with what they have learned, connecting the science knowledge presented in this section to larger ideas about animal behavior. Then, student groups dissect different flowers to observe the parts of the flower and to explore how their particular flower spreads pollen. Students present their observations and claims to the class. Then class discusses how the flowers they dissected spread their pollen.

Targeted Concepts, Skills, and Nature of Science	Performance Expectations
Scientists often work together and then share their findings. Sharing findings makes new information available and helps scientists refine their ideas and build on others' ideas. When another person's or group's idea is used, credit needs to be given.	Students should work in groups investigating the reproductive system of flowers as they dissect flowers. The class should work together to understand the mutualistic relationship between bees and flowers and they should update their *Project Board* with claims about what they have learned and the evidence that supports these claims.
Scientists must keep clear, accurate, and descriptive records of what they do so they can share their work with others and consider what they did, why they did it, and what they want to do next.	Students keep descriptive and accurate records of their observations as they dissect flowers.

Targeted Concepts, Skills, and Nature of Science	Performance Expectations
Scientific knowledge is developed through observations, recording and analysis of data, and development of explanations based on evidence.	Students put claims on the *Project Board* based on what they have read about the mutualistic relationship between bees and flowers.
Flowers are the reproductive organs of plants as well as food suppliers for many animals.	Students dissect flowers and think about how the parts of the flower help it in reproduction.

Materials	
1 roll per group	Invisible tape
1 per group	Flower (azalea, lily, gladiolus, tulip, daffodil, geranium, snapdragon, or sweet pea)
1 pair per group	Scissors
1 per group	Tweezers
1 per group	Magnifying glass
1 per student	*Flower Dissection Guide* page
1 per student	*Flower Dissection Observations* page
per group,enough to cover a desk	Newspapers or paper towels
1 per group	Poster paper and markers

Activity Setup and Preparation

Have a variety of flowers (azalea, lily, gladiolus, tulip, daffodil, geranium, snapdragon, or sweet pea) for each class. Each group should be given a different type of flower to dissect.

Keep flower stems in water and, if possible, refrigerate until students are ready to use them.

Assemble trays or bags of materials for each group.

Homework Options

Reflection

- **Science Content**: If you found an orchid with its nectar and pollen located at the end of a narrow twelve-inch tube, what could you determine? *(Students should realize that there must be a pollinator that can get to the nectar and pollen. The pollinator either needs some way to reach down the narrow 12-inch tube. For this particular orchid, there is a moth with a 12-inch proboscis (a special tongue) that it uses to sip the nectar out of the orchid.)*
- **Science Content**: Look for flowers outside, in your backyard, or at the grocery store. Draw the flowers you find, labeling the parts, and try to predict how they are designed to spread their pollen. Explain why you made the predictions you did. *(Students should use what they learned from dissections to label their flowers, and their predictions should be based on what they know about how pollinators and flowers work together.)*

Preparation for 2.7

- **Science Content**: What lessons did you learn from studying bees and flowers that you can apply to your study of other animals? *(Students should look for ways that other organisms in an animal's environment affect its behavior. They should be aware that different organisms can develop mutualistic relationships.)*

NOTES

NOTES

SECTION 2.6 IMPLEMENTATION

◀ $1\frac{1}{2}$ class periods*

2.6 Read

What Adaptations Do Bees Have that Affect Their Feeding Behavior?

reproduce: to produce offspring.

pollination: the transfer of pollen (male sex cells) from an anther to a stigma of a flower.

fertilize: in biology, to unite male and female sex cells to form a new organism.

In the last section, you learned about how bees' eyes work. Bees' eyes help bees forage because they have adapted to finding pollen and nectar in flowers. Bees and flowering plants also work together in other ways. In this section, you will read about how the relationship between bees and flowering plants helps both organisms survive.

How Bees and Flowers Help Each Other

Bees forage by looking for food in flowers. As they move from one flower to another, some of the pollen they find sticks to them and then falls off onto other flowers they visit. Flowering plants benefit from this process because bees move pollen from flower to flower. Combining pollen with the egg cell of a flowering plant makes it possible for flowers to develop seeds. Seeds are the way a flowering plant **reproduces** (produces more flowering plants).

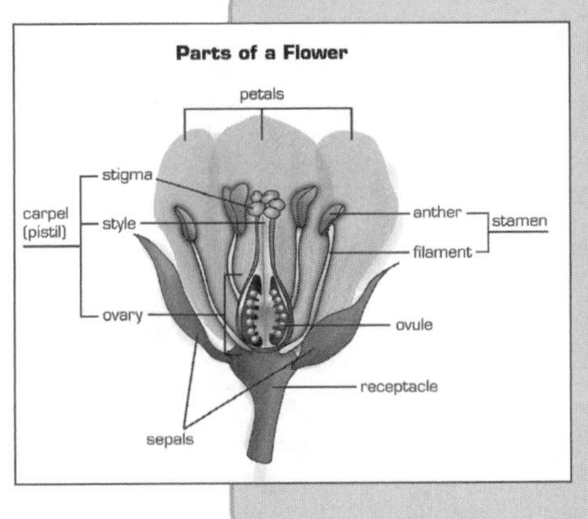

The diagram shows the parts of a flower. The production of *pollen* (male sex cells) takes place on an *anther*. The pollen is transferred to the *stigma*—a sticky surface at the top of the *pistil*. This process is called **pollination**. The male sex cells travel down the *style* to the *ovary*. In the ovary, the male sex cells **fertilize** the *ovules* (egg cells). The fertilized egg grows into an *embryo* (the early stage of an organism) that is surrounded by a supply of food and a protective layer. Together, the embryo, food, and protective layer form a seed. The seed can eventually produce another plant. You will learn more about the parts of a flower if you do the flower dissection that follows this section.

2.6 Read

What Adaptations Do Bees Have that Affect Their Feeding Behavior?

5 min.

Guide students' reading about the adaptations of bees.

Motivate students by eliciting their ideas on how flowers and insects are dependent on each other and have adapted mutually.

Record students' responses. Then, discuss how flowers depend on insects.

TEACHER TALK

"Flowers and bees have adapted to each other to survive. Flowers have nectar guides only visible in ultraviolet light and bees have the ability to see in the ultraviolet light. What other ways do you think flowers and insects have adapted to help each other?"

*A class period is considered to be one 40 to 50 minute class.

TEACHER TALK

"Many flowers depend on insects to help them reproduce through pollination and many insects depend on the flowers, nectar and pollen for food. It is amazing how these two have adapted. Flowers often lure insects by color, scent, or shape. The color, shape, scent of the flower, and the location of the nectar require the insect to have special features. You will learn about these features as we read about the relationship between pollinating insects and flowers."

Consider discussing how pollination affects humans. You may want to use the pollination of crops as an example. People in agriculture become concerned when bee populations are threatened. There is a fear for the occurrence of the Colony Collapse Disorder, a phenomenon which causes the worker bees in a colony to disappear and for the colony to collapse.

How Bees and Flowers Help Each Other

15 min.

Guide students' reading of how insects pollinate flowers and the mutualistic relationship between the two.

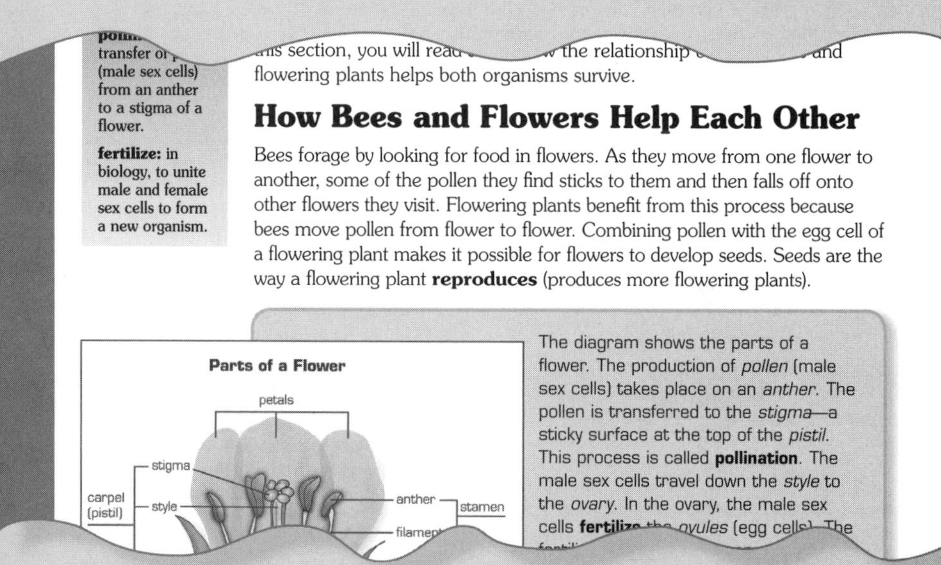
transfer of
(male sex cells)
from an anther
to a stigma of a
flower.
fertilize: in biology, to unite male and female sex cells to form a new organism.
section, you will read the relationship and
flowering plants helps both organisms survive.
How Bees and Flowers Help Each Other
Bees forage by looking for food in flowers. As they move from one flower to another, some of the pollen they find sticks to them and then falls off onto other flowers they visit. Flowering plants benefit from this process because bees move pollen from flower to flower. Combining pollen with the egg cell of a flowering plant makes it possible for flowers to develop seeds. Seeds are the way a flowering plant reproduces (produces more flowering plants).
Parts of a Flower
petals
stigma
carpel (pistil)
style
anther
stamen
The diagram shows the parts of a flower. The production of pollen (male sex cells) takes place on an anther. The pollen is transferred to the stigma—a sticky surface at the top of the pistil. This process is called pollination. The male sex cells travel down the style to the ovary. In the ovary, the male sex cells

△ Guide

Discuss how pollinators, such as bees, help flowers reproduce by carrying pollen from flower to flower. The student text has examples of other pollinators that help flowers reproduce. Describe these pollinators, emphasizing the pollinators students mentioned at the start of this section.

Describe how flowering plants reproduce when pollen from one flower is carried to another flower of the same species. Read the section on the parts of the flower on with the class and review the diagram. Describe how fertilization takes place using the image in the student text: when pollen lands on the stigma, it fertilizes the eggs, which are in the ovary at the base of the pistil. The eggs then develop into seeds.

Most flowering plants rely on animals such as bees, other insects, birds, and bats to transfer pollen from the anther of one flower to the stigma of another flower. These animals are called **pollinators**. When a pollinator lands on a flower to eat the nectar, it brushes some of the pollen onto its body. When it lands on another flower, some of that pollen may drop onto that flower. For this reason, attracting pollinators is very important to flowering plants.

pollinator: an insect or other animal that carries pollen from one flower to another.

mutualism: a relationship between organisms of two different species in which each member benefits.

Flowering plants have adaptations that help them attract these animals. Adaptations help animals and plants survive in their habitat. Flowering plants have flowers with bright colors and fragrant smells to attract animals. The shape of a flower is also an adaptation. Sometimes the nectar of a flower is located deep inside the flower. The pollinator has to reach deep into the flower to gather the nectar. When it moves to another flower, it goes deep into that flower, too, and leaves some of the pollen behind—deep in the flower where it needs to be for reproduction.

Animals that are pollinators have also developed adaptations that let them reach the nectar inside certain flowers. Insects such as moths and butterflies have long, slender feeding tubes that can reach deep inside a flower.

Both pollinators and flowering plants benefit from their relationship with each other. The flowers provide the animals with food. The animals help to transfer pollen from one flowering plant to the next. Biologists call this type of relationship **mutualism**.

The buttercup has open petals. The petals have lines, like the ones you saw in your simulation, that direct insects to the sacs containing nectar in the center. Many different lightweight insects with short mouthparts are adapted to pollinate the flowers of these plants.

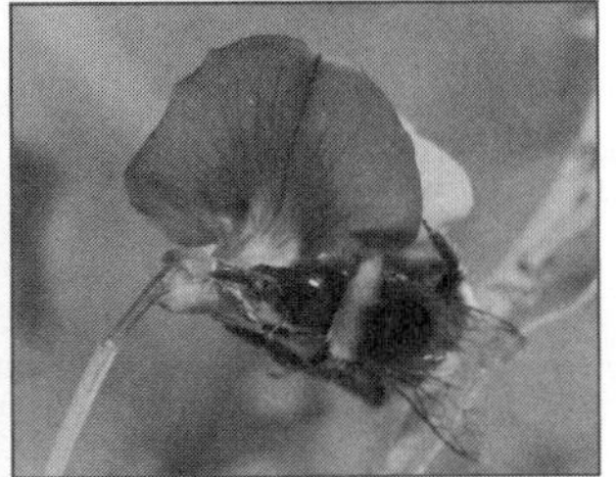

The sweet pea is adapted for pollination by insects like bumblebees. The nectar is inside a closed "cup" formed by a large petal and two smaller ones. The weight of a bumblebee landing on the large petal opens the cup. The bee can get the nectar inside. The pollen is transferred when the bee moves to another flower.

TEACHER TALK

"When bees fly from one flower to another, they take pollen with them and drop some of it on the other flower. Some of this pollen may land on the stigma and fertilize the flower's eggs. When this happens, seeds form which can begin the growth of a new plant under appropriate conditions."

Wind-Pollinated Flowering Plants

Not all flowering plants are pollinated by animals. Some are wind pollinated. The flowers of plants that depend on the wind to transfer their pollen from one flower to the next are different from animal-pollinated flowers. The flowers of wind-pollinated plants are usually small and have no scent. Beacuse they do not have to attract animals, they have no nectar. They are also not very colorful. Most trees and grasses are pollinated by the wind.

Wind pollination is not as efficient as animal pollination. Therefore, wind-pollinated flowers need to produce large amounts of pollen to ensure that some of the pollen reaches the stigmas of other flowers. They must also rely on the weather. On rainy days and days without wind, very little pollen can get transferred from one flower to another.

The flowers on this tree hang down so that the pollen can be easily caught and blown by the wind.

The flowers of grasses are located at the top of the plant where the pollen is exposed to the wind.

Ask students how flowers can get more bees to spread their pollen. Students may suggest that the colors of flowers' petals attract bees. They may also suggest that the fragrance of the flowers and the nectar in the flowers attract bees and other pollinators. The student text provides examples of how flowers' structures function to attract pollinators and to get pollen on them (e.g., color, shape, location of the nectar). Emphasize that these are adaptations of the flowers that help them attract pollinators.

Stop and Think

1. Adaptations are important for animals. Describe one adaptation of an insect and why it is important to that insect.
2. Sometimes the behavior of insects helps flowering plants. Without thinking about it, the bee moves the flower's pollen from one flower to another. Describe two ways that a flowering plant and an insect can have a mutualistic relationship.

Update the *Project Board*

You can now add what you have learned to the *What are we learning?* column of the *Project Board*. As you do this remember that you must support your learning with evidence. Put evidence from your reading and investigations in the *What is our evidence?* column.

What's the Point?

An animal's survival depends on its ability to gather and collect food. Adaptations are features of plants and animals that help them survive in their environment. Some adaptations help animals find and collect food. Sometimes adaptations of different organisms help the organisms work together to benefit each other. This is called mutualism.

For example, many insects rely on the nectar and pollen produced by flowering plants. At the same time, flowering plants rely on insects for reproduction. Flowers have adaptations, such as shape, color, and smell, that attract certain insects. The insects have adaptations, such as special vision or physical structures, that make it possible for them to get their food from the plant and help the

Butterflies feed on the nectar of a flower through long feeding tubes.

Stop and Think

10 min.

Guide and assess students understanding of mutualistic relationships between pollinators and flowers and the adaptation of animals.

You might want to share with students that flowers that attract bees tend to reflect light in the greens, blues, and ultraviolet lights. Flowers that attract moths tend to be light colored (e.g. white) which are more easily seen at night. Explain that this is an example of a mutualistic relationship.

Discuss the other examples of a mutualistic relationship using the images and image captions in the student text.

Consider providing other examples of mutualistic relationships. Cows, for instance, have a special sac located above the stomach that is filled with microorganisms that help the cow digest plant matter. The microorganisms benefit from the supply of plant materials. They are provided protection, and the warmth from the cow. The cow benefits from the microorganisms' help in digesting its food.

Before moving on to the *Stop and Think*, point out that not all flowers rely on pollinators: there are also wind-pollinated flowers.

△ Guide

Decide how you want students to answer the questions,. These *Stop and Think* questions are a good opportunity to assess students' understanding of the *Learning Set*. You can have students answer these individually and hand their answers in or you can discuss the questions with the class.

△ Guide and Assess

If students' answers do not have the information listed, guide them to the following:

1. Students may recount something they read about in this section, such as the long slender feeding tubes moths and butterflies have adapted to the shape of certain flowers in which the flower's nectar and pollen are deep inside the flower.
2. Again, students may use an example that they read about or one of their own. Answers should specify a benefit to the insect and a benefit to the plant. Student responses could include the insect that gets food (nectar) and the plant that gets pollinated so it can reproduce itself.

Update the *Project Board*

10 min.

Have a class discussion of what students have learned.

another. Describe two ways that a flowering plant and an insect can have a mutualistic relationship.

Update the *Project Board*

You can now add what you have learned to the *What are we learning?* column of the *Project Board*. As you do this remember that you must support your learning with evidence. Put evidence from your reading and investigations in the *What is our evidence?* column.

What's the Point?

An animal's survival depends on its ability to gather and collect food.

△ Guide and Evaluate

Ask the class, how what they have learned helps them to understand animal behavior. Students should be able to describe that animals are adaptable and that animals' behaviors are affected by the functioning of their bodies, such as the eyesight of bees. Students should be able to describe the mutualistic relationship between flowers and pollinators in which pollinators, such as bees, assist flowers by spreading pollen and flowers assist pollinators by providing nectar.

△ Guide

Remind students that the class Project Board is where they are keeping track of what they are learning. Ask the class what they want to include on their Project Board from what they have learned so far. Remind students to update their project board pages too.

As students discuss what they have learned, update the class Project Board, putting claims in the What have we learned? column (column 3) and recording evidence for each claim in the What is our evidence? column (column 4).

Example:

Claim: An animal's physical features affect its behaviors, including the way that animal gets food.

Evidence: Butterflies and moths tend to gather nectar from flowers with nectar in tubes because their long and slender mouth parts can reach the nectar.

and inv... the *What is our evi*...

What's the Point?

An animal's survival depends on its ability to gather and collect food. Adaptations are features of plants and animals that help them survive in their environment. Some adaptations help animals find and collect food. Sometimes adaptations of different organisms help the organisms work together to benefit each other. This is called mutualism.

For example, many insects rely on the nectar and pollen produced by flowering plants. At the same time, flowering plants rely on insects for reproduction. Flowers have adaptations, such as shape, color, and smell, that attract certain insects. The insects have adaptations, such as special vision or physical structures, that make it possible for them to get their food from the plant and help the

What's the Point?

1 to 5 min.

Make sure students understand that animals need to collect food for survival and that animals adapt to find and collect their food.

More to Learn: Flower Dissection

40 min.

Discuss the purpose of flowers and their anatomy, then have groups dissect flowers. Record their observations and have a discussion with the class.

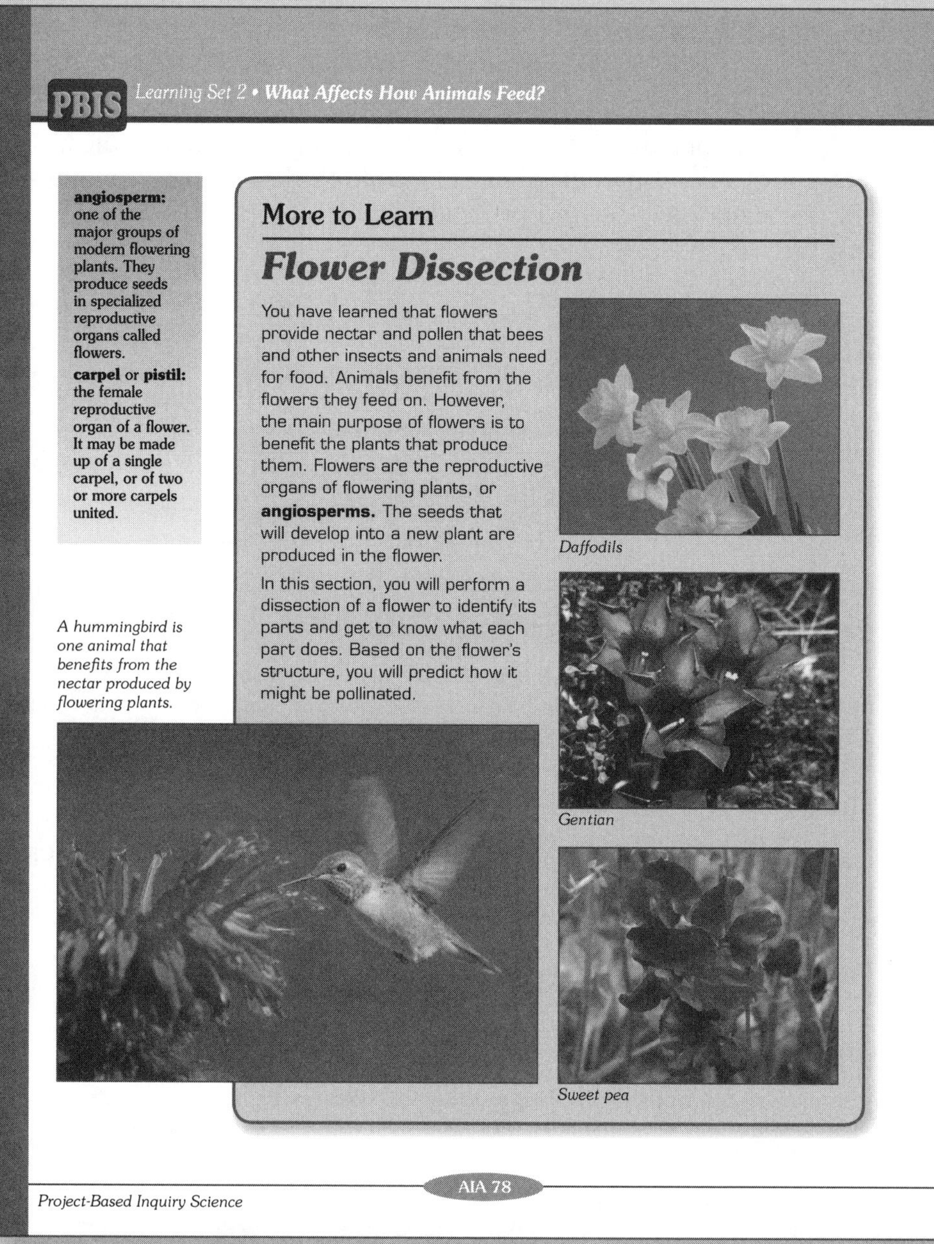

angiosperm: one of the major groups of modern flowering plants. They produce seeds in specialized reproductive organs called flowers.

carpel or **pistil:** the female reproductive organ of a flower. It may be made up of a single carpel, or of two or more carpels united.

More to Learn

Flower Dissection

You have learned that flowers provide nectar and pollen that bees and other insects and animals need for food. Animals benefit from the flowers they feed on. However, the main purpose of flowers is to benefit the plants that produce them. Flowers are the reproductive organs of flowering plants, or **angiosperms.** The seeds that will develop into a new plant are produced in the flower.

In this section, you will perform a dissection of a flower to identify its parts and get to know what each part does. Based on the flower's structure, you will predict how it might be pollinated.

Daffodils

Gentian

A hummingbird is one animal that benefits from the nectar produced by flowering plants.

Sweet pea

Describe that the purpose of the flower is to reproduce the plant. Point out that flowers are the reproductive organs of plants and have adapted to lure pollinating animals to help them reproduce. Discuss the fertilization process of flowers. Point out that in most flowers, fertilization takes place when pollen lands on the egg cell of the plant.

More to Learn

Anatomy of a Flower

Although flowers come in many sizes, colors, and shapes, they all have three major parts.

Parts of a Flower

petals, stigma, carpel (pistil), style, ovary, anther, stamen, filament, ovule, receptacle, sepals

1. The **carpel**, or **pistil**, is the female part of the flower. It is normally shaped like a bowling pin. The carpel includes the **stigma**, on top, a **style**, shaped like a tube, and the **ovary**, at the bottom. The ovary contains the **ovules** that will develop into seeds after fertilization.
2. The **stamen** is the male part of the flower. It includes the **anthers**, where pollen is produced, and a filament that attaches each anther to the base of the flower. Each flower has several anthers arranged around the carpel.
3. The outer protective cover includes the petals, which are larger and colored, and the **sepals**, which are shorter and usually green. The colors of the petals serve to attract insects and birds that help to pollinate the flower. The petals form the **corolla**.

The flower parts are attached to the **receptacle** at the base of the flower. Not all flowers look the same. Some flowers have both female and male parts. Other flowers have only the female, or only the male, parts. The colors and shape of the petals and corolla also vary from flower to flower.

The basic function of the flower is reproduction. Pollination takes place when pollen from the anthers of a flower is delivered to the stigma of a flower. The stigma is covered with a sticky substance to which pollen grains attach. Once on the stigma, the pollen grain germinates and forms a **pollen tube** that grows down the style. The male sex cells, contained within the pollen, travel down the tube to the ovary. Fertilization takes place when the male sex cells unite with the ovules, or eggs. The fertilized ovule develops into a seed.

stigma: the top part of the carpel where the pollen is deposited.

style: in plants, the slender, tube-like part f the carpel.

ovary: in a plant, the enlarged part at the base of the carpel that contains ovules, or eggs.

ovule: a tiny, egg-like structure in flowering plants that develops into a seed after fertilization.

stamen: the male reproductive structure of a flower. It consists of the anthers, which produce pollen, and of a slender filament.

anthers: the parts on the stamen of a flower where pollen is produced.

sepals: green leaf-like parts of a flower found outside the petals.

corolla: the collection of the colored petals of a flower.

receptacle: the base of a flower to which the flower parts are attached.

pollen tube: the slender tube formed by pollen grains to reach and fertilize the ovules.

AIA 79 ANIMALS IN ACTION

Anatomy of a Flower

10 min.

Discuss the anatomy of flowers before dissecting one.

Introduce the anatomy of a flower using the diagram and the student text. Emphasize that there are three major parts to a flower: the carpel, the stamens, and the corolla. Use the diagram in the student text to help guide students to where the various parts are in a flower and what they look like.

Emphasize that students will need to be able to identify and label the parts of the flower they observe during the dissection.

Procedure: Observe the Inside of a Flower

10 min.

Have groups dissect a flower.

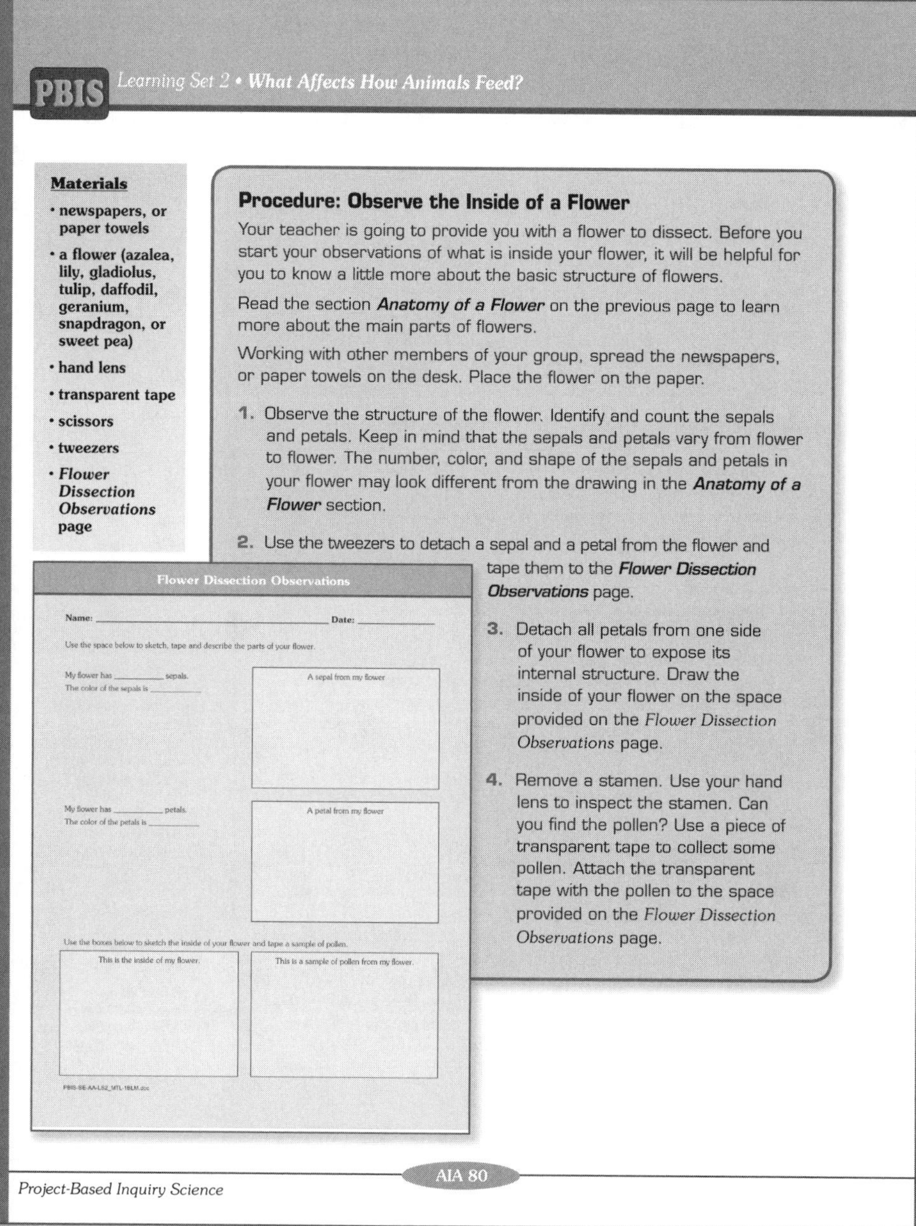

PBIS Learning Set 2 • *What Affects How Animals Feed?*

Materials

- newspapers, or paper towels
- a flower (azalea, lily, gladiolus, tulip, daffodil, geranium, snapdragon, or sweet pea)
- hand lens
- transparent tape
- scissors
- tweezers
- *Flower Dissection Observations* page

Procedure: Observe the Inside of a Flower

Your teacher is going to provide you with a flower to dissect. Before you start your observations of what is inside your flower, it will be helpful for you to know a little more about the basic structure of flowers.

Read the section ***Anatomy of a Flower*** on the previous page to learn more about the main parts of flowers.

Working with other members of your group, spread the newspapers, or paper towels on the desk. Place the flower on the paper.

1. Observe the structure of the flower. Identify and count the sepals and petals. Keep in mind that the sepals and petals vary from flower to flower. The number, color, and shape of the sepals and petals in your flower may look different from the drawing in the ***Anatomy of a Flower*** section.
2. Use the tweezers to detach a sepal and a petal from the flower and tape them to the ***Flower Dissection Observations*** page.
3. Detach all petals from one side of your flower to expose its internal structure. Draw the inside of your flower on the space provided on the *Flower Dissection Observations* page.
4. Remove a stamen. Use your hand lens to inspect the stamen. Can you find the pollen? Use a piece of transparent tape to collect some pollen. Attach the transparent tape with the pollen to the space provided on the *Flower Dissection Observations* page.

Flower Dissection Observations

Name: ____________ Date: ________

Use the space below to sketch, tape and describe the parts of your flower.

My flower has ________ sepals.
The color of the sepals is ________

A sepal from my flower

My flower has ________ petals.
The color of the petals is ________

A petal from my flower

Use the boxes below to sketch the inside of your flower and tape a sample of pollen.

This is the inside of my flower.

This is a sample of pollen from my flower.

PBIS-SE-AA-LS2_MTL-1BLM.doc

Project-Based Inquiry Science AIA 80

△ Guide

Briefly go over the procedure with students. Emphasize that they should start by recording the number of sepals and the number of petals and taping one of each to their *Flower Dissection Observations* pages. They should remove all the sepals and petals from one side of the flower and draw the inside of the flower. They should remove a stamen and collect its pollen on a piece of tape, fixing it to the page, and examine the stigma with their magnifying glasses and see what happens when they rub a stamen on it. Finally, they should open the ovary and draw a diagram of what they see.

5. Use your hand lens to look at the end of the pistil. This is the stigma. Rub a stamen onto the stigma and observe it, again, with the hand lens. Record your observations on the ***Flower Dissection Observations*** page.

6. Use the scissors or your fingernails to open the ovary. Use the hand lens to locate and count the ovules inside. Sketch a diagram of the inside of the pistil on the ***Flower Dissection Observations*** page.

Communicate Your Findings

Investigation Expo

Once you complete your flower dissection, you will share your observations with the rest of your class. You will prepare a poster with the results of your flower dissection to present to your classmates. Answer the following questions to prepare for your classroom discussion.

- Identify and describe the female parts your flower has. Identify and describe the male parts your flower has.
- Based on the structure of your flower and what you know about how animals pollinate flowers, predict how it might spread its pollen around. Do you think the pollen is spread by the wind or insects? Why do you think so? Explain your answer with evidence from your observations of the flower's structure.

Prepare a poster with the results of your classroom discussion. Include the drawings and parts of your flower. Make sure you label the parts and describe their functions. Using what you know about the flower parts, describe how you think your flower is pollinated.

Present to your class your drawings of your flower parts and your observations of the stamen and pistil.

Share your prediction about how your flower is pollinated. Tell why you think it is pollinated that way, using evidence from your observations of the structure of your flower.

Listen carefully as others present their results and observations. Be prepared to discuss the evidence you and your classmates present to justify your ideas about how your flowers are pollinated.

META NOTES

Each group should have a different kind of flower.

Get Going

Tell students to get their materials and begin dissecting their flowers. Also, let them know how much time they have.

Guide and Assess

Monitor groups' progress making sure they are following the procedure

Communicate Your Findings: *Investigation* Expo

20 min.

Have students' present to the class a poster of their observations and predictions of how their flower is pollinated.

△ Guide

First, let students know they will prepare a poster presentation of their observations and emphasize that they will need to address the bulleted items in the student text.

- describing the male and female parts of their flower
- how they think pollen is spread based on their observations

⬡ Get Going

Then, distribute poster materials and have students make their posters. Let them know how much time they have.

□ Assess

As students work on their poster, assess how students are using their observations to justify their predictions of how pollen is spread for the flower they dissected.

△ Guide

Have students present their posters and lead a class discussion.

Review how to use *Investigation Expo* to present results. There are two parts: poster presentations and discussions to share their observations. Describe how students will get to visit everyone's poster for about one minute and then how each group will present their results and a class discussion will follow.

When students have displayed their posters around the room and groups are visiting their classmates' posters, emphasize that students should be paying attention to groups' predictions of how the flower is pollinated (e.g., by a moth or the wind) and the supporting observations for their predictions.

Next, have groups present their posters and results to the class. Encourage students to ask questions, and model the kinds of questions you expect by asking groups how their evidence supports their predictions.

NOTES

Assessment Options

Targeted Concepts, Skills, and Nature of Science	How do I know if students got it?
Scientists must keep clear, accurate, and descriptive records of what they do so they can share their work with others, consider what they did, why they did it, and what they want to do next.	**ASK:** What is the point of writing what we learned about bees and flowers on the *Project Board?* **LISTEN:** Students should recognize that keeping a record of what they learned helps them see how their understanding has progressed and how they build on the information they learned.
Flowers are the reproductive organs of plants as well as food suppliers for many animals.	**ASK:** What two functions of some flowers are involved in the mutualistic relationship between them and animals **LISTEN:** The flowers are reproductive organs and animals assist them in their reproductive role. Flowers produce nectar, which provides food for the animals.

Teacher Reflection Questions

- What evidence do you have that students were able to make connections between animal anatomy, flower anatomy, and animal behavior?
- What ideas do you have for connecting the material in this section with your students' lives?
- How did you handle the variety of reading levels in your classroom? How would you deal with this in the future?

NOTES

NOTES

SECTION 2.7 INTRODUCTION

2.7 Explore

What Are the Feeding Behaviors of Some Other Carnivores?

◀ *2 class periods**

*A class period is considered to be one 40 to 50 minute class.

Overview

Students explore the feeding behavior of carnivores by observing lions and crocodiles in the video. Groups look for trends in their data by analyzing their observations and interpretations. Through group discussions, they categorize and separate observations and interpretations that were agreed upon from those that were not. Students use *Investigation Expos* to share their procedures, observations, and categories with the class. Groups use their observations and interpretations to create explanations for feeding behaviors. Then, students study the physical features of predators, learning that the joints in predators' bodies are the fulcrums of levers that help predators capture and overpower their prey. Using this new science information, students revise their explanations and share their revised explanations with the class. As a class, students decide on what they have learned and the supporting evidence and record this on their *Project Board.*

Targeted Concepts, Skills, and Nature of Science	Performance Expectations
Scientists often work together and then share their findings. Sharing findings makes new information available and helps scientists refine their ideas and build on others' ideas. When a person's of group's idea is used, credit needs to be given.	Students should work in groups to discuss and analyze their observations of animal behavior from video. Groups should present their analysis to the class during an *Investigation Expo* and, together, the class discusses the observations and interpretations made. Students should then work in groups to construct explanations of feeding behaviors and revise these based on science knowledge obtained later.

Targeted Concepts, Skills, and Nature of Science	Performance Expectations
Scientists must keep clear, accurate, and descriptive records of what they do so they can share their work with others, consider what they did, why they did it, and what they want to do next.	Students should be able to point out reliabl observations as those that group members agreed upon and that other groups also observed.
Observations and measurements are considered reliable if the results are repeatable by other scientists using the same procedures.	Students should be able to point out reliabl observations as those that group members agreed upon and that other group also obs
The structure and function of animals' bodies are complementary and affect animal behavior.	Students should describe how the general structure and form affect an animal's feedir behavior. They should also be able to descri how the joints of an animal affect its feedin behavior.
Animal's bodies have similarities to simple machines. These machines help the animal survive	Students should describe how moveable joi an animal act as the fulcrum point in a leve
Scientists differentiate between observations and interpretations. They use their observations and interpretations to explain animal behavior	Students should separate their interpretatio from their observations.
Scientific knowledge is developed through observations, recording and analysis of data, and development of explanations based on evidence.	Students should create explanations founde on evidence and add their explanations to t *Project Board.*
Explanations are claims supported by evidence. Evidence can be experimental results, observational data, and other accepted scientific knowledge.	Students should include observations and scientific knowledge in their explanations, lir their claims to their evidence.

Materials	
1 roll per class	*Animals in Action* DVD, to show the *Carnivore* video, and a way to show the video.
2 to 4 per student	*Observing and Interpreting Animal Behavior* page
1 to 2 per student	*Create Your Explanation* page
1 to 2 per student	*Project Board* page

Activity Setup and Preparation

Preview the *Carnivore* video before showing it to your class.

Homework Options

Reflection

- **Science Content**: What are some ways that crocodiles and lions conserve energy in hunting? *(The crocodile waits for prey to come to it. Lions hunt in groups, allowing them to conserve energy.)*
- **Science Content**: What are some predators you are familiar with that use the "lying in wait" hunting technique like crocodiles? What are some predators you are familiar with that chase down their prey like lions? *(Examples of predators that "lie in wait" might include: spiders, frogs, and certain lizards. Examples of predators that chase down their prey might include wolves, cats, hawks, and eagles. Many predators engage in both hunting behaviors. Polar bears will wait by seals' breathing holes and capture them when they come out, but they are also capable of chasing down prey.)*

Preparation for BBC

- **Science Content**: What do you think you might use from this Learning Set in your zoo enclosure design? *(The purpose of this is to prepare students for the next section in which they make a recommendation on a design for a zoo enclosure to study the feeding behavior of one of the animals discussed in the* Learning Set.*)*

NOTES

SECTION 2.7 IMPLEMENTATION

◀2 class periods*

2.7 Explore

What Are the Feeding Behaviors of Some Other Carnivores?

The animals you observed so far are foragers. They gather food by moving from place to place. Herbivores eat plants. Their bodies are adapted to eating and digesting plant material. Omnivores, such as chimpanzees, eat plants and animals. While chimpanzees are capable of hunting small prey and eating their meat, most of their diet is made up of fruit. Another group of animals eat only other animals. They are carnivores.

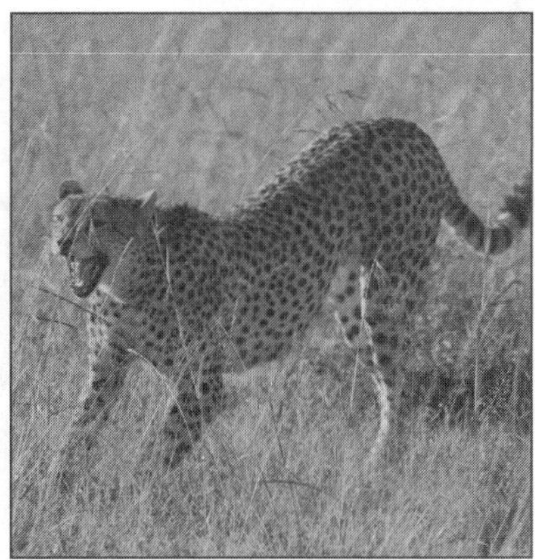

The cheetah is the perfect hunting machine with sharp eyesight, legs built for speed, teeth for tearing, and strong jaws to bring down its prey.

Some carnivores are predators that hunt, kill, and eat other animals. These carnivores must work hard for their food. The prey of some animals, such as lions or cheetahs, can move. This means the predator must be able to move faster. It must find and chase its prey in order to eat. Other predators wait until the prey comes to them. For example, alligators hide until their prey is near, and then they pounce. They use their powerful jaws to capture and consume their meal. Regardless of how predators get their food, their bodies must work like a well-designed machine, giving them the energy they need to overcome their prey. These animals have body parts that enable them to find, capture, kill, and eat other animals.

Conference

As an omnivore, you may have some experience eating meat. You may not hunt for your food. However, you have probably "hunted" for a meal at the grocery store and school cafeteria. Before you start this section, get together with your group and discuss what you already know about how a carnivore feeds. You can help get your discussion started by having one person in your group pretend to be a carnivore. As they pretend to "capture" and eat a piece of meat, you should watch their every move. Observe, and carefully

AIA 82

Project-Based Inquiry Science

2.7 Explore

What Are the Feeding Behaviors of Some Other Carnivores?

5 min.

Discuss several large carnivorous predators.

○ Engage

Briefly review the foragers students have learned about. Introduce the idea that not all animals are foragers and that many animals hunt for food. Then, ask students to identify several carnivorous predators and what they eat.

META NOTES

During this discussion, students may require clarification of what they have seen. They may request to watch the video again, especially the parts they were uncertain about. Provide time for students to check their observations and interpretations in this way.

TEACHER TALK

"What foragers have you learned about so far?

There are many different foragers that eat both plants and animals. But foraging is only one type of feeding behavior. Other animals feed differently. Some animals hunt. What animals do you know that hunt? *(Let students list some animals and record the list)*.

What do these animals eat? *(Students should suggest that these animals eat other animals).*

Animals that eat other animals are called carnivores and often carnivores hunt for their food"

META NOTES

The goal of this section is to connect the structure of an animals' body to the way in which it feeds. This section specifically looks at carnivorous predators. Carnivores eat animals but not all carnivores are predators. Some carnivores, such as vultures, eat dead animals (carrion). These animals are called scavengers.

△ Guide

Let students know that there are many different kinds of predators whose bodies and behaviors are adapted to enable them to capture their prey. Guide students to think about the physical characteristics of a predator by asking questions such as what the cheetah shown in the picture might eat and how it might capture its food.

TEACHER TALK

"One example of a carnivorous predator is a cheetah. What do you think the cheetah in the picture preys on? How does it capture its prey? Look at the cheetah's body. What parts of its body make it more able to hunt?"

Show students the video of the cheetah hunting. Students should observe the cheetah's body and see if they can find different ways the cheetah's body helps it become a better hunter.

Then provide students with some other examples of predators.

TEACHER TALK

"Another example of a predator is the polar bear. Polar bears are very large predators who usually hunt seals and other arctic animals, capturing their prey at holes in the arctic ice and using their size and power to kill.

Eagles are large birds that capture prey in their powerful talons and use their beaks to tear apart their prey."

Elicit from students other predators they know about and ask them to describe how they might hunt. For each predator, ask about the kind of animals they prey on and how they capture and kill their prey.

record, the details as they go to get the food, put it in their mouth, and eat it. When you are finished, use the following questions to guide your discussion.

- What actions and body parts were involved in
  - a) obtaining the food?
  - b) picking up the food?
  - c) putting the food in the mouth?
  - d) chewing and swallowing the food?
- What would be different if you were observing an animal in its natural environment hunting, chasing, capturing, and eating food?
- Describe any body parts or other characteristics that you think would make a predator a good hunter.

Predict

Use your group discussion to predict what characteristics and behaviors of carnivores help them meet their needs for obtaining food. Take into consideration that there are all kinds of carnivores—even a few plant species! What are some different strategies a predator might use to catch its next meal? Think about what forces it would take to overcome different kinds of prey.

Observe

As in other sections in this *Learning Set*, it is not possible to do direct field observations of a predator in action. In this case, it would be dangerous, as you could be viewed as prey by some animals! You may have seen some carnivores in the zoo. However, they do not have to hunt and capture their prey as animals in the wild do. The best way to observe the feeding behavior of carnivores is to watch a video. You will see three video clips. The first one shows a cheetah chasing and taking down a gazelle. The second one shows

Alligators use their powerful jaws to catch prey.

Conference

10 min.

Have groups discuss what they know about how predators feed.

META NOTES

The bulleted items require thinking about different aspects of the feeding process. These will be used to predict the different strategies that predators might use so it is important that students record their responses.

△ Guide

Using the list the students just created, let students know that they should act out and/or discuss how one predator captures and eats its prey. They should then describe how this would take place.

△ Guide and Assess

Monitor groups' progress and discussions. Check that students are recording their responses to the bulleted items.

TEACHER TALK

"Each group should pick out one of the animals from the list we created to act out and/or discuss how the predator captures its prey. While one member acts out how the predator captures and eats its prey, the rest of the group should carefully observe the carnivore and record their observations. Once the observers have finished making their observations, then the group should discuss how the animal hunts, using the bulleted items in the student text as a guide."

△ Guide

Ask groups to predict what characteristics and behaviors predators have that help them catch and kill prey. Have students list the different features as they make generalizations about the animals' bodies in obtaining food, picking up food, getting food in their mouths, and chewing and swallowing.

If students do not include plant species in their list of carnivorous predators, remind them that there are many different types of predators, including plants such as the Venus flytrap that catches and eats insects. Let students know that thinking about different strategies and body characteristics a predator might use will help them when they observe some video of predators.

When groups have finished making their predictions, consider having groups briefly share their ideas with the class.

Predict

5 min.

Have groups make predictions about the characteristics of predators.

- Describe ... parts or other characte... you think would make a predator a good hunter.

Predict

Use your group discussion to predict what characteristics and behaviors of carnivores help them meet their needs for obtaining food. Take into consideration that there are all kinds of carnivores—even a few plant species! What are some different strategies a predator might use to catch its next meal? Think about what forces it would take to overcome different kinds of prey.

Observe

As in other sections in this *Learning* ...

Pre…

Use your group discussion to predict what characteristics and behaviors of carnivores help them meet their needs for obtaining food. Take into consideration that there are all kinds of carnivores—even a few plant species! What are some different strategies a predator might use to catch its next meal? Think about what forces it would take to overcome different kinds of prey.

Observe

As in other sections in this *Learning Set*, it is not possible to do direct field observations of a predator in action. In this case, it would be dangerous, as you could be viewed as prey by some animals! You may have seen some carnivores in the zoo. However, they do not have to hunt and capture their prey as animals in the wild do. The best way to observe the feeding behavior of carnivores is to watch a video. You will see three video clips. The first one shows a cheetah chasing and taking down a gazelle. The second one shows

Alligators use their powerful jaws to catch prey.

AIA 83

ANIMALS IN ACTION

Observe

5 min.

Show students the video of cheetahs, lions, and crocodile hunting, and have them make observations.

△ Guide

Let students know that they will watch a video of cheetahs hunting a gazelle, lions subduing and eating a wildebeest and a crocodile capturing and subduing its prey. They will watch the video twice. The first time, they should focus on how the bodies are adapted for capturing and eating their prey and record what they see. Then they will have a chance to make observation plans with their groups and view the video again.

NOTES

two lions taking down and eating a wildebeest. The third video clip shows an alligator laying in wait and then capturing a wildebeest. As before, when you watched the other videos, you will see each video twice. The first time, you will be watching it to come up with an observation plan. The second time, you will watch more carefully, using your plan, and then work with your group to identify all the different behaviors.

As you watch the first video clip, pay attention to how the cheetah's body is adapted for chasing and capturing prey. While watching the second clip, pay attention to how the lions' bodies and mouths are adapted for taking down, killing, and eating its prey. In the third clip, look carefully at what parts of the alligator are adapted for catching prey that is much larger than it is.

Plan

Discuss with your group what you observed about the way each animal feeds. Then identify what you need to observe more closely to be able to write a detailed description of how lions and alligators feed. Plan the way you will observe the videos the next time they are played so you will be able to gather more detail about the ways lions and alligators feed. What details will you pay special attention to in each video? Do you want to divide up the observations among your group? Should you each focus on different details? Remember that you want to be able to describe the following three things:

- how each animal feeds
- what is special about the body of each animal that allows it to feed that way, and
- what is special about the environment that affects the way the animal feeds.

Observe

Watch each video clip again, this time using your observation plan. Record what you are observing, paying special attention to your own assignment. Watch carefully to see what body parts help the cheetah chase its prey. Observe what body parts help the lions take down and eat their prey. Watch carefully to see how the alligator uses its great strength to overcome its prey.

Analyze Your Data

After observing the video, analyze the behavior of the animals as you have done for the chimpanzee and the bee. Share with each other what you observed. Record the behaviors you observed on sticky notes. Arrange

Plan

5 min.

Have students plan observations of cheetahs', lions' and crocodiles' feeding behavior.

△ Guide

First, let students share their observations briefly. Then focus students to find how to make their observations better during the second viewing of the video clips. Ask them to think about the details they should watch for in the second viewing. Let students know that they should identify what they will attend to and who will make the observations. Emphasize that students should be able to describe the three bulleted items after viewing the video again:

- how each animal feeds
- what is special about the body of each animal that allows it to feed that way
- what is special about the environment that affects the way animals feed

how each ani...

- what is special about the body of each animal that allows it to feed that way, and
- what is special about the environment that affects the way the animal feeds.

Observe

Watch each video clip again, this time using your observation plan. Record what you are observing, paying special attention to your own assignment. Watch carefully to see what body parts help the cheetah chase its prey. Observe what body parts help the lions take down and eat their prey. Watch carefully to see how the alligator uses its great strength to overcome its prey.

Analyze Your Data

After observing the video, analyze the behavior of the animals as you have done for the chimpanzee and the bee. Share with each other what you observed. Record the behaviors you observed on sticky notes. Arrange

AIA 84

Project-Based Inquiry Science

Assess

Assess that groups' plans specify whether they will divide the work and what they will focus on. Check that groups incorporated their ideas about answering the bulleted items.

Get Going

Remind groups that they should be prepared to defend their explanations. Distribute the *Observing and Interpreting Animal Behavior* pages and show the video again. Remind students to record their observations in detail. They should record their interpretations in the Interpretations column.

Guide

After the video, have groups analyze their data using sticky notes as they've done before. Let students know that they should arrange the sticky notes in groups that allow them to interpret the behavior of each animal, and then they should use this information to fill out a new *Observing and Interpreting Behavior* page for each animal.

Observe

5 min.

Have students make observations of cheetahs', lions', and crocodiles' feeding behavior, following their plans.

Analyze Your Data

10 min.

Have groups discuss and categorize their observations

META NOTES

Often in classroom discussions, students will only present ideas that the group agrees on. Encourage students to share all their ideas, even those that cause dissent in the group discussion.

Communicate Your Interpretations: Investigation Expo

30 min.

Have a discussion on students' observations and categories in Investigation Expos.

the sticky notes in groups that allow you to interpret the behavior of each animal. Using a separate *Observing and Interpreting Animal Behavior* page for each animal, record the behaviors you observed, your interpretations of those behaviors, and what is special about each animal and its environment that allows it to feed the way it does.

You may ask your teacher to replay any parts of the videos that you did not understand or that you and your group members may have disagreements about.

Ask yourself how the bodies of the cheetah, the lions, and the alligator are different. Also, think about any similarities that you see.

Communicate Your Interpretations

Investigation Expo

In this *Investigation Expo*, some groups in the class will present their interpretations of the cheetah's behavior, others will present their interpretations of the lions' behavior, and some will present their interpretation of the alligator's behavior.

As you present, be clear about each of the following points:

- what you were observing
- your group's observation plan—what you were looking for and how you divided the work
- the behaviors you observed
- your ways of grouping the behaviors
- your interpretations of the behaviors, and
- what is special about your animal's body and the animal's environment that affects the way your animal feeds.

When you listen to other groups, be sure you understand their presentations. Ask questions if you need to know more. Notice observations and interpretations other groups have made that are different from yours. If you don't understand the reasons for the differences, or if you do not agree with their observations or interpretations, ask questions and state your disagreements. Any time you disagree, be prepared to present evidence that will allow you to support your claim. Remember to always be respectful, even when you are disagreeing with others.

Let students know that they should be thinking about the similarities and differences in the bodies of the cheetah, lion, and crocodile and how this might provide clues to how their bodies affect their feeding behaviors.

☐ Assess

As groups discuss and organize their data, assess whether they are having trouble categorizing their data. Use probing questions about the best way to choose categories, for example: What trends are you seeing? What would happen if you sorted your observations in a different way?

△ Guide

Let students know that they will be presenting their results to the class in another *Investigation Expo.* This time they will not use posters, but will use their *Observing and Interpreting Animal Behaviors* pages to prepare. Let students know that all three animals' behaviors (cheetahs, lions, and crocodiles) will be presented.

⬡ Get Going

Assign each animal behavior to one third of the groups. Remind groups that their presentations will have to respond to each of the bulleted items.

Let students know how much time they have to prepare their presentations and let them begin.

□ Assess

As groups prepare for their presentations, check to make sure that their presentations address the items bulleted in the student text.

△ Guide Presentations and Discussions

Begin a short discussion after each presentation. Ask the class if they had questions or comments. Encourage discussion between groups, noting areas of difficulty students are having. If there are any areas of strong disagreement, consider showing the video again.

Students should have an easier time leading the discussion now that they have experienced a few class discussions. Assist students with some of the language needed by modeling it for them or asking questions that guide them:

- How does your evidence or observation compare to what ... said?
- I agree with ... because ...
- I don't understand the reasoning behind your interpretation of your results. Tell me more about how the data supports your claim.

After the set of groups presenting the same animal have presented, ask questions to compare and contrast their observations, interpretations, and procedures.

TEACHER TALK

"These groups all focused on the same things. Do the results indicate the same thing? Do they show the same trend? How do their procedures differ and how are they the same? How did their observations differ and how are they the same? How did their interpretations compare?"

META NOTES

Students will learn more about the culture you are setting up by how you model being part of the audience during the presentations. Encourage students to ask questions of the presenting group and the presenting group to respond to the student asking the question. Students should not make you the focal point during the discussion. Try sitting with the audience during the presentations.

META NOTES

Questions should require more than a one word answers and should focus on a critical part of the investigation. Students will often want to question other students about color choice or layout. By modeling questions about content, you will assist students in asking better questions.

Discuss the claims made by the groups. Ask how confident they are of their claims. Students' claims should be based on the trends in their data; other groups that observed similar trends should have similar resulting claims.

TEACHER TALK

"Let's consider the claims. How do we determine if the claim is good? Do you trust the claims the groups made? Why or why not? Do you think we can make a claim based on all the presentations for this animal?"

Point out that scientists call claims valid if many different groups see similar trends when they are investigating the same thing.

◇ Evaluate

As groups present, note any interpretations included among the observations. Also note whether groups include observations about the forms and functions of the animals they observed.

Repeat this type of discussion for groups presenting the same animal.

After all the groups have presented ask questions to compare observations of the predators to see if students can find trends or make generalizations about how these predators eat. This should help students in the next segment where they construct explanations.

TEACHER TALK

"Remember that we make claims based on the trends in the data. Groups observed and gave presentations about the feeding behaviors of cheetahs, lions, and crocodiles. These animals are predators. What trends did you see in these predators that might help you make a claim about predators in general? "

△ Guide

Let students know that they will be constructing an explanation of the feeding behavior of predators. As with the explanation of bees' foraging behavior, they should use their interpretations as a claim or to develop a claim and use their observations and science knowledge to support the claim. Emphasize that every explanation is comprised of a claim (a conclusion or interpretation), evidence (observations) and science knowledge (ideas accepted in the scientific community) such that the evidence and science knowledge support the claim in a logical way.

Review the structure of an explanation using the example structure in the student text.

Explain

You have just developed interpretations of how lions and alligators feed. Now it is time to develop an explanation of what affects how carnivores feed. Work with your group to do that. Your explanation should make clear why lions and alligators feed differently from each other. You probably predicted and observed that some carnivorous predators need to have good eyesight, strong legs for running and pouncing, sharp teeth, and strong jaws. Other predators may not be built for speed, but they have other adaptations, such as the ability to blend in with their environment until prey comes close enough to capture. Working with your group and using a *Create Your Explanation* page, develop an explanation of how a carnivore's body affects its particular feeding behaviors.

You might want to structure your explanation like this.

Many carnivores feed by [*tell how*]. Others feed by [*tell how*].
We can predict how a carnivore feeds by looking at its body parts.
When a carnivore's body has [*tell what*], it feeds by [*tell how*].
When its body is [*tell what*], it feeds by [*tell how*].

machine: a device that makes work easier.

Incredible Predators

Predators have many adaptations that make them good hunters. They may have keen eyesight and hearing, the ability to blend in with their environment, specialized teeth for ripping and tearing flesh, and digestive systems that can break down meat efficiently.

Predators have to work to get their food. They must be able to chase down fast animals, capture prey that is sometimes larger than their own bodies, and kill and consume their food. To do this successfully, they must use force. Humans use **machines** to multiply force and make work easier.

Explain

10 min.

Have students create and discuss explanations of the behaviors they observed.

△ Guide and Assess

As groups are working on their explanations, monitor their progress and ask students to read their claims and the evidence they are using to support their claims. Their claims should be valid (based on the evidence) and their evidence should not include any opinions.

Example:

> Many carnivores feed by chasing their prey and using their teeth and claws to tear flesh. Others feed by lying in wait, pouncing, and overpowering their prey. We can predict how a carnivore feeds by looking at its body parts. When a carnivore's body has long, powerful legs, it hunts for food by chasing its prey. We know this because we observed that the cheetahs and lions were able to use their powerful legs to hold onto the gazelle and wildebeest. When the predator's body can be hidden and when it has great power, the predator lies in wait and then pounces. We know this because we observed a crocodile hiding in the water and then pouncing when it's prey got close.

☐ Assess

As groups are working on their explanations, ask students what their claims are and what evidence they are using to support their claims. At this point, they have limited science knowledge they can use, but they should use their observations from the video to support their claims. Note any difficulties students are having.

△ Guide

Hold a brief class discussion. Have a few groups present their explanations and spend the remainder of the class discussing similarities and differences in their explanations.

Let students know that they will be revising their explanations after reading the information box, *Incredible Predators* in the student text.

Incredible Predators

15 min.

Discuss how animal's bodies have parts that act like machines, such as moveable joints that act as levers. Introduce and describe the three classes of levers.

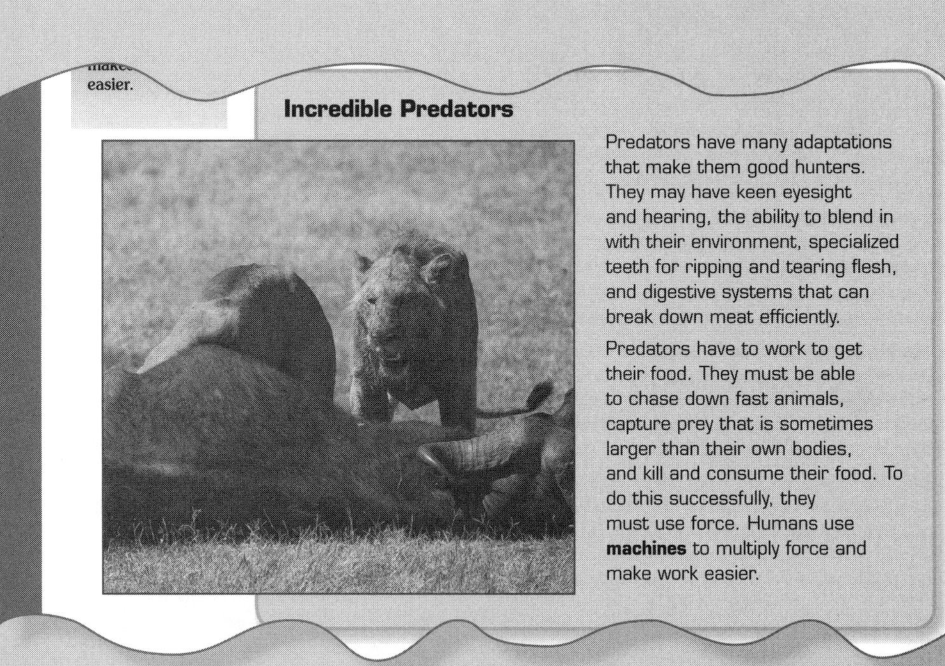
easier.

Incredible Predators

Predators have many adaptations that make them good hunters. They may have keen eyesight and hearing, the ability to blend in with their environment, specialized teeth for ripping and tearing flesh, and digestive systems that can break down meat efficiently.

Predators have to work to get their food. They must be able to chase down fast animals, capture prey that is sometimes larger than their own bodies, and kill and consume their food. To do this successfully, they must use force. Humans use **machines** to multiply force and make work easier.

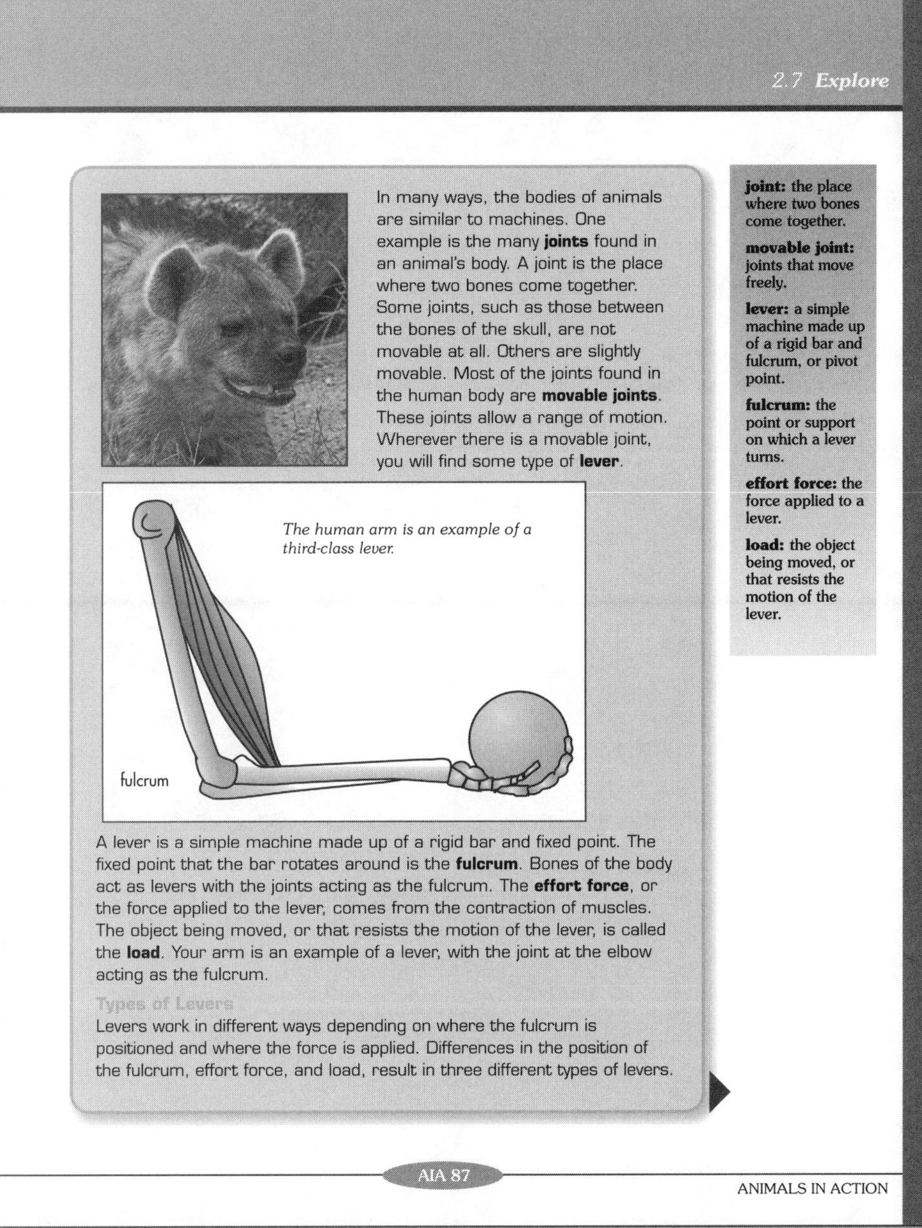

2.7 Explore

In many ways, the bodies of animals are similar to machines. One example is the many **joints** found in an animal's body. A joint is the place where two bones come together. Some joints, such as those between the bones of the skull, are not movable at all. Others are slightly movable. Most of the joints found in the human body are **movable joints**. These joints allow a range of motion. Wherever there is a movable joint, you will find some type of **lever**.

The human arm is an example of a third-class lever.

A lever is a simple machine made up of a rigid bar and fixed point. The fixed point that the bar rotates around is the **fulcrum**. Bones of the body act as levers with the joints acting as the fulcrum. The **effort force**, or the force applied to the lever, comes from the contraction of muscles. The object being moved, or that resists the motion of the lever, is called the **load**. Your arm is an example of a lever, with the joint at the elbow acting as the fulcrum.

Types of Levers

Levers work in different ways depending on where the fulcrum is positioned and where the force is applied. Differences in the position of the fulcrum, effort force, and load, result in three different types of levers.

joint: the place where two bones come together.

movable joint: joints that move freely.

lever: a simple machine made up of a rigid bar and fulcrum, or pivot point.

fulcrum: the point or support on which a lever turns.

effort force: the force applied to a lever.

load: the object being moved, or that resists the motion of the lever.

AIA 87

ANIMALS IN ACTION

△ Guide

Once students have made their explanations, discuss the variety of ways predators' bodies are adapted to the hunting task. Connect the hunting needs of animals to the ways the animal's bodies are built. Explain that any time animals need to hunt they must use force (a push or a pull), to bring down the animal, to kill an animal, or to eat the animal. To make sure they have enough force, animals' bodies are built from simple machines. Simple machines can change the amount of an applied force and the direction of the force.

This section is an introduction to the bodies of predators and a comparison of predator's bodies to simple machines. The content included in this part is important for addressing the Big *Challenge* of the Unit. The depth of inclusion of the simple machines discussion should be dictated by the science standards in your area. If needed, use the examples of simple machines included in this section to assist students in understanding simple machines.

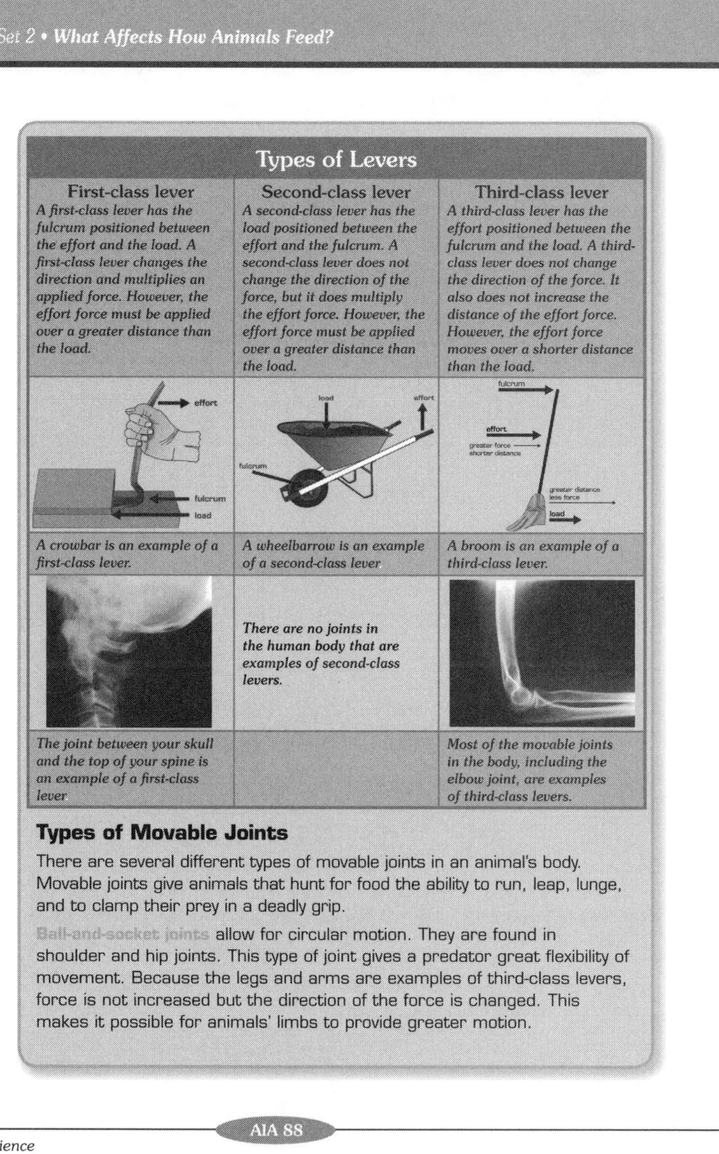

Types of Levers

First-class lever	Second-class lever	Third-class lever
A first-class lever has the fulcrum positioned between the effort and the load. A first-class lever changes the direction and multiplies an applied force. However, the effort force must be applied over a greater distance than the load.	*A second-class lever has the load positioned between the effort and the fulcrum. A second-class lever does not change the direction of the force, but it does multiply the effort force. However, the effort force must be applied over a greater distance than the load.*	*A third-class lever has the effort positioned between the fulcrum and the load. A third-class lever does not change the direction of the force. It also does not increase the distance of the effort force. However, the effort force moves over a shorter distance than the load.*
effort, fulcrum, load	load, effort, fulcrum	fulcrum, effort, greater force shorter distance, greater distance less force, load
A crowbar is an example of a first-class lever.	*A wheelbarrow is an example of a second-class lever.*	*A broom is an example of a third-class lever.*
	There are no joints in the human body that are examples of second-class levers.	
The joint between your skull and the top of your spine is an example of a first-class lever.		*Most of the movable joints in the body, including the elbow joint, are examples of third-class levers.*

Types of Movable Joints

There are several different types of movable joints in an animal's body. Movable joints give animals that hunt for food the ability to run, leap, lunge, and to clamp their prey in a deadly grip.

Ball-and-socket joints allow for circular motion. They are found in shoulder and hip joints. This type of joint gives a predator great flexibility of movement. Because the legs and arms are examples of third-class levers, force is not increased but the direction of the force is changed. This makes it possible for animals' limbs to provide greater motion.

One machine described in detail in the student text is a lever. Describe how most animals have multiple joints, including movable joints, which act as the fulcrums of levers. Explain to students that levers are simple machines that can multiply the effort force (the push or pull applied to it) to either increase or decrease that force, and they can change the direction of a load (what is resisting it). Let students know that the effort force comes from the contraction of muscles.

Then introduce the three types of levers: first class, second class, and third class. Describe how first-class levers have the fulcrum (pivot point) between the effort

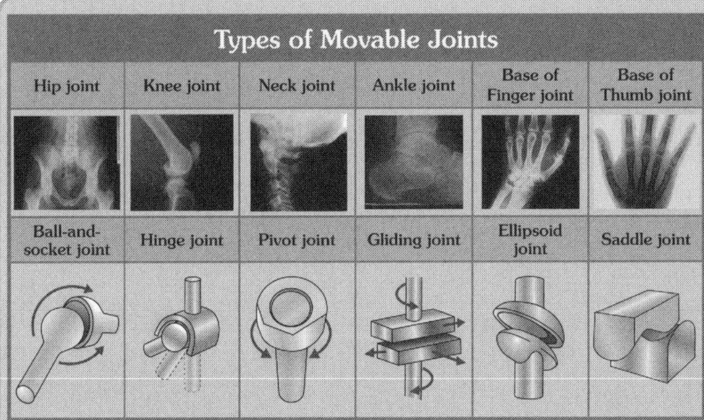

Types of Movable Joints
Hip joint
Knee joint
Neck joint
Ankle joint
Base of Finger joint
Base of Thumb joint
Ball-and-socket joint
Hinge joint
Pivot joint
Gliding joint
Ellipsoid joint
Saddle joint

Hinge joints permit a back-and-forth motion, similar to the movement of a gate swinging on hinges. They do not move from side to side. Your knee is an example of a hinge joint. Hinge joints are found in the jaws of many predators. Jaws are examples of two first-class levers with the point where they connect being the fulcrum. First-class levers multiply the force applied. This makes it possible for the animal to apply a lot of force with its jaws.

Pivot joints are characterized by one bone rotating around another. One bone is ring-shaped, and the other forms the pivot. This type of joint is found in the first two vertebrae of the neck. In birds of prey, such as owls, this type of joint lets them turn their heads and use their keen eyesight to spot prey.

Gliding joints occur where two bones with flattened or slightly curved surfaces come together. The two flattened surfaces can only slide past one another, giving this type of joint limited movement. Gliding joints are found in the wrists and ankles, as well as between the vertebrae that make up the spine. Having this type of joint in the backbone gives predators (as well as prey) an advantage. It gives four-legged animals the support they need while still allowing them to flex their backs. By flexing their backs, they gain power for movements such as running.

Ellipsoid and Saddle joints are commonly found in the bones of the hands and feet. Ellipsoid joints connect the fingers with the palms of your hands and toes to the feet. Saddle joints allow one bone to slide in two directions. The place where the thumb meets the wrist is a saddle joint. Both of these joints make hands and feet well adapted for climbing and grasping foods, which are behaviors typical of herbivores and omnivores.

and the load. First-class levers change the direction of the applied force (when pushed down, the lever pushes up on the load), and can change the amount of force as well. The force will be greater or smaller than the effort (applied force) depending on where the fulcrum is positioned. An example of this is a crowbar and the joint between the skull and the first vertebra of the spine.

Next, describe how second-class levers have the load (output force) between the fulcrum (pivot point) and the effort (input force). Nutcrackers and bottle openers are examples of second-class levers. Let students know that there are no known joints in the human body that are second-class levers.

However, the Achilles tendon pulls or pushes across the heel and acts as a second-class lever.

Then describe third-class levers where the effort is positioned between the fulcrum and the load. The direction of the effort (applied force) is the same as the direction on the load (output force), however, it acts so that the load moves faster than the effort because it has a greater distance to travel. Fishing poles, brooms, and hockey sticks are examples of third-class levers. In the human body, most joints are third-class levers.

Introduce the types of movable joints. Consider eliciting students' ideas on what part of their body fits each type of joint pictured. Then, read through the example for that particular joint in the student text. The shoulder and hip joints are ball-and-socket joints. Knee joints and jaw joints are hinge joints. The first two vertebrae in the neck contain pivot joints. The spine, wrists, and ankles contain gliding joints. The place where the thumb meets the wrist is a saddle joint.

Discuss how these joints can be advantageous for predators. A jaw joint with a lot of leverage, such as that of a crocodile, will help the animal capture and kill its food. Pivot joints in the neck allow a predator to scan the area for its prey.

NOTES

Stop and Think

1. What advantages does a hinge joint give carnivores?
2. Why are most of the joints in a predator's body third-class levers? How does this benefit a predator?
3. Make a sketch of a carnivore that you are familiar with. Label each feature, or characteristic that makes the animal well suited for capturing and eating prey.

Revise Your Explanation

Based on what you have just read, revisit your *Observing and Interpreting Animal Behavior* pages. Add more detail to the middle column. If you need to, revise your interpretations.

Then, revisit your *Create Your Explanation* pages. You now have more science knowledge about what affects the feeding behavior of carnivores. Revise your claim if you need to, add your new knowledge, and rewrite your explanation to add detail.

Share Your Explanation

Present your explanations to the class, and as a class, create your best explanation of what affects how carnivores feed.

Update the *Project Board*

Use the *Project Board* to record what you have learned about how carnivores feed. Be sure to link your learning statements with the evidence that supports them.

What's the Point?

Like all other organisms, carnivores are well-adapted for their feeding behaviors. You were able to observe and read about just a few different carnivores, but the world is full of many different meat eaters. Like lions and alligators, each has a different food preference, usually depending on its environment. Also, like lions and alligators, each is suited for capturing the particular prey they eat. Some have keen eyesight and sharp hearing, some have the ability to blend into their environment and lay in wait, and some have the ability to chase down fast prey.

The joints of all animals help the animal find and eat food. The joints of carnivores are particularly well-adapted for their feeding behaviors. Most joints are examples of third-class levers, meaning that the effort force is between the fulcrum (the joint) and the load. Muscles provide the effort force. Third-class levers increase distance over which the effort force is applied. This serves to increase the speed at which the animal can move. This is quite an advantage for any animal that must chase its dinner!

Stop and Think

15 min.

Have a class discussion on groups' responses.

META NOTES

The *Stop and Think* questions are very specific to connecting machines to different body parts. These questions will help students understand the content.

○ Get Going

Ask groups to answer the *Stop and Think* questions and let them know how much time they have.

△ Guide & Assess

Have a class discussion on groups' responses. During the discussion, guide students as needed. Listen for the following:

1. Hinge joints allow predators to run quickly. They also allow predators to capture their prey in their jaws.
2. Third-class levers allow the predator to move quickly and cover great distances in pursuit of their prey.
3. Drawings should be detailed, and all features should be labeled.

Revise Your Explanation

10 min.

Have students revise their explanations.

3. Make a sketch of a carnivore that you are familiar with. Label each feature, or characteristic that makes the animal well suited for capturing and eating prey.

Revise Your Explanation

Based on what you have just read, revisit your *Observing and Interpreting Animal Behavior* pages. Add more detail to the middle column. If you need to, revise your interpretations.

Then, revisit your *Create Your Explanation* pages. You now have more science knowledge about what affects the feeding behavior of carnivores. Revise your claim if you need to, add your new knowledge, and rewrite your explanation to add detail.

Sh… Your Ex…nation

△ Guide

Let students know that now that they have learned something new about animal's bodies, they should revise their explanations to include this new information. Have them add to the middle columns in their *Observing and Interpreting Animal Behavior* pages.

Let students know that they should revise their *Create Your Explanation* pages. Emphasize that they should use the new information as evidence to support their claims or, if their claims are inconsistent with what they have learned, they should revise their claims. Let them know that they will be sharing their explanations with the class.

Share Your Explanation

10 min.

Have students share their explanations with the class.

Share Your Explanation

Present your explanations to the class, and as a class, create your best explanation of what affects how carnivores feed.

△ Guide

Have groups present their explanations to the class. Groups should point out their claims, their evidence, and their reasons. They should also be prepared to defend their explanations, and to discuss how they revised them.

As groups present, students should ask questions to clarify and point out where something seems to be missing from an explanation. Model how you expect students to ask questions by asking about how the evidence in an explanation supports the claim.

Move the discussion directly into updating the *Project Board* after all groups have presented.

◇ Evaluate

Make sure students have included the new information they learned about joints, and which joints help the predators in their feeding behaviors.

△ Guide

Lead a discussion of what the class can add to the *Project Board*, focusing on the third and fourth columns—*What are we learning?* and *What is our evidence?*. Record students' ideas on the *Project Board*, and have them record their ideas on their own *Project Board* pages.

Students should include their claims about what affects how predators feed. Their evidence should be observations from the video and science knowledge presented in the readings.

Students should identify ways that many carnivores have adapted to hunt. Using the examples of cheetahs, lions, and crocodiles, students should be able to generalize some behaviors of all predators. Emphasize that all organisms are well adapted for their feeding behaviors.

Ask students to connect the feeding behaviors of animals, their body forms, and simple machines. Most predators have joints that are third-class levers that provide them with increased speed and distance to chase their dinner.

Update the *Project Board*

10 min.

Record the class's ideas on their Project Board.

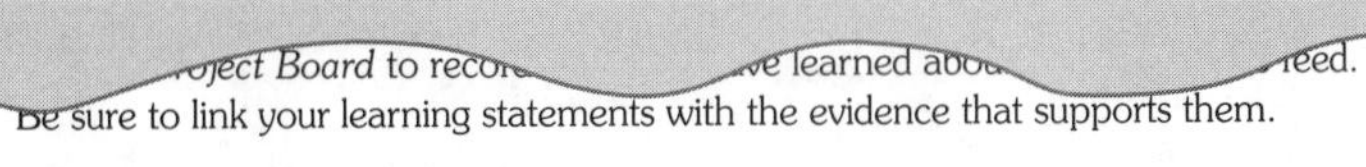

Be sure to link your learning statements with the evidence that supports them.

What's the Point?

Like all other organisms, carnivores are well-adapted for their feeding behaviors. You were able to observe and read about just a few different carnivores, but the world is full of many different meat eaters. Like lions and alligators, each has a different food preference, usually depending on its environment. Also, like lions and alligators, each is suited for capturing the particular prey they eat. Some have keen eyesight and sharp hearing, some have the ability to blend into their environment and lay in wait, and some have the ability to chase down fast prey.

The joints of all animals help the animal find and eat food. The joints of carnivores are particularly well-adapted for their feeding behaviors. Most joints are examples of third-class levers, meaning that the effort force is between the fulcrum (the joint) and the load. Muscles provide the effort force. Third-class levers increase distance over which the effort force is applied. This serves to increase the speed at which the animal can move. This is quite an advantage for any animal that must chase its dinner!

AIA 90

Project-Based Inquiry Science

What's the Point?

5 min.

Record the class's ideas on their Project Board.

Assessment Options

Targeted Concepts, Skills, and Nature of Science	How do I know if students got it?
Scientists must keep clear, accurate, and descriptive records of what they do so they can share their work with others, consider what they did, why they did it, and what they want to do next.	**ASK:** How can you use what you've learned in future explanations? **LISTEN:** Students should recognize that keeping records of their work allows them to use it for future work.
The structure and function of animals' bodies are complementary and affect animal behavior.	**ASK:** How do lions' powerful legs affect their behavior? How do crocodiles' short legs, flat bodies, and powerful jaws affect their behavior? **LISTEN:** Lions' powerful legs allow them to subdue and kill prey. Crocodiles' short legs and flat bodies allow them to lie in wait just below the water's surface, and their powerful jaws allow them to capture prey with their jaws alone.
Scientists differentiate between observations and interpretations. They use their observations and interpretations to explain animal behavior.	**ASK:** How did you use your interpretations to create explanations? **LISTEN:** Students should have combined their interpretations to create a claim.
Scientific knowledge is developed through observations, recording and analysis of data, and development of explanations based on evidence.	**ASK:** What steps did you have to use to develop your explanation? **LISTEN:** Students had to run a simulation and observe, record and analyze their observations, and formulate a claim.

Targeted Concepts, Skills, and Nature of Science	How do I know if students got it?
Explanations are claims supported by evidence. Evidence can be experimental results, observational data, and other accepted scientific knowledge.	**ASK:** How did you support your claims? **LISTEN:** Students should have used observations and scientific knowledge to support their claims.

Teacher Reflection Questions

- What difficulties did students have in making the connection between the physical features of crocodiles, lions, and cheetahs and their behaviors? How were you able to help them see how the levers in predators' bodies allow their hunting behaviors?
- How did students use comparisons within sets of groups and between sets of groups to help them interpret the feeding behavior of predators?
- Which discussions were students most involved in and why? What could you use from this discussion to assist future discussions?

NOTES

NOTES

BACK TO THE BIG CHALLENGE INTRODUCTION

Learning Set 2

Back to the Big Challenge

◀ ***1 class period****

*A class period is considered to be one 40 to 50 minute class.

Overview

Students create recommendations for enclosure designs that will allow an animal introduced in this *Learning Set* to feed as if in its natural environment. This is the first time students create recommendations, so time is spent on learning what a recommendation is. Groups present their recommendations to the class during a *Solution Briefing*, engaging in a common social practice among scientists—sharing and refining ideas within their community. Through class discussions, students will get ideas on how to revise and refine their recommendations. The class then updates its *Project Board*.

Targeted Concepts, Skills, and Nature of Science	Performance Expectations
The structure and function of animals' bodies are complementary and affect animal behavior.	Students' recommendations should include how the animal's structure affects its feeding behavior and the design of the enclosure.
Behavior is a type of response to internal or external stimuli. Behavior is determined by experience, physical characteristics, and environment.	Students' recommendations should be sensitive to how the animal's physical characteristics and environment affect its feeding behaviors.
Explanations are claims supported by evidence. Evidence can be experimental results, observational data, and other accepted scientific knowledge.	Students include observations and scientific knowledge in their explanations, to support their recommendations for designing an enclosure.
Scientists must keep clear, accurate, and descriptive records of what they do so they can share their work with others and consider what they did, why they did it, and what they want to do next.	Students should use their records from *Learning Set 1* and *Learning Set 2* to construct recommendations for an animal enclosure design.

Targeted Concepts, Skills, and Nature of Science	Performance Expectations
Scientists often work together and then share their findings. Sharing findings makes new information available and helps scientists refine their ideas and build on others' ideas. When another person's or group's idea is used, credit needs to be given.	Students share their ideas for designing an enclosure. By hearing the class's feedback on their ideas, they will be able to revise and refine their ideas during their next design iteration.
Criteria and constraints are important in determining effective scientific procedures and answering scientific questions..	Students' recommendations for designs should address all the class's criteria and constraints.

Materials	
3-4 per student	*Create Your Explanation* page.
1 per student	*Solution-Briefing Notes* page
1 per class	Class *Project Board*
1 per class	Class list of criteria and constraints

Homework Options

Reflection

- Science Content: Describe the best design enclosure recommendations you heard about in class for the animals you did not select. Why do you think they are best? *(This will help students recall others' ideas and consider features in their design that they might value in their own designs.)*
- Science Content: Answer the question of the Learning Set: What affects how animals feed? *(Students should describe how animals' feeding behavior depends on their environment, their physical structure, their inherent nature, and their learned behaviors.)*
- Science Content: Your enclosure design must also take into account how the animal you selected communicates. How do you think the animal you selected communicates? What do you think affects its communication? *(This elicitation will get students thinking about what and how animals communicate.)*

BACK TO THE BIG CHALLENGE IMPLEMENTATION

◀ *1 class period**

Learning Set 2

Back to the Big Challenge

15 min.

Introduce recommendations and have the class construct their recommendations.

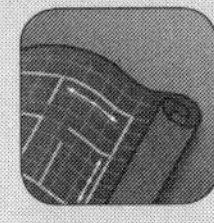

Learning Set 2

Back to the Big Challenge

You and your classmates have been developing an answer to the question, *What affects how animals feed?* Answering this question will help you complete the challenge of the Unit, to design an enclosure for an animal. Your enclosure should be designed so your animal can feed as naturally as possible. The enclosure should be as similar as possible to the animal's habitat.

To design your enclosure, you will need to think about some of the big ideas that affect how animals feed. Several big ideas were introduced in this *Learning Set*. You need to think about how those ideas are going to be built into your enclosure.

recommendation: a claim that suggests what to do in a described situation.

You are going to focus on one of the animals you learned about in this section. You will use observations, interpretations, and explanations of that animal's feeding behavior to help you develop a **recommendation** about designing an enclosure for that animal. You can also use information from the *Project Board*, especially the third and fourth columns, where you included information about each animal.

You will write a recommendation about how to provide the best environment for the animal so it can feed in the most natural way possible. Think about starting your recommendations with "If," "When," or "Because." For example, you might begin your recommendation by writing, "If the bees were going to feed in the most natural way possible, ..." Even better would be a recommendation of the form, "Because bees need to ..., ..." Then you need to complete the statement.

You will probably need to come up with more than one recommendation. Write as many recommendations as you need to write to make sure someone else will know everything that is important so that the animal's enclosure will allow your animal to feed naturally.

AIA 91 ANIMALS IN ACTION

△ Guide

Remind students that they have focused on answering the question: *What affects how animals feed?* in this *Learning Set*, and that this is one of the main issues to consider in their challenge to design an animal enclosure. Let students know that they will be presenting their ideas on how to design an animal enclosure based on the feeding behaviors of one of the animals presented in this *Learning Set*.

Give students a brief overview of what they will be doing: creating recommendations for the design of an enclosure for one of the animals they

TEACHER TALK

"During this *Learning Set*, you have focused on what affects how various animals feed. This is important in the design of an animal enclosure because you want the animals to feed as closely to how they would in their natural environment. Today, your group will construct a recommendation for an enclosure based solely on what affects how an animal feeds. First, you'll need to know more about what a recommendation is."

learned about, presenting their recommendations to the class and discussing them, and finally updating the class *Project Board*.

Then, discuss with the class what a recommendation is and that they will need to develop more than one recommendation for their enclosure design.

TEACHER TALK

"In our class, when we talk about a recommendation we mean a claim that suggests what to do in a situation based on evidence and science knowledge. This information is what we use when we make explanations and keep track of in columns 3 and 4 of our *Project Board*.

Recommendations may be of the forms:

- **If... then... because; or**
- **When ... occurs, then do, try, or expect ...**
- **Because ..., should....**

You will need more than one recommendation to describe what is important in the design of your enclosure."

Next, let students know that they should use a *Create Your Explanation* page for each recommendation their group makes and that they will be presenting their recommendations to the class.

Next, let students know that they should use a *Create Your Explanation* page for each recommendation their group makes and that they will be presenting their recommendations to the class.

Next, provide some examples of recommendations such as those in the student text in the *Making Recommendations* information box and those given below. Again, emphasize that recommendations require evidence and science knowledge to back them up.

Example of Recommendations:

If you want to build an enclosure for cats that takes into account how cats feed in the wild, then you should allow a way to introduce mice or other creatures it

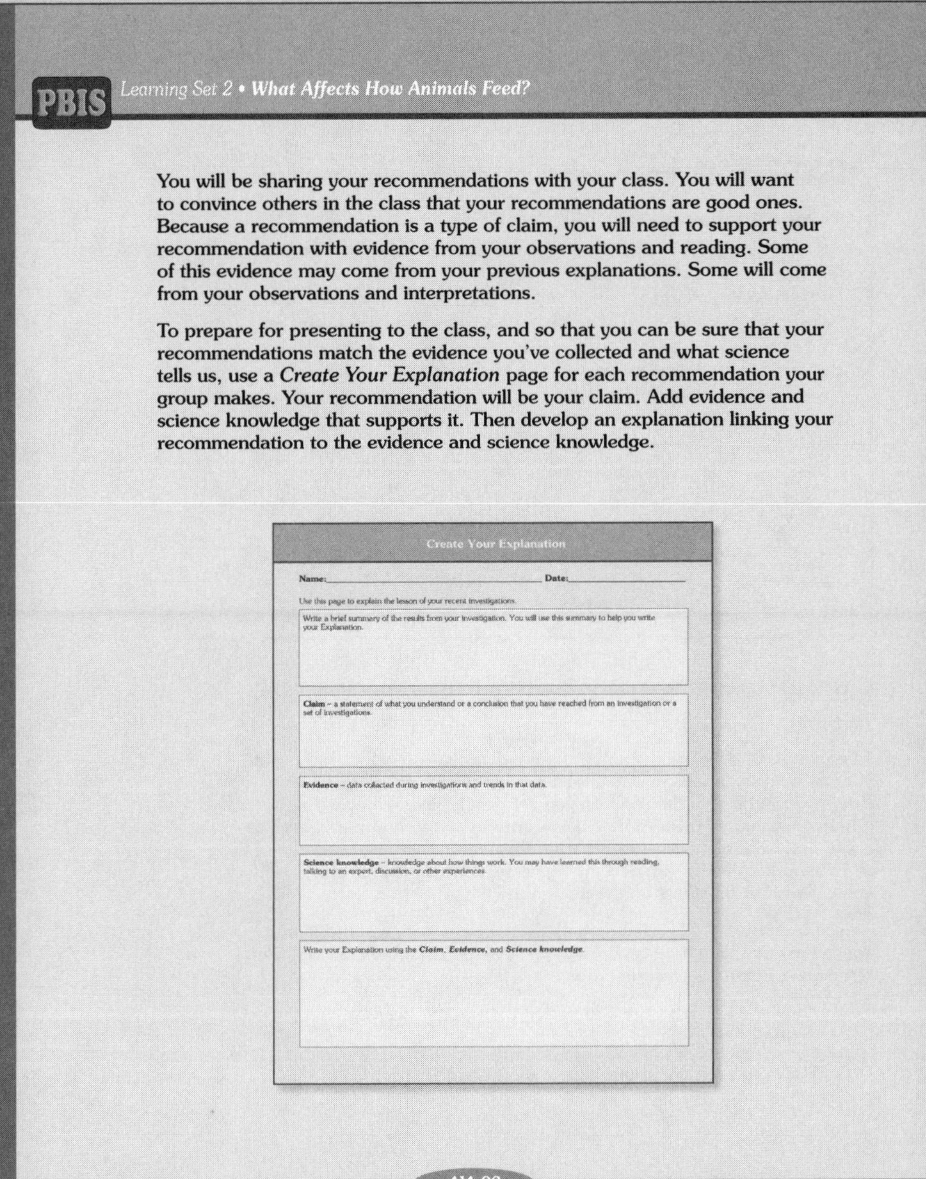

PBIS Learning Set 2 • What Affects How Animals Feed?

You will be sharing your recommendations with your class. You will want to convince others in the class that your recommendations are good ones. Because a recommendation is a type of claim, you will need to support your recommendation with evidence from your observations and reading. Some of this evidence may come from your previous explanations. Some will come from your observations and interpretations.

To prepare for presenting to the class, and so that you can be sure that your recommendations match the evidence you've collected and what science tells us, use a *Create Your Explanation* page for each recommendation your group makes. Your recommendation will be your claim. Add evidence and science knowledge that supports it. Then develop an explanation linking your recommendation to the evidence and science knowledge.

Create Your Explanation
Name:
Date:
Use this page to explain the lesson of your recent investigations.
Write a brief summary of the results from your investigation. You will use this summary to help you write your Explanation.
Claim – a statement of what you understand or a conclusion that you have reached from an investigation or a set of investigations.
Evidence – data collected during investigations and trends in that data.
Science knowledge – knowledge about how things work. You may have learned this through reading, talking to an expert, discussion, or other experiences.
Write your Explanation using the Claim, Evidence, and Science knowledge.

AIA 92

Project-Based Inquiry Science

ıunts into its enclosure. In the wild, cats are predators that hunt for their food. 3y having a way to introduce smaller animals into their enclosure, you can ·egulate how much they eat and allow them to hunt as they do in the wild.

We recommend having a small door outside the enclosure that leads to a tunnel ɔpening in the enclosure. This would allow rats or mice to be introduced into he enclosure. This way the cats will have to hunt for their food as they do in he wild.

Back to the Big Challenge

Be a Scientist

Making Recommendations

A recommendation is a type of claim that suggests to someone what to do when certain kinds of situations occur. It can have this form: When some situation occurs, do or try or expect something. For example, if you want to make a recommendation for crossing the street you might say the following:

When you have the right of way, expect that some cars will not have time to stop in time.

When you have the right of way, look both ways to make sure the traffic has stopped.

Recommendations might also begin with "if." For example, you might state another recommendation about crossing the street this way:

If you have the right of way, and the traffic has stopped, then you can cross the street.

You can also state a recommendation using a "because" statement.

Because some cars will not stop when the light turns red, make sure you look both ways carefully before you cross.

Communicate Your Solution

Solution Briefing

After you have developed your recommendations, you will communicate your recommendations to one another in a *Solution Briefing*. In a *Solution Briefing*, you present the solution you are developing in a way that will allow others to evaluate how well it achieves criteria and to make suggestions about how you might improve it. Your solution, in this case, consists of the recommendations you are developing. Before you start preparing, read more about *Solution Briefings* on the next page.

As you prepare for this briefing, make sure you revisit the criteria and constraints you identified in the beginning of the Unit. Use the following questions to plan your presentation:

- How is your enclosure addressing the feeding needs of your selected animal?
- How does the enclosure meet the criteria?

AIA 93

ANIMALS IN ACTION

META NOTES

The examples provided are not complete. The amount of space in the enclosure is not discussed or the type of environment in the enclosure. To simulate the wild, these things need to be taken into consideration and would be part of other recommendations for the enclosure's design.

⬡ Get Going

Distribute the *Create Your Explanation* pages to groups and let them know how much time they have.

△ Guide and Assess

Monitor students' progress and determine if students understand what is involve in making a recommendation by reviewing what they are doing. Guide them as needed by asking them to point out their claim, evidence, science knowledge, and how the evidence and science knowledge support their claim.

- How did the constraints affect your recommendations?
- What information did you use to help you make your recommendation?
- What other ideas did you think about along the way, and why did you not recommend them?
- What questions do you still have?

As you listen to your classmates' presentations, make sure you understand the answers to these questions. If you do not understand something, or if a group did not present something clearly enough, ask questions.

You can use the questions above as a guide. When you think something can be improved, make sure to contribute your ideas. Be careful to ask your questions and make your suggestions respectfully.

As you listen, record notes on a *Solution-Briefing Notes* page.

Solution-Briefing Notes

Name:________________ Date:________________

Design Iteration:________________

Design or group	How well it works	What I learned and useful ideas		
		Design ideas	Construction ideas	Science ideas

Plans for our next iteration

Communicate Your Solution: Solution Briefing

20 min.

Introduce Solution Briefings, *then have groups present their recommendations.*

META NOTES

A *Solution Briefing* is a common pedagogical tool used in *PBIS*. In a *Solution Briefing* students present their ideas and hear other students' ideas so that students can build on each other's ideas. In this *Solution Briefing,* students are sharing and building on their ideas about how to design an enclosure for an animal. In this and many other class presentations, students will learn how to communicate their ideas, ask questions, and sharpen their critical thinking skills. For more information please refer to the *Teacher's Resource Guide.*

△ Guide

Describe to students what a *Solution Briefing* is and how it works. Explain that they will be presenting their recommendations for designing an enclosure for an animal based on what affects how an animal feeds. Emphasize that these will be works in progress and that the point is to share their ideas and gather advice. The goal is for the larger group (the class) to help each small group to make their solution better.

TEACHER TALK

"You will all work for a while on your recommendations for the design of the enclosure and then you will get some advice from outside of your group. Designers and scientists get together often to share their ideas and get advice from each other on ways to improve their designs (or explanations). *PBIS* calls this a *Solution Briefing*. We are going to have our first *Solution Briefing* so you can share your animal enclosure design recommendations and get advice from the class. Everyone is encouraged to ask questions so they understand your recommendations for design and then offer suggestions on ways to improve it."

Guide students' through the *Be a Scientist* text box *Introducing a Solution Briefing* in the student text.

Describe how to prepare for a *Solution Briefing*. Explain that their presentations should describe their recommendations for the design of the enclosure, based on what they learned about what affects how an animal feeds. Emphasize that they will need to revisit the criteria and constraints for the challenge and that their presentations should address all the bulleted questions in the student text.

Also, explain that the audience should ask clarifying questions and offer suggestions. Everyone should voice their questions and ideas, in a polite and considerate manner, using language such as "I agree with ... because..." or "I disagree with... because."

TEACHER TALK

"It is important that you all think about what you are going to say when you present your design recommendations. In the student text, there is a list of questions that will help you think about what to talk about during your presentation. When you are in the audience or not presenting you will need to listen to make sure that you understand each design. You should also voice your opinions and ideas. Remember to always be polite and considerate whether you are presenting or are in the audience."

⬡ Get Going

Point out where you have displayed the class *Project Board* and their list of criteria and constraints for the enclosure.

Give groups about 10 minutes to prepare their presentation and have student groups present.

Back to the Big Challenge

Be a Scientist

Introducing a *Solution Briefing*

A *Solution Briefing* is useful when you have made one or more attempts to solve a problem or achieve a challenge and need some advice. It gives you a chance to share what you have tried and learned. It also provides an opportunity for you to learn from others. You can ask advice of others about difficulties you are having.

Real-life designers present their design plans and solutions to others several times as they work on design projects. A team of designers sets up their design solution or design plan, and everyone gathers around. They make sure everyone can see. The design team presents its solution to everyone. The other designers ask questions and give helpful advice about ways to improve the design.

You will do the same thing. In a *Solution Briefing*, each team presents their solution in progress for others to see. Then teams take turns presenting to the class. Other classmates ask questions and offer helpful advice.

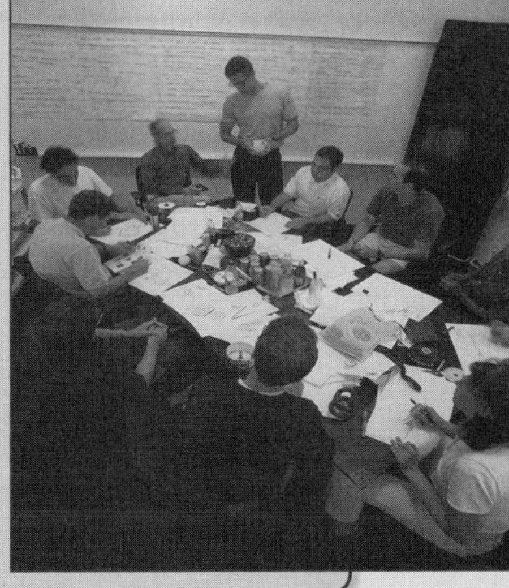

A *Solution Briefing* works best when everyone communicates well. Before you present your design or recommendation to the rest of the class, think about what might be important to share. What aspects of your solution should you present? What parts do you want to discuss with others? You need to be ready to justify to others what you decided to do and why.

When you are listening during a *Solution Briefing*, it is important to pay close attention. Look at each design or plan. Think about questions you would like to ask about each.

Each time you hold a briefing, you will take notes. You will fill out a *Solution-Briefing Notes* page as you listen to each group's presentation.

△ Guide

Describe the *Solution-Briefing Notes* page to students as a place to keep track of the presented design recommendations and the ideas they found useful. Remind students that each group will be presenting their recommendations and that the audience will be taking notes, asking questions and giving advice. Emphasize the importance of listening for reasoning that supports the recommendations and to prepare their questions and advice for the presenting group.

TEACHER TALK

"During this *Learning Set,* you have focused on what affects how various animals communicate. What affects how animals communicate? This is important in the design of an animal enclosure because you want the animals to communicate as closely to how they would in their natural environment. Today, your group will construct recommendations for an enclosure based solely on what affects how an animal communicates."

Distribute the *Solution-Briefing* Notes page and begin the presentations.

While groups are presenting, you and the class should be listening for how recommendations are supported, if the criteria and constraints are met, and if the questions in the student text have been addressed.

META NOTES

Remember this is the first time students have seen a *Solution Briefing*, and it will take time for them to understand how to participate as presenters and as the audience.

If groups get off track or stuck, you can prompt them with one or two of the questions listed in the student text. As the *Solution Briefing* progresses, encourage students in the audience to participate more in asking questions to help the presenters clarify their design recommendations.

Listen for the following types of response to the to student text questions.

- Students' recommendations should describe the feeding needs of the animal based on information provided in this *Learning Set* and address these in their recommended design.
- The recommendations for the enclosure must meet the class's list of criteria.
- Students should describe how their recommendations address the constraints of the enclosure, such as building the enclosure locally and dealing with the local weather.
- Students' recommendations should be supported by their observations and science knowledge in the student text and they should be able to describe how these support their recommendations.
- Students should describe ideas they decided against and their reasoning for choosing or not choosing an idea.
- Students should list questions they still have. One of these could be about how the animal they chose communicates. Others might involve more information about what affects how the animal they selected feeds in the wild such as: Do all bees feed the same way? Is there anything special about the UV patterns bees see on flowers that makes them want to go to that flower over another flower with UV patterns?

META NOTES

If students say that another group copied their idea, tell them this is an important issue and will be addressed later, continue with the presentations. After the presentations, discuss building on each other's ideas versus copying and the need to give credit to those who formulated the ideas being used.

Update the *Project Board*

The last column on the *Project Board* helps you pull together all that you have learned during the Unit. In this column, you record partial answers to the *Big Question* or *Big Challenge*. Each investigation you do and each reading you complete is like a piece of a puzzle. You fit the pieces together to help you address the challenge. Each piece provides you with a critical factor that must be addressed to answer the *Big Question*.

The *Big Challenge* is to create an enclosure for an animal that makes it possible for the animal to feed or communicate as naturally as possible. The last column of the *Project Board* is the place to record your recommendations. Recording your recommendations on the *Project Board* as you go along will allow you to easily remember your recommendations when you return to the *Big Challenge* at the end of the Unit.

META NOTES

If students say that another group copied their idea, tell them this is an important issue and will be addressed later, continue with the presentations. After the presentations, discuss building on each other's ideas versus copying and the need to give credit to those who formulated the ideas being used.

Update the Project Board

5 min.

Record the class's ideas on their Project Board.

After all groups have presented, ask students if they heard any interesting ideas or thought of new ideas they would like to use to improve their own design recommendations. Guide students to think about how sharing ideas will help them improve their own ideas.

Finally, compare this session with what scientists and designers do to improve their solutions. Reflect on the usefulness of iterations and record keeping. Let students know that scientists are always building on each others' ideas. Bring up the idea of giving credit to others when you build from their ideas. This is not copying. Emphasize that the difference between this and copying is that copying does not give credit to the people who thought of the ideas you used.

META NOTES

This *Learning Set* has been focused on investigating what affects how animals feed. Today you considered how that might affect the challenge by constructing recommendations for designing an enclosure that allows animals to feed as they would in their natural environment. What do we know about what affects how animals feed now and how it will affect the enclosure designs? Think about all of the recommendations you heard and our class discussion on them to help you decide what to put up on the class *Project Board*.

△ Guide

Have a class discussion on what to add to the *Project Board*, focusing on the last column—*What does it mean for the challenge or question?* Record students' ideas on their class *Project Board*, and have them record their ideas on their own *Project Board* pages.

Example:

Recommendation: Termite mounds and grass should be included in chimpanzee enclosures because we and other scientists observed that chimpanzees use grass blades to collect termites from their mound to eat.

◇ Evaluate

Make sure the class *Project Board* contains recommendations for enclosures only pertaining to animals discussed in *Learning Set 2* and focuses on feeding.

Teacher Reflection Questions

- What difficulties did students have in making recommendations? What ideas do you have to help them?
- What evidence do you have that students understand that the sharing of ideas and building on other's ideas are practices of scientists? How can you further their understanding of the importance of these practices?
- What difficulties do your classes still have in participating in class discussions? What ideas do you have to assist your classes?

NOTES

LEARNING SET 3 INTRODUCTION

Learning Set 3

What Affects How Animals Communicate?

◀ *12 class periods**

*A class period is considered to be one 40 to 50 minute class.

Student ethologists make observations of animals communicating and focus on how their physical characteristics and how environment affects their communication.

Overview

Student groups work as ethologists studying the communication behaviors of animals and how the animals' physical characteristics and environment affect their communication. They begin by considering how humans communicate and how the environment affects how humans communicate. They will explore the differences between humans' verbal and nonverbal communication to recognize that the constraints of an animal's body and environment determine the form of communication that the animal uses. Students observe the waggle dance of bees, learn the history of scientists' understanding of the waggle dance, and learn how bees communicate the location of nectar to the hive. Students then observe elephant communication and they develop explanations for how elephants communicate, learning that elephants' range of hearing allows them to communicate using pitches that humans cannot hear. Sound waves are then introduced, along with the parts of the wave, loudness, and the structure and function of the human ear. Students then observe dolphin communication and develop explanations of how dolphins communicate. Students are introduced to echolocation and how sound is used to "image" objects by various animals, including humans.

LOOKING AHEAD

- **Students will observe animals using videos during this *Learning Set*. This requires appropriate equipment.**
- **The *Big Challenge* at the end of this Unit requires students to refer back to their *Observing and Interpreting Animal Behavior* pages. Students will benefit if they have these pages organized.**

Targeted Concepts, Skills, and Nature of Science	Section
Scientists often work together and then share their findings. Sharing findings makes new information available and helps scientists refine their ideas and build on others' ideas. When another person's or group's idea is used, credit needs to be given.	3.1, 3.2, 3.3, 3.4, 3.5, 3.6, 3.7, BBC
Scientists must keep clear, accurate, and descriptive records of what they do so they can share their work with others, consider what they did, why they did it, and what they want to do next.	3.1, 3.2, 3.3, 3.4, 3.5, 3.6, 3.7, BBC

Targeted Concepts, Skills, and Nature of Science	Section
Tables are an effective way to communicate results of a scientific investigation.	3.2, 3.4, 3.6
Scientists differentiate between observations and interpretations. They use their observations and interpretations to explain animal behavior.	3.4, 3.6
Studying the work of different scientists provides understanding of scientific inquiry and reminds students that science is a human endeavor.	3.3
Scientific knowledge is developed through observations, recording and analysis of data, and development of explanations based on evidence.	3.3, 3.5, 3.6, 3.7
Scientists make claims (conclusions) based on evidence obtained (trends in data) from reliable investigations.	3.2, 3.3, 3.4, 3.6
Explanations are claims supported by evidence. Evidence can be experimental results, observational data, and other accepted scientific knowledge.	3.4, 3.5, 3.6, 3.7, BBC
Criteria and constraints are important in determining effective scientific procedures and answering scientific questions.	BBC
Behavior is a type of response to internal or external stimulus. Behavior is determined by experience, physical characteristics, and environment.	3.2, 3.3, 3.4, 3.6, BBC
The structure and function of animals' bodies are complementary and affect animal behavior.	3.3, 3.4, 3.5, 3.6, 3.7, BBC
Animals communicate with other animals using sound. The sounds they can make and the sounds they can hear are adaptations to their environment.	3.1, 3.2, 3.4, 3.5, 3.6, 3.7
Vibrations of molecules produce sound. Sound is a compression wave that can be described by its amplitude, frequency, and wavelength. Sound moves differently through different matter.	3.5, 3.6, 3.7
Animals' ears are adapted to hearing sounds in their environment. Some animals use sound that is out of the range of human hearing.	3.4, 3.5, 3.6, 3.7
Animals' sense of sight is adapted to their environment. Some animals see things that humans cannot see.	3.7

Students' Initial Conceptions and Capabilities

- Students may have difficulties understanding the difference between observations and inferences. (Allen, Statkiewitz, & Donovan, 1983; Kuhn 1991, 1992; Roseberry, Warrant, & Conant, 1992.)

Understanding for Teachers

Animal behavior is affected by many different factors. In *Animals in Action,* students consider animals' body structure and their environment as influences on their communication. Students use these two factors to develop explanations for an animal's communication in this *Learning Set.*

Animals communicate to feed, to protect themselves, and to reproduce. Some have very complicated communication behaviors (such as speech) and some have very simple communication behaviors. When animals communicate, they are constrained by their body structure. Animals use the features of their bodies to communicate. Bees, elephants, and dolphins all communicate, but they communicate in very different ways.

NOTES

NOTES

LEARNING SET 3 IMPLEMENTATION

Learning Set 3

What Affects How Animals Communicate?

When you were a baby, you may have cried a lot. You cried to let someone know you wanted to eat, be held, or have your diaper changed. Your parents learned very quickly what your cries meant. By the time you were a year old, you were able to communicate in many new ways. You may have pulled on someone's sleeve, pointed to what you wanted, or even used a few words like "eat" or "more." By the time you were two, you could talk to others in two- or three-word sentences. These sentences communicated your wants and needs to your parents.

A baby cries to communicate.

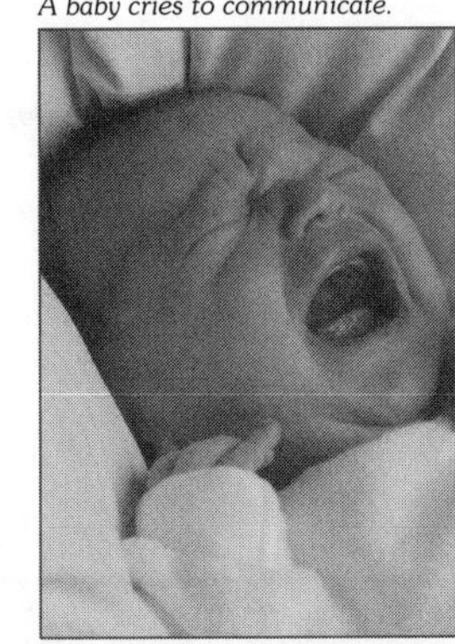

Communication is a behavior common to many animals. Baby animals often need to communicate to others their need for food or help. Communication can allow adult animals to live more comfortably. Once animals are able to communicate with each other, they can form groups and help each other find food and defend themselves from predators.

To answer the *Big Question: How do scientists answer big questions and solve big problems?* you need to break it into smaller questions. In this *Learning Set*, you will investigate animal communication and answer the question *What affects how animals communicate?* You will read about and observe several different animals. You will examine how the form of the animal and the environment it lives in affects the ways it communicates.

A mother polar bear and her cub exchange greetings.

AIA 97 — ANIMALS IN ACTION

Learning Set 3

What Affects How Animals Communicate?

5 min.

Introduce the question: What Affects How Animals Communicate?

○ Engage

Begin by eliciting students' ideas about how animals communicate.

TEACHER TALK

"You've seen different animals, including people, communicate. How do babies communicate?

How do pet dogs or cats communicate (for instance, when they want to be fed)? Do they always use the same parts of their bodies or the same senses to communicate?

Why do animals communicate in such different ways?"

Then, point out that communication is a common behavior for these animals and that many animals need to communicate to survive. Even amoebas communicate! Let students know that this *Learning Set* focuses on answering the question, *What Affects How Animals Communicate?* and that they should be able to answer this question by the end of the *Learning Set*. Emphasize that this will help students address the *Big Challenge* — to design an enclosure that will encourage the feeding or communication of one of the animals studied in this Unit — and it will also help students answer the *Big Question* of the Unit: *How do scientists answer big questions and solve big problems?*

Let students know that they will be considering various animals in this *Learning Set*.

NOTES

SECTION 3.1 INTRODUCTION

3.1 Understand the Question

Thinking about What Affects How Animals Communicate

◀ ***1 class period****

*A class period is considered to be one 40 to 50 minute class.

Overview

Students list different ways humans communicate and they record situations, noting when and why people use each way. This enables them to begin thinking about what affects communication and the purposes of communication. Then, students categorize the ways humans communicate and they share their categories and lists with the class. Together, the class looks for similarities and differences in the ways groups categorized their lists and chose final categories. Having already thought about what affects the way people communicate, students develop questions about what affects how animals communicate. Each group selects two questions to put on the *Project Board.*

Targeted Concepts, Skills, and Nature of Science	Performance Expectations
Scientists often work together and then share their findings. Sharing findings makes new information available and helps scientists refine their ideas and build on others' ideas. When another person's or group's idea is used, credit needs to be given.	Students should work with their groups to develop categories of communication and to develop questions about ways of communicating.
Scientists must keep clear, accurate, and descriptive records of what they do so they can share their work with others, consider what they did, why they did it, and what they want to do next.	Students should add to their class's *Project Board,* which organizes their class's ideas. Students record their observations and knowledge of human communication behaviors.
Animals communicate with other animals using sound. The sounds they can make and the sounds they can hear are adaptations to their environment.	Students should include the ways humans use their voice to communicate in the lists of ways of communicating.

Materials	
1 per class	*Project Board*
1 per student	*Project Board* pages
1 per group	Sticky notepads
10 to 15 per group	Index cards

Homework Options

Reflection

- **Science Content:** What are some of the ways your environment can change how you communicate? Are things in the environment used for any of the ways of communicating you identified? *(Some environments are too loud to communicate by speaking, while some are too dark for communicating by gestures. We often use parts of our environment to make sounds or project lights to communicate. For example, while driving a car we might honk the horn or flash the headlights; in other instances we might ring a bell or knock on a door.)*
- **Science Content:** Think about an animal other than a human that you are familiar with, such as a cat or dog. What are some ways this animal uses its body to communicate? How does it use its body to communicate in ways that are different from how humans use their bodies? In what ways are they similar? *(This question is meant to engage students in thinking about how other animals communicate and what affects the way they communicate. For example, a dog wags its tail when it is happy, bears its teeth when angry, and may put its tail between its legs when scared.)*

Preparation for 3.2

- **Science Content:** Which of the ways of communicating that you identified are best for communicating complex ideas? *(Speaking, writing, and drawing pictures are very effective ways of communicating complex ideas.)*

SECTION 3.1 IMPLEMENTATION

3.1 Understand the Question

Thinking about What Affects How Animals Communicate

The energy that food provides is a necessity of life. All animals find some way to feed. Another necessity of life is communication. Animals communicate when they need to affect the behavior of other animals. Human communication may include discussions about big and important ideas that affect the world. However, communication does not have to be complex. For most animals, communication is much more simple.

In this *Learning Set*, you will learn about different ways animals communicate. Some animals use sound, some use body characteristics like tails or coloring, some use gestures or facial expressions, and some use smell. More complex animals, including many mammals, use a combination of methods to communicate.

Animals communicate with other animals of the same species and sometimes with animals of different species. For example, dogs might bark at other dogs that enter their territory. Dogs also bark at other species of animals, like squirrels and people.

Communication is important to the survival of animals. Without the ability to communicate with others, animals would not be able to protect themselves, hunt for food, reproduce, find other animals in their group, or identify their territory.

A peacock communicates by spreading its tail feathers.

Wolves communicate with facial expressions.

◀ *1 class period**

3.1 Understand the Question

Thinking about What Affects How Animals Communicate

5 min.

Have a brief discussion on why animals communicate.

△ Guide

Using the information in the student text, briefly discuss why animals communicate. Emphasize that animals communicate when they need to affect the behavior of other animals. Many animals communicate to protect themselves, hunt for food, reproduce, find their group, or identify their territory.

Ask students what examples of animal communication they have experienced. Record their responses to refer to later.

*A class period is considered to be one 40 to 50 minute class.

Get Started

10 min.

Have students consider human communications and the situations in which they are used.

META NOTES

Students first consider human communications, which they are familiar with. This should help them make more connections to other animal communications.

Get Started

Humans communicate in many different ways. Their communication can be simple, like a smile or a frown, or more complicated, like a wink or a hand gesture. Humans also use language to communicate. Create a list of the different ways people communicate with one another, with and without using their voices, and with or without any tools. Record the kinds of situations and the reasons they communicate in each way. Use a chart like the one shown. One example is already recorded on the chart. Make your chart as detailed as possible. List at least five ways of communicating.

Observation Notes		
Ways of Communicating	Situations when people communicate this way	Why people communicate this way
Jumping up and down and waving	The person you are communicating with is far away.	The jumping and waving can be seen from a distance. Sound might nt carry far enough and it might be rude to yell.

Conference

Work with your small group to classify the different ways of communicating. Begin by reading to your group the different ways of communicating that you identified. On separate sticky notes, write each situation when people communicate that way. Then group the different ways of communicating under each situation. When you are done with your groupings, name each group. Make a list of your categories and the ways of communicating.

△ Guide

Let students know that they will first be considering an animal they are very familiar with, humans. Let them know that they will be filling in their observations in a chart like that shown in the student text. Describe the chart to the students. Emphasize that they will need to think of at least five ways of communicating and that these may consist of verbal and nonverbal forms of communication.

○ Get Going

Distribute the *Observation Notes* page and let students know how much time they have to complete their charts.

Students should be working individually. They will meet with their groups later to share their ideas.

□ Assess

Monitor students and check what kinds of communication and communication situations they are identifying. If all of the ways of communicating they identify occur in the same situation, suggest that they think about the different situations in which they have communicated and if that affected the communication.

Conference

Work with your small group to classify the different ways of communicating. Begin by reading to your group the different ways of communicating that you identified. On separate sticky notes, write each situation when people communicate that way. Then group the different ways of communicating under each situation. When you are done with your groupings, name each group. Make a list of your categories and the ways of communicating.

AIA 99

ANIMALS IN ACTION

Conference

10 min.

Groups share and categorize the ways of communicating that they listed.

△ Guide

Let students know that they will be meeting with their group members to share their lists and to make a list of categories of ways of communicating, and that they will share this list with the class. Let them know that they will be using sticky notes and index cards like they have in past *Learning Sets* to construct their categories. On each sticky note, they should write the ways of communicating and on the index cards they should write the situations. Then they should group the ways of communication by situation.

Provide students with an example such as the one shown below:

Situations on index cards			
	The person you need to communicate with is inside and you're outside	**You cannot speak**	**You are driving a car and you need to communicate with other drivers**
Way of communicating on sticky note:	Knocking on the door	Writing	Honking your horn
Way of communicating on sticky note:	Ringing the doorbell	Gesturing	Using turn signals
Way of communicating on sticky note:	Waving through the window	Drawing pictures	Flashing your headlights

⬡ Get Going

Distribute index cards and sticky notes and let groups know how much time they have to create their categories.

☐ Assess

Monitor student discussions. If there are any disagreements about how to categorize these lists, take note and remember to bring up the opposed viewpoints in the class discussion.

△ Guide

Once groups have categorized the ways of communicating they identified, they should name their categories and write a list of the categories and ways of communicating on a sheet of paper. They will need to refer to this when they present their categorizations to the class and during the class discussions.

△ Guide

Let the class know that each group will be presenting their categories and that the class will also come up with a list of communication categories. Emphasize that as each group presents their categories, the audience should be comparing their group's categories and noting how they are similar or different.

Then, have groups present their lists to the class.

Communicate

10 min.

Have the class create a list of categories based on group categories.

META NOTES

The final categories for the class list will be used to help students update the *Project Board*.

PBIS *Learning Set 3* • ***What Affects How Animals Communicate?***

Communicate

Share your list with your class. Listen carefully as your classmates describe their lists. How are their categories different from or the same as your group's? Create a class list of communication categories.

Conference

You have considered communication in people. You have thought about different reasons why people communicate. Now think about questions you can ask about how other animals communicate.

AIA 100

Project-Based Inquiry Science

Conference

10 min.

Have groups develop two questions about how animals communicate.

☐ Assess

As groups present, keep track of the similarities and differences in the ways they have categorized their ideas and assess how comprehensive the categories are. You will want to make sure that the students discuss any differences in the ways groups categorized their ideas and that they discuss any ideas that they were unable to categorize. You can use the similarities you notice in the ways groups categorized their ideas to help them develop final categories for the class list.

META NOTES

While students should discuss the reasons for the differences in the categories groups developed, it is not necessary for them to reach complete agreement. The final categories should help students to see how the environment and other factors (such as an animal's anatomy and its need to eat) affect how it communicates.

△ Guide

When groups have presented their lists, lead a discussion to decide what categories and ways of communicating will go in the class list. If groups have categorized things very differently, you can lead them to think about what categories would be most useful to biologists studying behavior. (Categories that include subjective situations, such as "you are angry," might not be as useful as categories describing only objective situations, such as "the person you are trying to communicate with is far away.")

If groups have ways of communicating that they weren't able to fit into any categories, look for ways to combine categories or redefine categories to include them.

△ Guide

Let students know that based on what they have discussed, they should each develop two questions about how animals (not just humans) communicate and what affects animal communication. Let students know that the questions can be about ideas they discussed as they developed categories for their lists, but they do not have to be.

Emphasize to students that the questions they develop should not have yes or no answers, but should interest the student, require several resources to answer, relate to the *Big Question,* and require the student to collect and use data. Also, encourage students to think about the enclosure challenge as they develop their questions.

⬡ Get Going

Let students know how much time they have and that they will meet with their groups to discuss their questions afterward.

△ Guide

Let students know that they will meet with their group members to share and refine their questions and that each group will pick out the two most interesting questions to the group. Emphasize that they will be turning in all the questions.

META NOTES

Familiarize yourself with the extra questions. Some could be included on the *Project Board* or used later in this *Learning Set.*

META NOTES

Groups will share their questions as part of updating the class *Project Board.*

NOTES

Update the *Project Board*

10 min.

Have a class discussion of groups' questions and update the Project Board.

Develop two questions that might help you understand what affects how animals communicate. Make sure the questions are not simply yes/no questions or ones you can answer with a single word or sentence.

When you write your questions, keep in mind that your questions should

- be interesting to you;
- require several resources to answer;
- relate to the *Big Question* or designing a new enclosure that will encourage the feeding or communication of an animal; and
- require collecting and using data.

When you have completed your two questions, meet with your small group. Share all the questions with each other. Carefully consider each question and decide if it meets the criteria for a good question. With your group, refine the questions that do not meet the criteria. Choose the two most interesting questions to share with the class. Give your teacher the rest of the questions so they might be used later.

Update the *Project Board*

The *Project Board* helps you to organize your ideas as you answer the *Big Question* and address the *Big Challenge*. You will now share your group's two questions with your class. Be prepared to justify why yours are good questions. Your teacher will add your questions to the *Project Board*. Throughout this *Learning Set*, you will work to answer some of these questions.

What's the Point?

Like feeding, communication is important for all animals. Some types of communication are simple. Language is a complicated form of communication. Humans use words to communicate with one another. They also use different methods like gestures and facial expressions. Reasons for communicating depend on the animals' needs for survival. Animals that live in groups need to be able to communicate with one another. Hunted animals or ones that need a lot of food communicate to protect their territory from invaders.

△ Guide

Have each group share their questions with the class. Encourage students to think about and ask groups how their questions meet the criteria of good questions, emphasizing that questions should be polite.

During the discussion, encourage students to share the reasoning behind their ideas.

As groups share their questions and the class discusses them, record them for the class to see. As a class, combine questions that they think belong together and edit the wording as they suggest. Then include these questions on the class *Project Board*.

Share and discuss any of the extra questions that should be included on the class *Project Board.*

◇ Evaluate

The class should have questions that they can begin to answer through observation, such as "How do animals use their bodies to communicate?" Make sure that questions like this are in the class list before proceeding.

META NOTES

It is helpful to keep track of the date items were posted on the *Project Board.* That way, students can monitor changes in their ideas and their progress on the challenge.

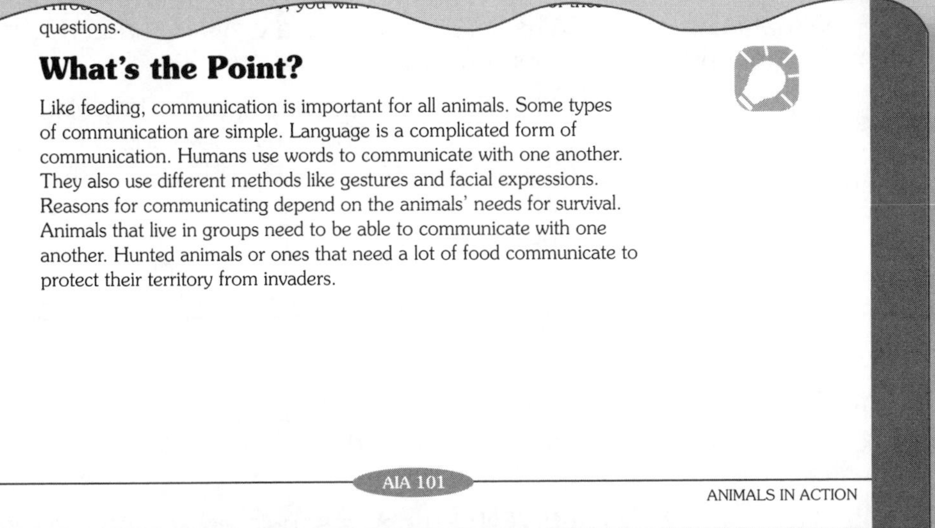

questions.

What's the Point?

Like feeding, communication is important for all animals. Some types of communication are simple. Language is a complicated form of communication. Humans use words to communicate with one another. They also use different methods like gestures and facial expressions. Reasons for communicating depend on the animals' needs for survival. Animals that live in groups need to be able to communicate with one another. Hunted animals or ones that need a lot of food communicate to protect their territory from invaders.

AIA 101

ANIMALS IN ACTION

What's the Point?

5 min.

◇ Evaluate

Make sure students understand that an animal's survival may depend on its ability to communicate.

Assessment Options

Targeted Concepts, Skills, and Nature of Science	How do I know if students got it?
Animals communicate with other animals using sound. The sounds they can make and the sounds they can hear are adaptations to their environment.	**ASK:** What are some of the ways that humans use sound to communicate? What are some of the ways humans might communicate without sound? **LISTEN:** Primarily, they speak. They also scream or cry, and they use objects in their environment to make sound (they honk horns, knock on doors, sound alarms, and ring bells). Humans often communicate using gestures (like waving, winking, hand signals). They may also communicate using sign language or writing.
Scientists often work together and then share their findings. Sharing findings makes new information available and helps scientists refine their ideas and build on others' ideas. When another person's or group's idea is used, credit needs to be given.	**ASK:** How did discussing your group's categories with the class help to develop a good final list? **LISTEN:** Students should have used the discussion to develop categories that include all of the ways of communicating.

Teacher Reflection Questions

- What challenges do you foresee for students as they try to answer the questions they developed? What can you do to help students meet those challenges?
- What difficulties did students have forming and selecting questions to investigate? How did the group and class discussions affect this?
- How were you able to help students develop their final questions while keeping the discussion student-centered? What ideas do you have for next time?

SECTION 3.2 INTRODUCTION

3.2 Investigate

How Do Humans Communicate?

◄ *2 class periods**

*A class period is considered to be one 40 to 50 minute class.

Overview

Students explore how humans communicate verbally and nonverbally through a puzzle-solving activity. In each group, a pair of students is provided with a tangram. One person knows the solution to the puzzle and must communicate the solution to the person with the unsolved puzzle. The remaining group members record behaviors they observe. Half of the groups should use words to communicate the solution, the others should not. After groups have solved the tangram, the class compares the time it took to solve. Reflecting on the different ways students communicated to solve the tangram, students consider how communication is determined by the constraints of an animal's physical features and/or their environment.

Targeted Concepts, Skills, and Nature of Science	Performance Expectations
Scientists often work together and then share their findings. Sharing findings makes new information available and helps scientists refine their ideas and build on others' ideas. When another person's or group's idea is used, credit needs to be given.	Students should work with their groups to analyze their observations of human communication. Then, they should share their results with the class and decide what to include on the class's *Project Board.*
Scientists must keep clear, accurate, and descriptive records of what they do so they can share their work with others, consider what they did, why they did it, and what they want to do next.	Students should record their observations and their interpretations of their observations from the puzzle-solving activity and refer to these records as they share their results with the class.
Tables are an effective way to communicate results of a scientific investigation.	Students may use tables to describe their findings. Students should recognize the value of using tables such as the *Project Board.*

Targeted Concepts, Skills, and Nature of Science	Performance Expectations
Scientists make claims (conclusions) based on evidence obtained (trends in data) from reliable investigations.	Students should make claims about human communications based on their observations.
Behavior is a type of response to internal or external stimulus. Behavior is determined by experience, physical characteristics, and environment.	Students should be able to describe how their communication behaviors were affected by the constraint of using or not using their voice. Students may recognize that their environment and experience affect the way they communicate.
Animals communicate with other animals using sound. The sounds they can make and the sounds they can hear are adaptations to their environment.	Students should explore and discuss how and why they use speech to communicate.

Materials	
2 per group	Tangrams (small puzzle, 12-24 pieces)
1 per group	Poster paper and markers
1 per group	*Observing and Interpreting Animal Behavior* page
1 per class	*Project Board*
1 per student	*Project Board* page

Homework Options

Reflection

- **Science Content:** Did the nonverbal groups' ways of communicating fit into any of the communication categories you created in the last section? How was solving the puzzle with nonverbal communication similar to situations you identified in the last section? *(Students should make connections between this section and the previous one.)*

- **Science Content:** What are some ways that technology extends humans' ability to communicate? What are the physical constraints on human communication that communication technology helps humans with? *(There are many communication technologies—telephones, radio, television, the Internet, or writing. Humans do not naturally have the ability to communicate over great distances—across continents, for instance—and many of these technologies help humans extend the range of their communication. They also extend humans' ability to communicate to others at future times—such as through writing or recording.)*

Preparation for 3.3

- **Science Content:** For an outside observer, it might be difficult to interpret what the nonverbal puzzle solvers were communicating. What clues would you look for in an animal's environment to help you interpret complicated communication between them? *(Animals often communicate in order to get food or to protect themselves. One strategy for interpreting animals' communications is to look in the environment for situations when communication might help an animal avoid dangers or obtain food.)*

Activity Setup and Preparation

Each group will need one assembled tangram and one unassembled tangram. Prepare these sets before class and have them ready to distribute. The tangrams will need to be cut, one placed in an envelope and the other assembled. The assembled tangram may be covered with heavy sheets of paper or placed in a large envelope.

Two of the students in each group will need to face each other as they work on the tangrams without being able to see each other's tangrams. You can accomplish this in a number of ways—you can put a stack of books between them or use carrels. Be aware that the remaining group members need to be able to observe the two group members with tangrams.

Decide where you want the students to display their posters for the *Investigation Expo*.

NOTES

SECTION 3.2 IMPLEMENTATION

3.2 Investigate

How Do Humans Communicate?

A mime communicates without words.

Your class list of forms of communication helped you begin to see many different ways animals might communicate. In this section, you will think about what affects communication. You will explore two different ways humans communicate.

Predict

The goal for this activity is to solve a puzzle in two different ways. Some groups will use words to solve the puzzle and some will not use words. The groups will then analyze their observational data and share their conclusions with the class.

Before you begin solving the puzzle, predict which groups you think will be more successful at solving the puzzle: the "with–words" groups or the "without–words" groups. Begin by using your knowledge and past experiences to think about what instructions you might need to communicate to someone as you solve a puzzle together. How would you communicate those instructions, and what differences might there be between communicating with words and communicating without words? Then write your prediction as a statement that answers the following question:

> Will the "with-words" groups or the "without-words" groups be more successful at solving a puzzle together, and why?

You might start your sentence with, "I think the (*with-words/without-words*) groups will be more successful because, when solving a puzzle, it is helpful to..."

Be a Scientist

Predicting

When you write a prediction, you connect what you think you know about something, along with past experiences, to what you are learning now. Using all the information you have, you form an educated guess.

◀ *2 class periods**

3.2 Investigate

How Do Humans Communicate?

5 min.

Introduce the section by eliciting students' ideas on how they could explore human communications using words and no words.

○ Engage

Remind students of the categories they came up with in the previous section for the ways humans communicate. Then, elicit their ideas on how they might observe verbal and nonverbal communication between humans.

*A class period is considered to be one 40 to 50 minute class.

Predict

10 min.

Introduce how they will investigate communication "with-words" versus communication "without-words" and discuss their predictions.

Predict

The goal for this activity is to solve a puzzle in two different ways. Some groups will use words to solve the puzzle and some will not use words. The groups will then analyze their observational data and share their conclusions with the class.

Before they begin solving the puzzle, students will predict which groups they think will be more successful at solving the puzzle: the "with–words" groups or the "without–words" groups.

Students might start their predictions with, "I think the (*with-words/without-words*) groups will be more successful because, when solving a puzzle, it is helpful to..."

△ Guide

Let students know that they will explore the differences between communicating "with-words" and "without-words" by working in groups to complete a puzzle and describe the situation to the class.

TEACHER TALK

"Today you will be exploring the differences between communicating with and without words. For each group, one person will know the puzzle's solution and one person will not; the other group members will be making observations of the communication behavior. For half of the groups, the solution can be communicated using words. In the rest of the groups, the solution must be communicated without any words. Without any words means that you cannot speak or write words. The "no words" groups will have to come up with another way to communicate, such as through gestures. After completing the puzzle, the class will compare the difficulty between the groups who used words, and those who did not, by comparing the time it took to complete the puzzle."

Then, describe what a prediction is using the *Be a Scientist* box in the student text. Emphasize that predictions are based on what you know about something and what you know from past experiences to form an educated guess.

Then, have each student make a prediction about which set of groups will be most successful (most efficient, or quickest) at solving the puzzle. Emphasize that they should specify the reasons for their predictions as well. Have a brief discussion of students' predictions and reasoning.

Procedure

For this investigation, you will work in groups of three. There will be two puzzle solvers and one observer in each group. The two puzzle solvers will solve the puzzle while the third group member carefully observes the puzzle solving and watches for successes and difficulties.

Some groups will follow the directions for "Solving a Puzzle with Words" and some for "Solving a Puzzle without Words." Read the directions very carefully because the two procedures are not the same. Notice the criteria and constraints that go with each set of instructions.

Materials

- **2 copies of a puzzle–one assembled, one unassembled**
- **paper**
- **pencils**

Solving the Puzzle without Words

1. The puzzle solvers sit facing each other.
2. One group member has an assembled puzzle, and one group member has an unassembled puzzle. The person with the unassembled puzzle should never see the assembled puzzle.
3. Using only facial expressions, hand gestures, or other means that do not involve speaking or writing, the two puzzle solvers should now solve the puzzle.
4. As the puzzle solvers are working, the observer should record observations about the work, paying attention to how the puzzle solving goes, watching the challenges and successes of solving the puzzle without words. The observer should also record how much time it took to solve the puzzle.
5. The time limit for solving the puzzle is 10 minutes.

Criteria	Constraints
Puzzle solvers should solve the puzzle together as quickly as they can.	Puzzle solvers have no more than 10 minutes to assemble the puzzle.
One puzzle solver can use an assembled puzzle for help.	The second puzzle solver cannot see the assembled puzzle.
Puzzle solvers can communicate only with facial expressions and hand gestures.	No words can be spoken or written.

Procedure

15 min.

Have groups solve the puzzle, following the procedure assigned to them.

△ Guide

Briefly go over each set of procedures *(Solving the Puzzle without Words* and *Solving the Puzzle with Words)* with the class. Discuss the criteria and constraints for each situation. Emphasize that all groups have ten minutes. Remind students that the group members observing must pay close attention and record their observations of the situation, including the challenges and successes of solving the puzzle and the time to solve the puzzle. Let students know that the difference between the two sets of directions is the use of words versus not using words.

Emphasize that groups cannot begin until you tell them to start.

Solving the Puzzle with Words

1. The puzzle solvers sit facing each other.
2. One group member has an assembled puzzle and one group member an unassembled puzzle. The person with the unassembled puzzle should never see the assembled puzzle.
3. Using spoken or written words, the two puzzle solvers will work together to solve the puzzle.
4. As the puzzle solvers are working, the observer should record observations about the work, paying attention to how the puzzle solving goes, watching the challenges and successes of solving the puzzle using words. The observer should also record how much time it took to solve the puzzle.
5. The time limit for solving the puzzle is 10 minutes.

Criteria	Constraints
Puzzle solvers should solve the puzzle together as quickly as they can.	Puzzle solvers have no more than 10 minutes to assemble the puzzle.
One puzzle solver can use an assembled puzzle for help.	The second puzzle solver cannot see the assembled puzzle.
Puzzle solvers can communicate with spoken and written words	

Get Going

Assign half the groups the procedure with words and the other half the procedures without words. Distribute tangrams to the groups (one assembled and one unassembled) and an *Observing and Interpreting Animal Behavior* page. Emphasize that they may not look at the tangrams yet. Make sure students understand which set of procedures they are to follow and their role. Remind observers to record the time when the puzzle solvers in their group finish.

Assess

Monitor groups to see what kinds of observations the observers are recording. They should be evaluating how efficiently the puzzle solvers are working and identifying what is difficult for them and what works for them.

Analyze Your Data

ethnographer: scientist who studies people.

The first task in analyzing your data is sharing information between the observer and puzzle solver. Make a chart with three columns. Label them Observations, Interpretations and Conclusions. Using his or her recorded notes, the observer should share his or her observations with the puzzle solvers, and then add these observations to the first column of the chart. Add the observers' interpretations of what was happening during each observation to the second column. The puzzle solvers will know what behaviors the observer is referring to only if the observer provides enough detail about what the puzzle solvers were doing during each communication. **Ethnographers** (scientists who study people) call the exchange of information between groups a "member check." This is the first part of the member check.

Now, working with one observation at a time, the puzzle solvers should add more detail to help support the interpretations. This information should include why the puzzle solvers communicated the way they did. Include as much detail as you can and add this to the second column.. Observers will be able to understand the reasons why puzzle solvers communicated the way they did only if puzzle solvers are specific about why they communicated as they did. This is the second part of the member check.

Next, develop some conclusions about the different forms of communication your group members used. Identify the reason for each form of communication. Also, identify the difficulties of each form of communication. Record this information in the third column. Think about and record the kind of communication that would have been easier.

Communicate

Investigation Expo

Make a poster with your observations and interpretations. Your poster should include the following information:

- whether you were a "with-words" or "without-words" group
- your group's observations and interpretations
- the forms of communication your group members used, and what was easy and difficult to communicate using each one
- any conclusions your group reached about efficient communication

Analyze Your Data

10 min.

Have groups analyze their observations, developing conclusions about how efficient the ways of communicating they observed are.

△ Guide

Discuss with the class how they should analyze their data.

TEACHER TALK

"To analyze your data, each group should make a chart with three columns. The first column should contain the observations that were made while the puzzle solvers were solving the puzzle. The second column should contain the interpretations of the observations. Discuss these as a group and decide what to put in this column. The third column should contain the conclusions decided upon by the group. Remember that your conclusions must be supported by the observations.

The conclusions you develop should be about the different forms of communications your group members used. You should identify the reason for each form of communication and the difficulties with each and put this in your conclusion column. You should also record what kind of communication would have been easier and why."

META NOTES

Groups may find that their observers made observations that included interpretations that the puzzle-solvers disagreed with. This is OK. They should determine what the actual observation was and record it in the chart.

⬡ Get Going

Have groups begin their discussion and analysis. Let them know how much time they have.

△ Guide and Assess

Monitor groups' discussions and assist them if they are having difficulties. Encourage puzzle solvers to add detail and explain the reasons for what they did. Check to see that their interpretations and their conclusions are supported by observations. Check that the conclusion column contains the way of communication, difficulties with the communications, and ways of improving it. Also check to see if students wrote down the times it took to solve the puzzle.

and reco[…]ommunication that w[…]en easier.

Communicate

Investigation Expo

Make a poster with your observations and interpretations. Your poster should include the following information:

- whether you were a "with-words" or "without-words" group
- your group's observations and interpretations
- the forms of communication your group members used, and what was easy and difficult to communicate using each one
- any conclusions your group reached about efficient communication

Communicate

40 min.

Have each group present their observations and interpretations in an Investigation Expo.

For this *Investigation Expo*, you will hang your posters on the wall. Everyone will have a chance to see all the posters. Then some groups will have a chance to present to the class.

As you look at the posters and listen to the presentations, think about the following questions:

- What similarities do you see in what each of the "with-words" groups observed?
- What differences do you see between what each of the "with-words" groups observed?
- What similarities do you see in what each of the "without-words" groups observed?
- What differences do you see between what each of the "without-words" groups observed?
- Which groups solved the puzzle more quickly—those who used words or those who did not?
- What conclusions did "with-words" groups develop?
- What conclusions did "without-words" groups develop?
- Which conclusions do you agree with, and which do you disagree with? Why?

△ Guide

Begin by having students create poster presentations that answer the questions provided in the student text. Emphasize that groups will need to inform the class whether they used words or not, how long it took them to solve the puzzle and what their observations were (including what kinds of communication were easy and what kinds were difficult), and what conclusions they reached about efficient communication.

META NOTES

It is important for the group and the class to notice differences in observations, conclusions, and procedures. The questions for the *Investigation Expo* will help to focus students on these points.

⬡ Get Going

Distribute poster materials and have groups make posters for the *Investigation Expo*. Remind them to address the bulleted list in the student text when constructing their posters and let them know how much time they have.

◇ Evaluate

Monitor groups to see if they are addressing all the bulleted items required for their poster.

Decide if you want all groups or some of the groups to present their results to the class after the posters have been viewed. At least two groups from each category (with-words and without-words) should present.

△ Guide

Then have students place their posters around the room for viewing. Remind the class that for each poster, they will need to look for what claims are being made and if they are believable. Emphasize that they should use the second list of bulleted items in the *Communicate* section of the student text when reviewing each poster and later when they hear some groups present. Let students know that they should record questions they have for each group. They will have an opportunity to ask questions when groups present their interpretations.

⬡ Get Going

Have each group display their posters around the classroom and allow everyone to visit each poster for a minute or so to become familiar with each group's work and to formulate questions they may have.

Transition the class to a whole-class discussion.

△ Guide

Remind students that during the presentations, they should be listening for the kinds of communication observed and conclusions (claims) they have made. Let them know that if they are unclear about the kinds of communication observed or the conclusions (claims) they have made, they should ask questions of the presenting group.

Inform the groups you selected that they will be presenting.

The "without-words" groups should present first, followed by the "with-words" groups. It is important to have a discussion comparing groups that followed the same procedures.

After the last groups present, initiate a discussion comparing the conclusions of the "without-words" groups to the conclusions of the "with-words" groups.

META NOTES

You should model what you expect from the students. Encourage students to ask questions of the presenting group and for the presenting group to respond to the student asking the question. Students should not make you the focal point during the discussion.

Questions should require more than one-word answers and should focus on a critical part of the investigation. Model questions about content to assist students in asking better questions.

It is important to let students determine what trends are in their observations. Be careful not to tell them what trends they should look for.

For this *Investigation Expo*, you will hang your posters on the wall. Everyone will have a chance to see all the posters. Then some groups will have a chance to present to the class.

As you look at the posters and listen to the presentations, think about the following questions:

- What similarities do you see in what each of the "with-words" groups observed?
- What differences do you see between what each of the "with-words" groups observed?
- What similarities do you see in what each of the "without-words" groups observed?
- What differences do you see between what each of the "without-words" groups observed?
- Which groups solved the puzzle more quickly—those who used words or those who did not?
- What conclusions did "with-words" groups develop?
- What conclusions did "without-words" groups develop?
- Which conclusions do you agree with, and which do you disagree with? Why?

Reflect

As you listened to other groups, you probably noticed the differences in puzzle-solving methods among the groups. You probably also noticed that some kinds of communication were easier with words and some were easier with gestures. With your group, discuss and record answers to the questions below. Be prepared to share your answers with the class.

1. Which form of communication was more effective in helping the solvers complete the puzzle? Why?
2. Describe the similarities and differences between the verbal and non-verbal (without words) forms of communication used in this investigation. What is non-verbal communication good for?
3. Some animals do not have the ability to communicate with words. How do you think they successfully communicate with one another?
4. There may be times when you need to communicate without words. Describe a time when that happened. How did you manage to communicate even without words?

Reflect

10 min.

Have a class discussion on verbal and non-verbal communication.

Get Going

Let students know that they should discuss and answer the questions on verbal and nonverbal communication with their group members. Let them know how much time they will have and that a class discussion will follow.

META NOTES

These questions should guide the students' understanding of verbal and nonverbal communication and prepare them for their future explorations of animal communication behaviors.

△ Guide and Assess

Have students answer the questions in small groups and record their answers.

1. Students should base their answers on the time it took groups to solve the puzzle.
2. Both groups may have resorted to nonverbal communication to signify shapes or directions, but the class probably found that verbal communication allowed them to communicate a variety of complex ideas. Students should realize that nonverbal communication is good for situations in which one cannot or should not speak (e.g., in a library) or when one cannot hear.
3. This question should help students connect what they have just experienced with the forms of animal communication they will investigate in this *Learning Set*. Students should be able to describe examples of animal communication such as: a dog may wag its tail when it is happy; a bird may dance when trying to woo a mate; a cat may rub up against a person's leg to show affection. Emphasize to students that they should keep in mind the importance of gestures or body language when they explore the communications of other animals during this *Learning Set*.
4. This question should help students connect what they have just experienced and the ideas of this *Learning Set* to past experiences in various situations. Emphasize the importance of being able to observe gestures when trying to understand animal communication.

Update the *Project Board*

While doing this investigation and discussing your findings, you may have thought of some new ideas or new questions. Use the *Project Board* to record your ideas about communicating with words and without words. Include in the second column of the *Project Board* any new investigation questions you might want to answer. Think about what you have learned and the evidence you have for your learning. Record your learning and the evidence on the *Project Board*. Think about how this learning will help you answer the Big Question, *How do scientists answer big questions and solve big problems?*

Update the *Project Board*

10 min.

Have a discussion focused on updating the class Project Board.

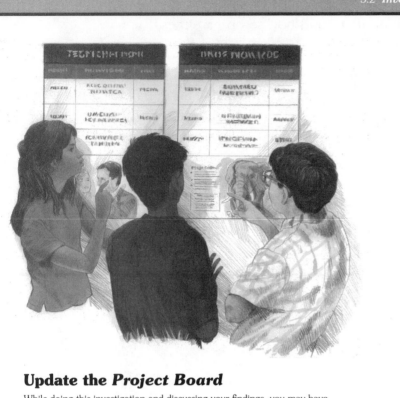

△ Guide

Ask students what their experiences in this activity suggest about the way animals communicate that they can use to design zoo enclosures that encourage communication. What information would help confirm their ideas or make them more useful?

Then draw students' attention to the *Project Board*. What can students put in the first column *(What do we think we know?)* and the second column *(What do we need to investigate?)* of the *Project Board?*

META NOTES

The emphasis should be on the second column. It is important that students formulate investigative questions about animal communication behavior.

◇ Evaluate

Students should include investigative questions pertaining to gestures in animal communications and how to observe these.

△ Guide

Ask if there is anything that students want to put into the third column *(What are we learning?)* and the fourth column *(What is our evidence?)* of their *Project Board*.

Remind students to write down the new entries onto their *Project Board* pages.

NOTES

What's the Point?

Communication is important for letting others know your wants and needs. You have seen through observations of humans that communication can happen without using spoken or written words. There may be times when it is necessary to communicate without words. There may have been times you experienced this yourself. However, humans do have the ability to use words. You were able to see in the investigation that this gives them a tremendous advantage when solving complex problems. All animals have the ability to communicate in some way. Sometimes constraints determine the form of communication.

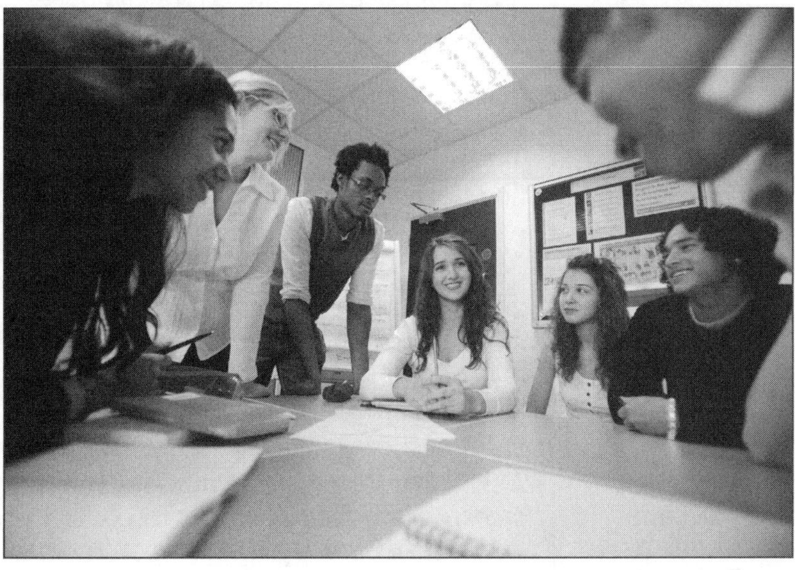

What's the Point?

5 min.

◇ Evaluate

Make sure students recognize that constraints determine the form of communication an animal uses.

Assessment Options

Targeted Concepts, Skills, and Nature of Science	How do I know if students got it?
Animals communicate with other animals using sound. The sounds they can make and the sounds they can hear are adaptations to their environment.	**ASK:** What physical features allow you to communicate with words? **LISTEN:** Humans are able to use language because they have vocal cords, tongues, and other body parts that allow them to produce speech. They also have ears, which allow them to hear.
Scientists must keep clear, accurate, and descriptive records of what they do so they can share their work with others, consider what they did, why they did it, and what they want to do next.	**ASK:** How did having an observer record what happened help your group to analyze and interpret what happened? **LISTEN:** Students should have used the time that the observer recorded to compare how efficiently different groups solved the puzzle. They also should have used the observer's notes to identify forms of communication that they used.
Scientists often work together and then share their findings. Sharing findings makes new information available and helps scientists refine their ideas and build on others' ideas. When another person's or group's idea is used, credit needs to be given.	**ASK:** How did sharing your ideas during the *Investigation Expo* help you to understand how constraints may affect communication? **LISTEN:** Students should describe how the constraints of the activity affected communication in different groups when they shared their results.

Teacher Reflection Questions

- How were you able to help students connect their experiences in this section with animal communication?
- How are students progressing in engaging in the social behavior of scientists? How can you assist students to reach this goal?
- What difficulties did the class encounter solving the tangrams? Is there anything you would do differently next time?

SECTION 3.3 INTRODUCTION

3.3 Explore

How Do Bees Communicate and Why?

◀ *1 class period**

*A class period is considered to be one 40 to 50 minute class.

Overview

Students explore how bees communicate the location of food to the hive. Students watch a video of bees doing the waggle dance, create plans for observing bees in the video, and use the plans to make detailed observations of the waggle dance. Then, groups work to develop interpretations of bees' behavior using what they know about bees' bodies and environments. Students then read about the history of scientific thinking on the waggle dance and learn that the meaning of the waggle dance is still being debated.

Targeted Concepts, Skills, and Nature of Science	Performance Expectations
Scientists often work together and then share their findings. Sharing findings makes new information available and helps scientists refine their ideas and build on others' ideas. When another person's or group's idea is used, credit needs to be given.	Students should work with their groups to analyze their observations and interpretations of bees' waggle dance. Groups should share their interpretations with the class and decide what to include on the class *Project Board.*
Scientists must keep clear, accurate, and descriptive records of what they do so they can share their work with others, consider what they did, why they did it, and what they want to do next.	Students should record what they are learning to construct new investigative questions about animal communication on the *Project Board.*
Studying the work of different scientists provides understanding of scientific inquiry and reminds students that science is a human endeavor.	Students should recognize that the meaning and purpose of the waggle dance are still being debated.
Scientific knowledge is developed through observations, recording and analysis of data, and development of explanations based on evidence.	Students should recognize that by recording and analyzing observations, they are developing interpretations of animal behavior, just as scientists do.

Targeted Concepts, Skills, and Nature of Science	Performance Expectations
Scientists make claims (conclusions) based on evidence obtained (trends in data) from reliable investigations.	Students should construct conclusions and interpretations based on their observations of the waggle dance from the video.
Behavior is a type of response to internal or external stimulus. Behavior is determined by experience, physical characteristics, and environment.	Students should consider how stimuli in the environment affect bees' communication behavior.
The structure and function of animals' bodies are complementary and affect animal behavior.	Students should consider the structure and function of animals' bodies as they interpret the waggle dance.

Materials	
1 per class	*Animals in Action* DVD (*Honeybee Waggle Dance* video) and a way to view the video
1 per student	*Observing and Interpreting Animal Behavior* page
1 per class	*Project Board*
1 per student	*Project Board* page

Activity Setup and Preparation

Set up the DVD equipment before class. View the *Honeybee Waggle Dance* video. Make sure you turn off the sound before showing it to students. Students should only watch the first 50 seconds of the video, with the sound off, during their first two viewings.

Consider doing an Internet search on bee communication and the waggle dance. There are a number of sources that show and describe bee communication.

Homework Options

Reflection

- **Science Process:** You based your interpretation of the waggle dance on observations of the bees doing the waggle dance. What other observations would have helped you to interpret the waggle dance? *(Students would probably have had a much easier time interpreting the waggle dance if they had been able to observe what the bees did after watching the waggle dance, like where and how they foraged.)*
- **Science Content:** Translate a waggle dance to our language. Consider if the bee doing the waggle dance could speak in English, how would it communicate how to get to flowers with nectar? *(Students should provide an example of how a waggle dance might be translated to words.)*

Preparation for 3.4

- **Science Content:** What about an animal's environment affects their communication? How might this information help you understand the feeding of the animal? *(Students should look for ways the environment, including the sources of food and danger, can be used to understand how and why animals communicate.)*

NOTES

NOTES

SECTION 3.3 IMPLEMENTATION

3.3 Explore

How Do Bees Communicate and Why?

In the last *Learning Set*, you learned that bees forage for their food. Bees need to forage efficiently, because they use a lot of energy to collect their food. Bees have a unique way of communicating to other bees where they have found the food.

The bee in the center of the comb is performing a waggle dance that tells the other bees where food for the hive can be found. The dancing bee is blurry because it is moving quickly.

The Mystery of the Waggle Dance

The waggle dance is one of the most amazing behaviors found in the animal world. Karl von Frisch described the dance in the 1960s. Through careful observation, he noticed that bees would return to the hive with nectar and pollen and then shake themselves and turn around as the other bees watched. Karl von Frisch named this the waggle dance. He observed and recorded many bees engaged in this behavior and wondered why they did this.

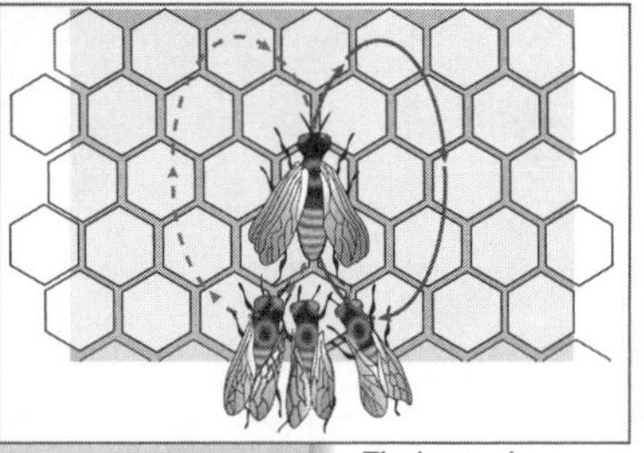

The bee in the center dances along the direction of the lines while the other bees watch.

◀ *1 class period**

3.3 Explore

How Do Bees Communicate and Why?

Elicit students' ideas on how bees communicate, and introduce the waggle dance.

○ Engage

Begin by asking students how they think bees communicate where nectar is to the hive. Record students' ideas.

Let students know that bees' bodies aren't structured to allow them to speak like humans, but they definitely communicate with each other.

Next, introduce the information box in the student text titled *The Mystery of the Waggle Dance*. Let students know that the waggle dance is a dance a bee does in the hive that tells the other bees where food is. Let students know that Karl von Frisch described the waggle dance in the 1960s after many careful observations he made of bees.

*A class period is considered to be one 40 to 50 minute class.

Observe

5 min.

Show students the video of the waggle dance without sound.

Observe

Think about what it might have been like to be von Frisch watching the bees and trying to develop an explanation of something other people had wondered about for centuries. You will watch a video of bees doing the waggle dance. To get an idea of what the bees are doing, watch the video once without taking notes. Think about what you will need to pay close attention to as you watch again. Remember that you will be trying to interpret the bees' behavior—what they are doing when they do the waggle dance, and why they are doing it.

Now watch the video again. This time, record your observations of the waggle dance. Record as much detail as you can. You might need to see the video several times before you feel ready to analyze what you saw.

Analyze Your Data

When you have completed your observations, work with your group to interpret the bees' behavior. Use an *Observing and Interpreting Animal Behavior* page to help you.

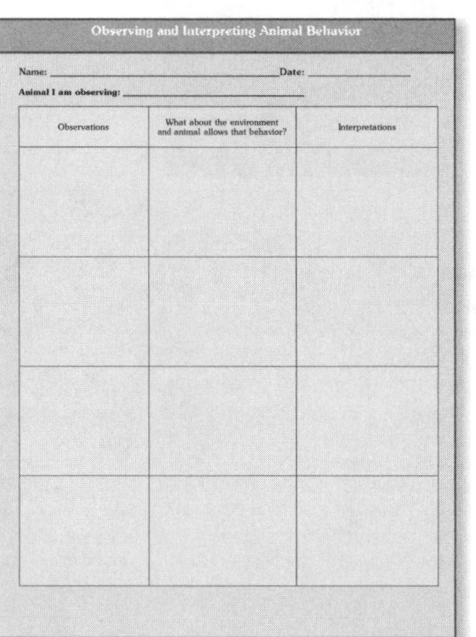

Observing and Interpreting Animal Behavior

Name: ______ Date: ______

Animal I am observing: ______

Observations	What about the environment and animal allows that behavior?	Interpretations

Decide which observations of behavior to record in the Behavior column. Include in the second column what you know about bees' bodies and their environment that might affect that behavior. Then develop some interpretations. Why do you think the bees were doing what they were doing? What does it look like the bees are trying to communicate? What messages do you think they are giving to other bees?

You may not know about bees yet, so your interpretations will be educated guesses. After you read more about bee communication, you will revisit your *Observing and Interpreting Animal Behavior* page and add more detail.

△ Guide

Let students know that they will be watching a video of bees to make observations. Emphasize that the first time they view the video they should not take any notes. They should think about what details they want to pay special attention to for when they watch it a second time. Let students know that the second time they view the video they should record their observations.

⬡ Get Going

Make sure the sound is off before showing the video. Students should not hear the narration until the end of this section. For these first two viewings students should only view the first 50-51 seconds of the video.

Start the video of the bees without narration. Only showing students the first 50 seconds of the video is enough to develop a sense of what the waggle dance is.

Emphasize to students that they should be ready to make observations and show the first 50 seconds of the video with the sound off again.

META NOTES

You should show the entire video with narration before students update their *Project Board*.

Analyze Your Data

When you have completed your observations, work with your group to interpret the bees' behavior. Use an *Observing and Interpreting Animal Behavior* page to help you.

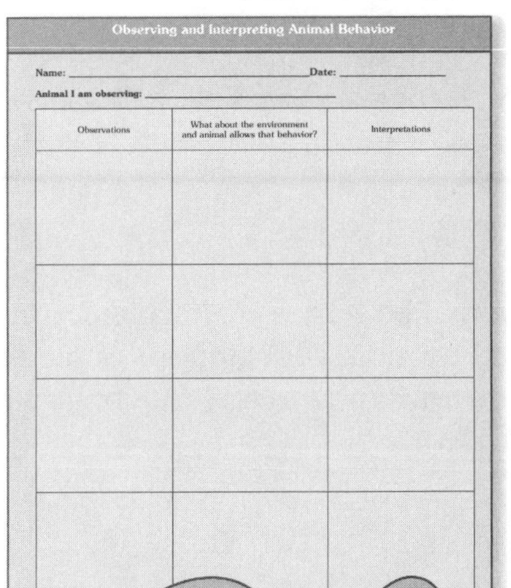

Observing and Interpreting Animal Behavior

Name: ______ Date: ______

Animal I am observing: ______

Observations	What about the environment and animal allows that behavior?	Interpretations

Decide which observations of behavior to record in the Behavior column. Include in the second column what you know about bees' bodies and their environment that might affect that behavior. Then develop some interpretations. Why do you think the bees were doing what they were doing? What does it look like the bees are trying to communicate? What messages do you think they are giving to other bees?

You may not know about bees yet, so your interpretations will be educated guesses. After you read more about bee communication, you will revisit your *Observing and Interpreting Animal Behavior* page and add more detail.

Analyze Your Data

10 min.

Have students analyze their observations and interpret what the waggle dance means.

△ Guide

Let the class know that groups should discuss their observations, decide upon interpretations about why the bees were doing what they observed, what they think was being communicated, and what it means.

TEACHER TALK

"Each of you made some observations about what you saw in the video, and they may be different. You should share your observations with your group members and come up with interpretations of what you saw. You should have interpretations to address the three questions in the student text: Why do you think the bees were doing what they were doing? What does it look like the bees are trying to communicate? What messages do you think they are giving to other bees?

You may also have interpretations that address other issues and you should list these on your *Observing and Interpreting Animal Behavior* page."

⬡ Get Going

Then distribute the *Observing and Interpreting Animal Behavior* pages and have groups begin their discussions. Let the class know how much time they have.

△ Guide and Assess

Monitor groups to see if each group member's observations are being discussed, and that they are discussing the reasons for any differences between observations. Also check that students include items supporting information in the column *What about the environment and animal allows that behavior?* This column should contain what they know about bees' bodies and environment, and they should record their ideas in the middle column.

Have a brief class discussion in which a few groups share some of their interpretations. Let the class know that they will be able to improve their interpretations with more observations or information from other scientists' observations.

TEACHER TALK

"If you had made many observations, like Karl von Frisch did, your interpretations would be supported by more observations. Since we will not be observing bees extensively, we will now read a bit about why bees communicate and Karl von Frisch's work."

Why Do Bees Communicate?

Bees are very social animals. They live in large groups in one hive. Their survival depends on cooperation in the hive. Each bee has a job to do. The job of the queen bee and drones (male bees) is to make more bees by reproducing. The rest of the bees are female worker bees. They do not reproduce. Recently hatched worker bees clean the hive and care for the queen's offspring. As the worker bees mature, they build parts of the hive. Later, they guard the hive from predators. The last job of a worker bee is to leave the hive and forage for food.

Worker bees feeding the queen bee.

If a beehive is to be successful, the bees need to make more bees. So they must have food. You already know that the worker bees have responsibility for getting food and bringing it back to the hive. And you know that each of them discovers new flowers with nectar or pollen as they forage. Remember that the worker bees need to be very successful at this task or they will not be able to supply enough food to the hive. It seems only natural, then, that they should have a way to communicate with the other bees. But remember that, up until recently, scientists thought the bees were dancing only to get attention.

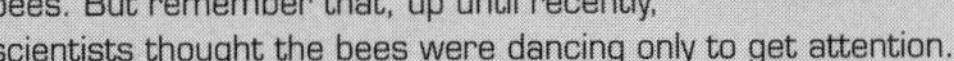

Karl von Frisch thought there must be other reasons for the dance. It seemed too complicated to be just for getting attention, and he had seen bees become more efficient at locating food. He observed honeybees and performed experiments to better understand the purpose of the dance. From his observations, he concluded that bees waggle to show other bees in the colony the location and distance of food from the hive. Other worker bees then use the information from the dance to quickly locate food. This saves a tremendous amount of energy, since the worker bees do not have to fly around randomly, looking for food sources. This discovery was very important. It earned Karl von Frisch a Nobel Prize in 1973.

Even though Karl von Frisch won an important prize for his discovery, some scientists still disagree with his idea. One scientist, Adrian Wenner, started out thinking the waggle dance was a form of bee communication but now thinks bees communicate through scents. Adrian Wenner and others think bees could be marking the flowers with scents from their bodies. Then the other bees could follow the odors to find the food.

Why Do Bees Communicate?

10 min.

Have a discussion on why bees communicate and the dynamic nature of science.

△ Guide

Have a class discussion on *How Do Bees Communicate?*

Point out that for the hive to survive, the worker bees that forage for food must find the food and let the rest of the worker bees know where that food is so they do not expend a lot of energy trying to find and gather food. Then, let students know that Karl von Frisch conducted many experiments and made numerous observations and concluded that the bees' waggle

META NOTES

Consider having students read aloud in groups, summarizing paragraphs, pulling out key ideas, or consider reading to them.

dance actually lets the bees know how far away and in what direction the food is located. Emphasize that this saves the hive a lot of energy. Let students know that Karl von Frisch won a Nobel Prize in 1973 for this work.

TEACHER TALK

"Scientists must be open to changing their ideas if there is evidence that does not support their ideas. However, if there is no evidence that goes against their views and their views are supported by reliable evidence, then their ideas are considered valid. This does not mean that there are not multiple ideas that could be valid.

For example, in the 1600s Sir Isaac Newton was able to describe how and why objects move with three rules. There was no contradiction to these rules for about 250 years, and they became known as scientific laws. Then, in the 1920s Albert Einstein introduced his theory of special relativity that negated Newton's Laws for special cases (as when an object travels at a speed near that of light). Einstein's theory has become accepted through experimental evidence and the scientific community now considers Newton's work as valid only for certain situations."

Next, discuss how science is a dynamic field. With the class, read about the scientific debate of the waggle dance and how new technology helped to support Karl von Frisch's conclusions. However, the controversy is still not settled.

Emphasize that scientists are still debating why bees do the waggle dance.

Then show students the entire *Honeybee Waggle Dance* video with the sound on. After the viewing, let groups update their *Observing and Interpreting Animal Behavior* page.

Worker bee with attached radio antenna.

Scientists have debated different theories and worked to explain why bees did the waggle dance. The problem they faced was that they could not follow individual bees after they left the hive.

New technologies are making it possible for scientists to make more detailed observations of bee behavior. In 2005, a group of British scientists glued tiny radar antennae to worker bees. The radar equipment made it possible to follow the path of individual worker bees. Scientists could also monitor worker bee movement over time. The researchers concluded that the waggle dance is indeed the way bees share information about food with one another. These scientists concluded that Karl von Frisch was right. Their interpretation of the observations leads them to agree with him that the dance shows other worker bees the location of a food source and its distance from the hive.

The controversy is not settled though. Many bee scientists think the evidence supports the waggle-dance conclusion. But Wenner and other scientists question the data and evidence presented in the radar studies. They still think the bees communicate by leaving an odor on the flowers and having other bees find the odor. Scientists are identifying new ways to investigate bee behavior so they can know for sure.

Stop and Think

1. Scientists are still debating the use of the waggle dance. Why do you think bees use the waggle dance? What makes the waggle dance an effective way for bees to communicate?
2. What would happen to human communication if the only way to communicate was through a human "waggle dance?" What things would we have a lot of trouble communicating? How would we be able to do science?
3. Karl von Frisch first wrote about the waggle dance in 1927. He won the Nobel Prize in 1973, and scientists continue to debate the use of the waggle dance. Everyone agrees that bees move in certain ways, but they still don't agree about why bees do the dance. Why do you think the debate has continued for so long?

Stop and Think

10 min.

Have a discussion on students' responses.

⬡ Get Going

Describe how you want students to answer the questions (individually or in groups) and let students know how much time they have.

△ Guide and Assess

Have a discussion of students' responses. Use the information below to guide and assess their understanding.

1. Students should support their answers with good reasons.
2. Students should draw on the results of the activity from the previous section to support their answers.
3. One reason that students might suggest is that the information we have available changes as our technology changes.

NOTES

Update the *Project Board*

As you observed the waggle dance and interpreted the bees' behavior with your group, you may have thought of new questions you want to ask about animal behavior. You also learned about how scientists interpret the waggle dance. Add your questions and what you learned to the *Project Board*. Make sure you add evidence to support any new science you learned. Also, think about how what you just learned can help you answer the *Big Question: How do scientists answer big questions and solve big problems?* As your teacher records this information on the big *Project Board*, add the information to your own *Project Board* page.

What's the Point?

One of the most important tools for animal survival is the ability to communicate. Animals can exchange information about food and water sources, possible dangers, and mating. Scientists think bees communicate using a waggle dance. Other scientists think bees use scent markers. Although bees do not have bodies that allow them to communicate through gestures, they have developed an interesting dance that only they fully understand and use to communicate with other bees.

Scientists debate different scientific ideas. They use scientific methods to investigate their ideas. They analyze the data and generate conclusions supported by the data. Sometimes, theories are so complicated and difficult to investigate that debate over them lasts for many years. The waggle dance is one theory that has been debated for decades. Scientists continue to question the scientific methods used to study the waggle dance.

Bees use the waggle dance to communicate the location of food to other bees in the hive.

Update the Project Board

15 min.

Have a class discussion of what students have learned.

△ Guide

Have a class discussion focused on what the class can add to the *Project Board*. Begin with column 2, *What do we need to investigate?* Ask students what they think they need to investigate that will help them design enclosures that allow animal communication as in their natural habitat.

Then ask the class to think about what they should put in columns 3 and 4 *(What are we learning?* and *What is our evidence?)* Students should suggest things about animal communication that does not involve sound, how survival may depend on communication, and how the environment affects communication.

Examples:

Column 3 (claim): Some animals' survival depends on how well they can communicate where food is located.

Column 4 (evidence): An example of this is bees. Bees depend on forager bees to communicate where the food is so that the worker bees can efficiently gather food for the entire hive. The bees perform the waggle dance in the hive to let members of the hive know the distance from where the nectar is and the direction to fly in.

Column 3 (claim): The environment can affect how or what animals communicate.

Column 4 (evidence): An example of this is the location of the Sun for the bee's waggle dance. The forager bee lets the hive know where the nectar is by doing the waggle dance to communicate the distance and direction to fly in for nectar. The direction to fly in is based on the location of the Sun in the sky. The waggle dance changes as the Sun moves across the sky.

Column 3 (claim): Some animals do not have the ability to communicate using vocal chords and find other means to communicate that are adapted to their body structures.

Column 4 (evidence): For example, bees do not have vocal chords but can communicate the location of food to the hive by doing a special dance known as the waggle dance.

information ... own *Project Board* page.

What's the Point?

One of the most important tools for animal survival is the ability to communicate. Animals can exchange information about food and water sources, possible dangers, and mating. Scientists think bees communicate using a waggle dance. Other scientists think bees use scent markers. Although bees do not have bodies that allow them to communicate through gestures, they have developed an interesting dance that only they fully understand and use to communicate with other bees.

Scientists debate different scientific ideas. They use scientific methods to investigate their ideas. They analyze the data and generate conclusions supported by the data. Sometimes, theories are so complicated and difficult to investigate that debate over them lasts for many years. The waggle dance is one theory that has been debated for decades. Scientists continue to question the scientific methods used to study the waggle dance.

What's the Point?

5 min.

◇ Evaluate

Make sure students realize that scientists debate different scientific ideas and that scientific ideas may change as more information becomes available. Scientific claims must be supported by evidence (observations).

Assessment Options

Targeted Concepts, Skills, and Nature of Science	How do I know if students got it?
Behavior is a type of response to internal or external stimulus. Behavior is determined by experience, physical characteristics, and environment.	**ASK:** What in bees' environment prompts them to communicate? **LISTEN:** The need for food for survival. Bees must be able to efficiently communicate where nectar can be found in order for the hive to survive.
The structure and function of animals' bodies are complementary and affect animal behavior.	**ASK:** What are the constraints on bees' communication from the structure and function of their bodies? **LISTEN:** Students should recognize that bees do not have vocal cords, so they cannot speak. Bees use their bodies in a waggle dance to communicate the location of food to the hive.

Targeted Concepts, Skills, and Nature of Science	How do I know if students got it?
Scientists must keep clear, accurate, and descriptive records of what they do so they can share their work with others, consider what they did, why they did it, and what they want to do next.	**ASK:** Why was it important to make detailed and descriptive observations of the waggle dance? **LISTEN:** The clues to possible meanings of the waggle dance are in the details, such as the way the dancing bee moves in the same direction each time it dances.
Studying the work of different scientists provides understanding of scientific inquiry and reminds students that science is a human endeavor.	**ASK:** Do you think it's surprising that Karl von Frisch won a Nobel Prize even though there is still debate about his conclusions of the waggle dance? Do you think that he should have been able to prove his idea once and for all? **LISTEN:** Students should recognize that scientific theories are open to revision as new information becomes available. In some cases, the phenomenon a theory tries to explain may be so difficult to investigate that the theory will be very hard to test.
Scientific knowledge is developed through observations, recording and analysis of data, and development of explanations based on evidence.	**ASK:** How did Karl von Frisch develop his theory about bees' waggle dance? **LISTEN:** Karl von Frisch started by observing bees' waggle dances and how they located food. He then analyzed and interpreted his observations and developed an explanation. Later, other scientists observed bees search for food, dance, and then collect food. The collection of observations from these various scientists leads scientists to currently believe von Frisch's theory of the waggle dance.

Teacher Reflection Questions

- What difficulties did students have understanding the theory and debate surrounding the waggle dance? What could you do to make this clearer next time?
- What difficulties did students have in understanding the importance of making careful observations, keeping records, and sharing information? Where will your students need the most help with these ideas in the next section?
- How were you able to keep students organized and focused during group discussions of the video?

NOTES

NOTES

SECTION 3.4 INTRODUCTION

3.4 Explore

What Affects How Elephants Communicate?

◀ *1 class period**

*A class period is considered to be one 40 to 50 minute class.

Overview

Students observe elephants' communication behavior and develop explanations of how elephants communicate. Before making their observations, students predict how elephants communicate, considering the physical characteristics of elephants and the needs of elephants that might affect how they communicate. Students observe elephant communication in a video, develop observation plans, and then view the video again, making detailed observations following their plans. Student groups then analyze their observations, separating observations and interpretations, and present their observations and analyses to the class. After listening to the ideas of their classmates, they build on the collective observations and interpretations of the class to develop explanations of how elephants communicate, working collaboratively as scientists work.

Targeted Concepts, Skills, and Nature of Science	Performance Expectations
Scientists often work together and then share their findings. Sharing findings makes new information available and helps scientists refine their ideas and build on others' ideas. When another person's or group's idea is used, credit must be given.	Students should share their groups' observations and interpretations with the class during an *Investigation Expo*. Using every group's information students develop explanations to share with the class and refine these based on class discussions.
Scientists must keep clear, accurate, and descriptive records of what they do so they can share their work with others, consider what they did, why they did it, and what they want to do next.	Students should use *Observing and Interpreting Animal Behavior* pages to record and analyze their observations. Then they create and record explanations of elephants' communication behaviors. Students should refer to these records in future sections.
Tables are an effective way to communicate results of a scientific investigation.	Students should use tables where appropriate in their *Investigation Expos* and to share their explanations.

Targeted Concepts, Skills, and Nature of Science	Performance Expectations
Scientists differentiate between observations and interpretations. They use their observations and interpretations to explain animal behavior.	Students should separate their observations from their interpretations.
Scientists make claims (conclusions) based on evidence obtained (trends in data) from reliable investigations.	Students should construct claims and interpretations based on trends in their observations of elephants in the video they watched.
Explanations are claims supported by evidence. Evidence can be experimental results, observational data, and other accepted scientific knowledge.	Students should use their observations to support their interpretations and other claims they make about elephant communication.
Behavior is a type of response to internal or external stimulus. Behavior is determined by experience, physical characteristics, and environment.	Students should consider how stimuli in the environment affect elephants' communication behavior based on their observations of elephants in the video.
The structure and function of animals' bodies are complementary and affect animal behavior.	Students should recognize that elephants' physical features allow for the kinds of communication they observed.
Animals communicate with other animals using sound. The sounds they can make and the sounds they can hear are adaptations to their environment.	Students should recognize that elephants use sound as one means of communication.
Animals' ears are adapted to hearing sounds in their environment. Some animals use sound that is out of the range of human hearing.	Students should recognize that elephants use their ears to hear communications, but making and detecting sound are not their only form of communication.

Materials	
2 per student and 1 per group	*Observing and Interpreting Animal Behavior* pages
1 per student	*Create Your Explanation* pages
1 per class	*Animals in Action* DVD (*What Affects How Elephants Communicate?*) and a way to view the video.
1 per group	Sticky notepad
10 to 15 per group	Index cards
1 per group	Presentation materials

Activity Setup and Preparation

Set up the DVD equipment before class. View the *What Affects How Elephants Communicate?* video and make sure students will be able to see and hear the video from all viewing locations.

Homework Options

Reflection

- **Science Process:** What were some differences between your observations and interpretations and other groups' observations and interpretations? What were some possible reasons for the different observations and interpretations? Why was it important to share your results? *(Groups probably used different observation plans; this may have led to different observations. Other groups may have carried out their plans more conscientiously. Although groups are likely to notice different things in any case, this is one reason to have groups share their observations.)*
- **Science Process:** What evidence did you use in your explanation? How confident are you of your interpretation of the evidence? What are some other possible explanations of your observations? *(Students should evaluate their interpretations and think critically about the implications of the evidence.)*

Preparation for 3.5

- **Science Content:** What information about elephants' behaviors and physical features could help you to better explain how they communicate? *(This question is meant to engage students in thinking about what they need to learn.)*

NOTES

SECTION 3.4 IMPLEMENTATION

3.4 Explore

What Affects How Elephants Communicate?

You found when you were solving the puzzle that communicating with words is more effective in many ways than communicating without words. Spoken words can be used when you can see the person you are communicating with and when you cannot see the person you are trying to communicate with. For example, if you were in a large store, you could call for your parents even if you could not see them. But often, verbal communication works better when someone is making eye contact with the person they are speaking to. Then facial expressions and body language help send the message.

Humans and bees use different methods to communicate. Communication methods are affected by an animal's body structure. Humans use different parts of their bodies to communicate than bees do. Humans can use their hands, faces, and entire bodies. Bees do not have all these body parts to use. Their bodies do not allow for the range of communication that humans have. Some of the ways animals communicate are not visible. Bees communicate through odors and dances. Odors are not visible, but they still provide information for communication.

An elephant can communicate using its eyes. An elephant's eyes can show that it is aggressive or is paying attention.

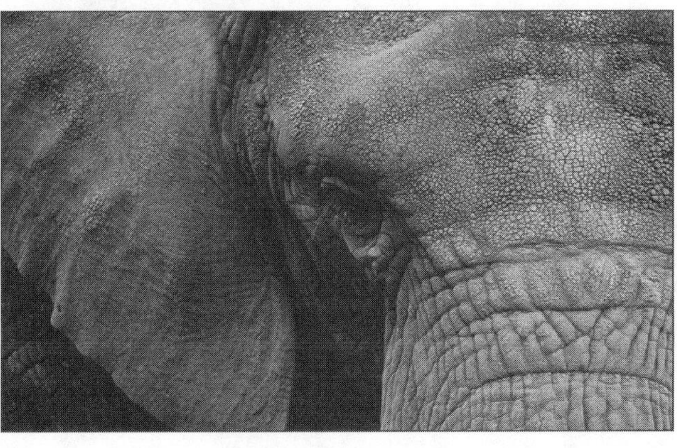

◀ *2 class periods**

3.4 Explore

What Affects How Elephants Communicate?

5 min.

Introduce and elicit students' ideas about elephant communication.

Engage

Begin by reviewing what students have observed so far about the communication behavior of humans and bees. Then elicit students' ideas about how elephants communicate and record these ideas.

*A class period is considered to be one 40 to 50 minute class.

Predict

5 min.

Have groups make predictions about how elephants communicate.

Elephants communicate in some ways that are similar to bees and humans. They also communicate in ways that are different from these two animals. As you learn more about elephant communication, you should pay attention to three different things that affect how elephants communicate. First, think about how an elephant's body affects how it communicates. Next, pay attention to ways elephants might communicate that cannot be seen, such as using odor. Finally, think about how an elephant's environment affects the way it communicates.

savanna: a grassland area with scattered trees, usually found in tropical or subtropical regions.

Predict

The picture shows some African elephants as they walk in a group. These elephants live on the **savanna**, a large, mostly grassy area of Africa. Just like people and bees, elephants live in groups. They need to communicate with other members of the group. Predict how these animals communicate with one another. What tools do you think elephants use to help them communicate? Think about what you learned about how humans communicate and the tools humans use. Identify any similarities and differences between humans and elephants. You can use these questions to help you think about elephant communication.

- What are some reasons animals need to communicate?
- What are some different ways animals communicate with one another?
- Think about the way an elephant's body works. How do you think elephants use their bodies to communicate with one another?
- What type of elephant sounds do you think an elephant uses to communicate with another elephant?

A group of African elephants walking on the savanna.

Observe Elephant Communication

It would be difficult for your class to make field observations in the African savanna. Although a safari to Africa to observe elephants would be exciting, it is probably not possible. You may have observed elephants communicating at a zoo. However, the zoo is not like the savanna. Once again, the best way to observe elephant communication is to watch a video.

TEACHER TALK

"In this section, you will be observing elephant communication behavior. How do you think elephants use their bodies to communicate?

How do you think elephants use communication that cannot be seen, such as odor?

How do you think an elephant's environment affects its communication?"

Guide

Discuss how African elephants live in groups on the savanna (a large, grassy area with scattered trees). Let them know that they should think about when and how elephants might need to communicate.

Let students know that they should address the four bulleted questions in their predictions.

Get Going

Let students know how much time they have and let them begin.

Guide

Have some groups share their predictions and briefly discuss what they might need to observe to support these predictions.

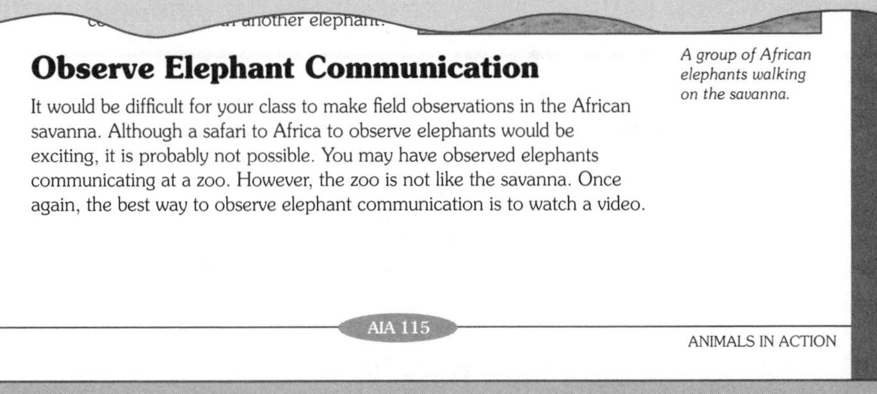

Observe Elephant Communication

It would be difficult for your class to make field observations in the African savanna. Although a safari to Africa to observe elephants would be exciting, it is probably not possible. You may have observed elephants communicating at a zoo. However, the zoo is not like the savanna. Once again, the best way to observe elephant communication is to watch a video.

A group of African elephants walking on the savanna.

AIA 115 ANIMALS IN ACTION

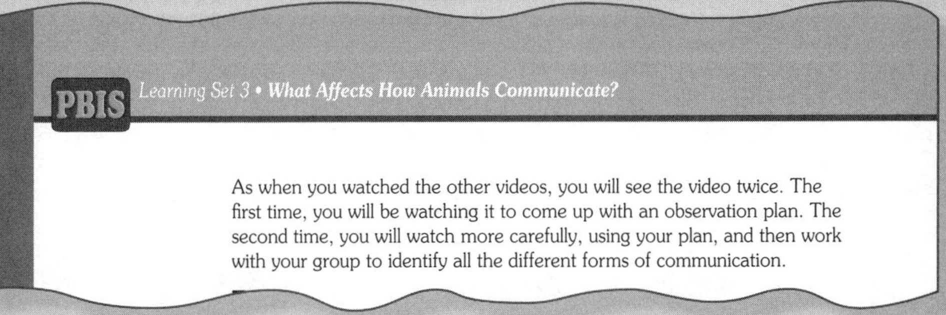

PBIS Learning Set 3 • What Affects How Animals Communicate?

As when you watched the other videos, you will see the video twice. The first time, you will be watching it to come up with an observation plan. The second time, you will watch more carefully, using your plan, and then work with your group to identify all the different forms of communication.

Observe Elephant Communication

5 min.

Have students watch the elephant video and think about how to make detailed observations.

Guide

Let students know that they will observe elephants communicating by watching a video of elephants in the savanna. Inform them that they will watch the video twice. First, they will watch it to plan their observations, and the second time to make observations following their plan.

Get Going

Show the video.

Plan

5 min.

Have students construct observation plans.

As when you watched the other videos, you will see the video twice. The first time, you will be watching it to come up with an observation plan. The second time, you will watch more carefully, using your plan, and then work with your group to identify all the different forms of communication.

Plan

Discuss with your group what you observed about the way the elephants communicate. Then identify what you need to observe more closely to be able to write a detailed description of how they communicate. Plan the way you will observe the video the next time so that you can make detailed observations. What details will you pay special attention to in each video? Do you want to divide up the observations among your group? Should you each focus on different details?

Remember that you want to be able to describe three things.

- how elephants communicate
- what is special about the elephant's body that allows it to communicate that way
- what is special about the environment that affects the way elephants communicate

Observe

Remember to follow the plan your group decided on. Record your observations so you will be able to share them with your group and your class. Once you have your plan written and all the members of your group understand the plan, you will watch the video again. Remember to take notes about the elephants' behavior and habitat. Pay attention to the elephants' bodies. Think about how their body structure helps them to be better at communicating. Think also about how an elephant's habitat affects how it communicates.

Analyze Your Data

The video you watched showed elephants communicating. How did you know the elephants were communicating? You may have predicted and observed that elephants make different sounds. These sounds communicate a variety of things to other elephants. This is one way other mammals, such as people, communicate.

Get Going

tudents should develop their observation plans as a group.

fter students view the entire video once, ask them to write their bservation plans. Remind them of the three bulleted items: how elephants ommunicate, what is special about the elephant's body that allows it to ommunicate that way, and what is special about the environment that ffects the way elephants communicate. Then, let students know how much ime they have.

Guide and Assess

Aonitor students' progress and check to see if they are addressing the ulleted items. Guide students as needed, asking them questions about revious observation plans involving communication that they wrote. Help tudents identify ways they might tighten up their plans so that they can ollect accurate, detailed data.

META NOTES

Students will use their observations to create an explanation of elephants' communication behavior based on the form and function of the elephants' bodies and on their habitat. Remember that explanations in *PBIS* have three parts: claims, evidence, and science knowledge. Students will write several explanations in this Unit and will revise them as they learn more to support their claims.

Observe

Remember to follow the plan your group decided on. Record your observations so you will be able to share them with your group and your class. Once you have your plan written and all the members of your group understand the plan, you will watch the video again. Remember to take notes about the elephants' behavior and habitat. Pay attention to the elephants' bodies. Think about how their body structure helps them to be better at communicating. Think also about how an elephant's habitat affects how it communicates.

Observe

5 min.

Have students view the video and make observations according to their plans.

Get Going

Vhen groups have finished their plans, have them view the video again and ecord their observations following their observation plans.

META NOTES

These observations are difficult—elephants communicate in more ways than may be immediately apparent in the video. Help students by encouraging them to consider all of the ways elephants may be communicating.

If needed, show the video again.

Analyze Your Data

15 min.

Have groups discuss, analyze, and categorize their observations and interpretations.

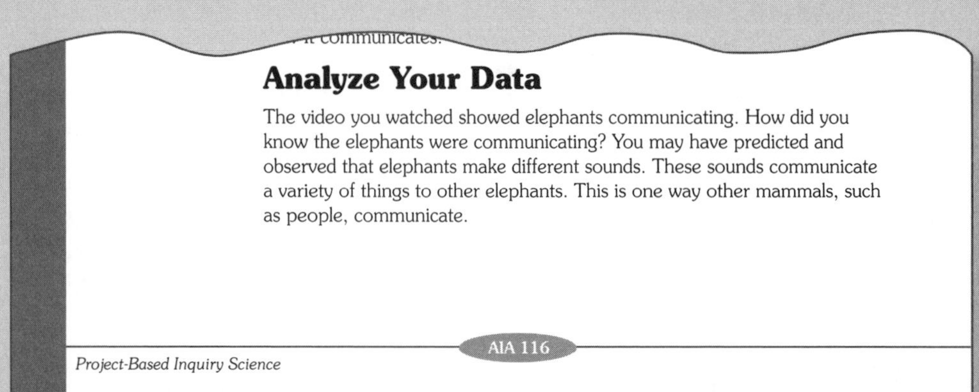

it communicates.
Analyze Your Data
The video you watched showed elephants communicating. How did you know the elephants were communicating? You may have predicted and observed that elephants make different sounds. These sounds communicate a variety of things to other elephants. This is one way other mammals, such as people, communicate.
AIA 116
Project-Based Inquiry Science

△ Guide

Let students know that while they share their observations and interpretations with their group members, they should be thinking about how they knew the elephants were communicating.

Then, describe for students how they will analyze the data using the information in the student text.

TEACHER TALK

"To analyze your data, you will first be sharing your observations with your group members. Remember that you may not have observed the same things.

Then, you should create a list of all the observations your group agrees upon. Using this list, create a sticky note for each observation and group the observations into categories. For example, one category might be how the elephant uses body movements.These categories should identify types of communication.

Then, with your group, record the observations and interpretations you agree upon in an *Observing and Interpreting Animal Behavior* page. Be prepared to share these with the class."

⬡ Get Going

Distribute sticky pads, index cards, and *Observing and Interpreting Animal Behavior* pages and let students know how much time they have for their analysis.

△ Guide and Assess

Monitor students and identify places where students have confused observations and interpretations, are in disagreement about observations, or have many observations in common (which may indicate that they have created and implemented a good observation plan).

Meet with your group to share your observations. Each member of the group should have a chance to share all their observations. Remember that, depending on your observation plan, you may not have been watching what the other group members were watching, and they may have a lot of information to share with you. Listen carefully for observations you might not have seen. Create a group list of observations that everyone agrees on. Make another list of observations that were not agreed on. Save the observations you could not agree on for the *Investigation Expo*.

Using the same procedure you used in *Learning Set 1*, create a sticky note for each observation on which you all agreed. Organize your observations into groups based on the type of communication that was shown. For example, how an elephant uses body movements to communicate might be a group of observations. If you think that some forms of communication might fall into two groups, you can list those twice.

After identifying the types of communication you observed, it will be time to think about what your observations mean. Think about when elephants make different sounds and what the different sounds might mean. Think about the different ways elephants use their bodies to send information to each other and what those body movements might mean. These interpretations will help you as you begin to develop explanations.

Using an *Observing and Interpreting Animal Behavior* page, record what you know about how elephants communicate. Record your observations in the left column. Include one behavior in each row of the chart. Then add what you know about what allows these behaviors in the middle column and your interpretations of these behaviors in the right column. Collaborate with the members of your group to create one chart for your group. Be prepared to share your observations and interpretations with the class.

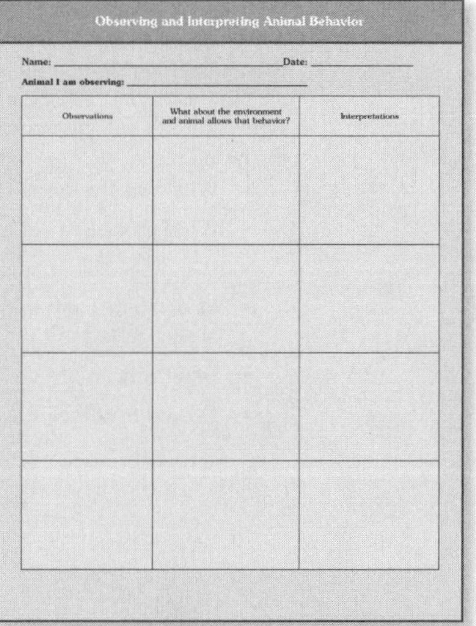
Observing and Interpreting Animal Behavior

Name: ______ Date: ______

Animal I am observing: ______

Observations	What about the environment and animal allows that behavior?	Interpretations

Discuss the groups' lists and help students identify interpretations versus observations, faults in the plan that may have affected the data collection, and difficulties they had with making their observations.

As groups are categorizing their observations, note the different categories groups are using. Check whether groups are identifying any inconsistencies in their results and problems with their procedures. Keep track of the differences in groups' interpretations and categories so that they can be discussed during the *Investigation Expo*.

META NOTES

During this discussion, students may require clarification of what they have seen. They may request to watch the video again, especially the parts they were uncertain about. Provide time for students to check their observations and interpretations in this way.

Communicate

35 min.

Have a discussion on groups' presentations of their observations and interpretations.

Communicate

Investigation Expo

You will share your observations and data analysis in an *Investigation Expo*. In this *Expo*, each group presents their poster to the class. In your presentation, you will share what you have seen with the class. Remember that each group created their own plan for making their observations and analyzing their data. Groups' observations may have been affected by their plan, and the analyses of other groups may be different from yours.

For your presentation, create a poster that describes each of the following:

- the questions you were trying to answer in your observations
- your observation procedure and how it helped you make observations
- your analysis of your observations and how confident you are about the analysis

At the bottom of your poster, make a list of the observations your group did not agree on.

Before the presentations, read the questions below. Plan your group's presentation so that you answer all of the questions. Then as each group is presenting their poster, listen for the answers. If a group does not answer all these questions, ask them to help you understand what they found out. When you ask the questions, focus on better understanding the observations the group made.

- What was the group trying to find out?
- What procedure did they use to collect their data?
- Were they able to make clear observations that were detailed?
- How did they group their observations? What did their groupings allow you to see?
- What conclusions do their results suggest?
- Do you trust their observations? Why or why not?

△ Guide

Let students know that they will be presenting their results in an *Investigation Expo* and that each group will need to create a poster of their results and will be presenting it to the class. Emphasize that student posters and presentations should address the first set of bulleted items on page 118 of the student text. Also let students know that they should list the observations they did not agree on at the bottom of their posters.

⬡ Get Going

Distribute poster materials and have groups make posters for the *Investigation Expo*. Let students know how much time they have.

△ Guide

As groups are working on their posters and preparing for their presentations, assist them as needed. Some issues that may arise are:

- Students might include opinions or interpretations in their observations. Ask them if anyone could argue with their observations. If any of the observations are debatable, they probably include opinions or interpretations.
- Determining confidence levels may still be difficult for students. Several issues can affect their confidence level. They may think they were unable to record their observations quickly enough or with enough detail. They may also have confused observation and interpretation. This can either raise or lower their confidence level. Finally, students may suggest that their confidence is high, even though it is clear that the observation plan was not effective or was not followed. Be cautious of students' stating more confidence than they should have in their results. Ask students if the observations on their posters give a clear picture of what they observed and how they observed it.

◇ Evaluate

Make sure groups have included responses to the bulleted points in the student text on their posters.

△ Guide

Let students know that before anyone presents, they will have the chance to familiarize themselves with each groups' work and formulate questions. Emphasize to the class that they will need to be able to answer each of the bulleted items in the second set of bullets for each group's results.

⬡ Get Going

Have students display their posters around the room and let students visit each poster for no more than a few minutes.

META NOTES

It is important for the group and the class to notice differences in observations, interpretations, and procedures. The questions for the *Investigation Expo* will help to focus students on these points.

△ Guide

Emphasize to the class that they will need to be able to answer each of the bulleted items in the second set of bullets for each presentation. Then let groups present. Have a brief discussion after each presentation. Encourage students to lead the discussion.

META NOTES

When students are presenting their posters, they should be presenting to each other, not to the teacher. By moving away from the presenters or sitting with the audience, you will encourage students to speak to the entire class.

Provide sufficient time for students to ask questions about the data analysis and conclusions (interpretations). If students are struggling, use some model questions to support thinking about the posters in a different way. By modeling questions about content, you will also assist students in asking better questions.

☐ Assess

As you listen to groups' presentations, focus on the ways students have used their observations to support their claims (interpretations). Identify areas of strength and weakness in students' observational plans. Note any interpretations mixed in with the observations and look for interpretations that are supported by observations.

△ Guide

After all the groups have presented, have a class discussion looking for similarities in the observations and interpretations from all the presentations. Discuss major differences. Ask students to consider what these similarities and differences might indicate.

META NOTES

If needed, show the video again to resolve disputes in observations and/or interpretations.

NOTES

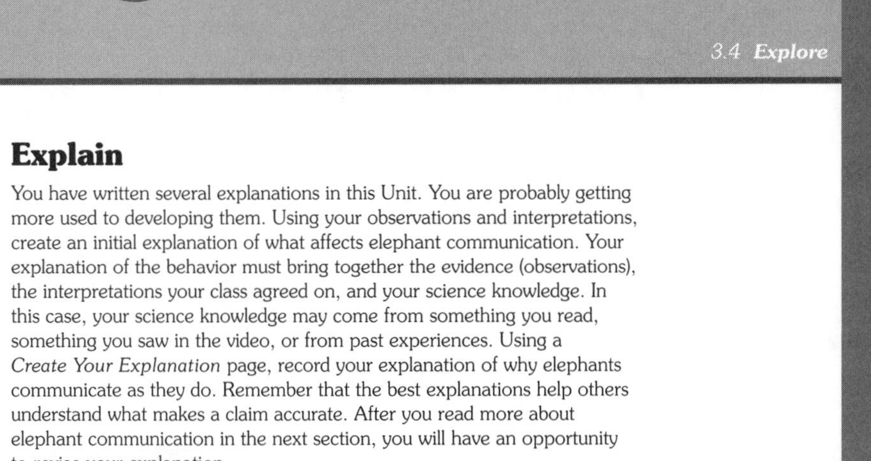

3.4 Explore

Explain

You have written several explanations in this Unit. You are probably getting more used to developing them. Using your observations and interpretations, create an initial explanation of what affects elephant communication. Your explanation of the behavior must bring together the evidence (observations), the interpretations your class agreed on, and your science knowledge. In this case, your science knowledge may come from something you read, something you saw in the video, or from past experiences. Using a *Create Your Explanation* page, record your explanation of why elephants communicate as they do. Remember that the best explanations help others understand what makes a claim accurate. After you read more about elephant communication in the next section, you will have an opportunity to revise your explanation.

Explain

10 min.

Have students create and discuss explanations of the behaviors they observed.

△ Guide

Let students know that they will be constructing another explanation. This explanation will be an answer to the question, *What Affects How Elephants Communicate?* As in previous explanations, they should construct a claim (which may be one of their interpretations or a combination of their interpretations) that is supported by their observations and science knowledge. Emphasize that explanations are comprised of a claim (a conclusion or interpretation) supported by evidence (observations), and science knowledge (ideas accepted in the scientific community) in a logical way.

Example:

Elephants make rumbling sounds to communicate. We know this because we saw elephants standing apart or walking separately and making rumbling sounds. We know elephants live in groups in the savanna, and they need to communicate across wide-open plains. We think these rumbling sounds will be heard across the wide-open plains.

□ Assess

As groups are working on their explanations, ask students what their claims are and what evidence they are using to support their claims. Their claims should be valid (based on evidence) and their evidence should not include any opinions. At this point, their evidence is slim and they don't have a lot of science knowledge they can use, but they should use their observations from the video to support their claims. Note any difficulties students are having with this.

NOTES

Explain

You have written several explanations in this Unit. You are probably getting more used to developing them. Using your observations and interpretations, create an initial explanation of what affects elephant communication. Your explanation of the behavior must bring together the evidence (observations), the interpretations your class agreed on, and your science knowledge. In this case, your science knowledge may come from something you read, something you saw in the video, or from past experiences. Using a *Create Your Explanation* page, record your explanation of why elephants communicate as they do. Remember that the best explanations help others understand what makes a claim accurate. After you read more about elephant communication in the next section, you will have an opportunity to revise your explanation.

Communicate

Share Your Explanation

Share your explanation with the class. As each group shares their explanation, pay special attention to how the other groups have supported their claims with science knowledge. Ask questions or make suggestions if you think a group's claim is not as accurate as it should be or if the group has not supported their claim well enough with observations and science knowledge.

What's the Point?

Humans and bees use different methods to communicate. Communication methods are affected by an animal's body structure. Some animals, like humans, have more body parts to use for communication than other animals. Some methods of communication, such as odors, are not visible.

Elephants communicate in some ways that are similar to bees and humans. They also communicate in ways that are different from these two animals. Elephants use body movements, sounds, touch, and odors for communication. Their environment also affects the way elephants communicate.

In science, observations are usually done before interpretations. First, scientists observe without making interpretations. Then, they use their observations and other information to make interpretations. Lastly, using the interpretations of their observations and all the evidence they have gathered, scientists develop explanations.

Communicate

10 min.

Have students share their explanations.

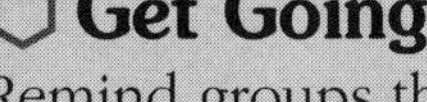

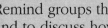

Get Going

Remind groups that they should be prepared to defend their explanations and to discuss how they revised them after hearing everyone's explanations.

Have groups present their explanations to the class about elephant communication. Ask the class to help pick out the claims, their evidence, and their reasons.

What's the Point?

5 min.

△ Guide

As groups present, students should ask questions to clarify and point out where something seems to be missing from an explanation. Model how you expect students to ask questions by asking how the evidence in an explanation supports the claim.

science k...

What's the Point?

Humans and bees use different methods to communicate. Communication methods are affected by an animal's body structure. Some animals, like humans, have more body parts to use for communication than other animals. Some methods of communication, such as odors, are not visible.

Elephants communicate in some ways that are similar to bees and humans. They also communicate in ways that are different from these two animals. Elephants use body movements, sounds, touch, and odors for communication. Their environment also affects the way elephants communicate.

In science, observations are usually done before interpretations. First, scientists observe without making interpretations. Then, they use their observations and other information to make interpretations. Lastly, using the interpretations of their observations and all the evidence they have gathered, scientists develop explanations.

AIA 119 ANIMALS IN ACTION

Students should realize that an animal's structure and environment affect its communication. Students should recognize that elephants use sounds and movement for communication.

Students should understand the importance of observations, which are needed to support scientific claims and explanations.

Assessment Options

Targeted Concepts, Skills, and Nature of Science	How do I know if students got it?
Behavior is a type of response to internal or external stimulus. Behavior is determined by experience, physical characteristics, and environment.	**ASK:** What things in the environment might evoke communication between elephants? **LISTEN:** Students should recognize that communication in all animals is often a response to danger or to the opportunity for food in the environment.

Targeted Concepts, Skills, and Nature of Science	How do I know if students got it?
The structure and function of animals' bodies are complementary and affect animal behavior.	**ASK:** What features of elephants' bodies affect how they communicate? **LISTEN:** Students should identify the parts of elephants' bodies, such as the trunk, that are used to produce sound. They should also identify the ears which allow elephants to hear, but which can also be moved to communicate. Students may question if elephants have vocal chords like humans. They do, they use their lungs, larynx, and vocal track.
Animals' ears are adapted to hearing sounds in their environment. Some animals use sound that is out of the range of human hearing.	**ASK:** What features of elephants' bodies allow them to detect communication? **LISTEN:** Students should identify elephants' hearing as an important adaptation that allows them to communicate.
Scientists must keep clear, accurate, and descriptive records of what they do so they can share their work with others, consider what they did, why they did it, and what they want to do next.	**ASK:** How did you use your observations to create explanations? **LISTEN:** Students should have used their observations as evidence to support their claims.
Scientists often work together and then share their findings. Sharing findings makes new information available and helps scientists refine their ideas and build on others' ideas. When another person's or group's idea is used, credit needs to be given.	**ASK:** How did sharing your ideas in the *Investigation Expo* help you create explanations? **LISTEN:** Students should have used the observations and interpretations of the class to create their explanations.

Targeted Concepts, Skills, and Nature of Science	How do I know if students got it?
Tables are an effective way to communicate results of a scientific investigation.	**ASK:** Did you use tables in your presentations? If not, could you have used tables? What advantages do tables have? **LISTEN:** Students should describe how tables make it easy to see relationships between the ideas or observations presented.
Scientists differentiate between observations and interpretations. They use their observations and interpretations to explain animal behavior.	**ASK:** Did you discover that there were interpretations mixed in your observations? How did you identify them? **LISTEN:** Students should have recognized that anything that was debatable was probably an interpretation.
Scientific knowledge is developed through observations, recording and analysis of data, and development of explanations based on evidence.	**ASK:** What were the important steps you took to develop your explanations? How are these steps similar to what scientists do? **LISTEN:** Students observed elephants in the video, recorded observations, analyzed and interpreted their observations, and developed explanations. Similarly, scientists make observations, analyze and interpret their data, and develop explanations.
Explanations are claims supported by evidence. Evidence can be experimental results, observational data, and other accepted scientific knowledge.	**ASK:** How did you support the claims in your explanations? **LISTEN:** Students should have used evidence from their observations to support their claims.

Teacher Reflection Questions

- What difficulties did students have observing physical and environmental factors that may affect elephant communication?
- What difficulties did students have in identifying and discussing the similarities and differences in groups' results?
- How were you able to keep students engaged in presenting and discussing each others' ideas?

SECTION 3.5 INTRODUCTION

3.5 Read

How Are Elephants Adapted for Communication?

◀ *2 class periods**

*A class period is considered to be one 40 to 50 minute class.

Overview

Students read about the ways elephants communicate, and how these forms of communication are determined by elephants' physical characteristics and by the environment. Students study how elephants' bodies allow them to produce low-pitched rumbles, which they can hear far across an open savanna, and a variety of other sounds. They learn how elephants also communicate through gestures, by touch and by using chemicals with their keen sense of smell. Using what they have learned, students revise their explanations, supporting their claims with new science knowledge and/or revising claims based on the new information they have. Then, they share their revised explanations with the class, create a class explanation, and update the *Project Board*. Students also learn about how sound is created, sound waves and their characteristics, and how the human ear hears.

Targeted Concepts, Skills, and Nature of Science	Performance Expectations
Scientists often work together and then share their findings. Sharing findings makes new information available and helps scientists refine their ideas and build on others' ideas. When another person's or group's idea is used, credit needs to be given.	Students should work in small groups to revise their explanations about elephant communication. They should share these and then work with the rest of the class to create a final explanation.
Scientists must keep clear, accurate, and descriptive records of what they do so they can share their work with others, consider what they did, why they did it, and what they want to do next.	Students should use their previous records to create and record their revised explanation. They then create and record a class explanation. These records should be used for making recommendations on a zoo enclosure design.
Scientific knowledge is developed through observations, recording and analysis of data, and development of explanations based on evidence.	Students should recognize that their *Project Board* is a record of how they develop ideas and information — through observation, reading, analysis, and creating explanations.

Targeted Concepts, Skills, and Nature of Science	Performance Expectations
Explanations are claims supported by evidence. Evidence can be experimental results, observational data, and other accepted scientific knowledge.	Students should construct claims that use observations and science knowledge to support valid claims.
The structure and function of animals' bodies are complementary and affect animal behavior.	Students should recognize that elephants' physical features allow for the kinds of communication they explored in *Section 3.4* and this section.
Animals communicate with other animals using sound. The sounds they can make and the sounds they can hear are adaptations to their environment.	Students should describe how the elephant uses sound as one means of communication, and that the sound an elephant makes is an adaptation to the environment where it lives.
Vibrations of molecules produce sound. Sound is a compression wave that can be described by its amplitude, frequency, and wavelength. Sound moves differently through different matter.	Students should be able to describe what causes sound (a compression wave) and that it moves differently in various materials. Students should also be able to describe the parts of a wave.
Animals' ears are adapted to hearing sounds in their environment. Some animals use sound that is out of the range of human hearing.	Students should describe that elephant communication uses, in part, sounds that are outside the range of human hearing, because elephants' ears are adapted to hear very low sounds that humans cannot hear. Elephants also have large ears that can pinpoint sounds from distant sources. Students should also be able to describe the human ear and how it functions.

Materials	
2 per student	*Create Your Explanation* pages
1 per class	*Project Board*
1 per student	*Project Board* pages
1 per class (optional)	Tuning fork
1 per class, if an additional viewing is needed	*Animals in Action* DVD *(What Affects How Elephants Communicate?)* and a way to view the video

Activity Setup and Preparation

Consider doing an Internet search on elephant communication. It may be helpful to search for organizations that have researched elephant communication. Look for video and sound clips using keywords like elephant voices, sounds, or communications.

Homework Options

Reflection

- **Science Content:** Why is an elephant's rumble an effective way to communicate with other members of the elephant's group on the savanna? *(The elephant's low-pitch sounds can be detected from distances as great as 10 km. This allows elephants to communicate with each other in the wide-open spaces of the savanna.)*
- **Science Content:** What have you learned about elephant communication that you could apply to designing an enclosure that would encourage elephants to communicate? *(This question should help students connect what they are learning to the* Big Challenge.)

Preparation for 3.6

- **Science Process:** How did you and the class improve your explanations when you revised them? Could you have made any of these revisions without the science knowledge you learned in this section? *(This question is meant to get students to look for lessons they can draw from their revisions about how to create good explanations.)*

NOTES

SECTION 3.5 IMPLEMENTATION

3.5 Read

How Are Elephants Adapted for Communication?

Sound is a very important form of communication for elephants. One truly amazing thing about elephants is the number of different sounds they can produce. Researchers have found that elephants can produce at least 25 different sounds. These sounds include rumbles, trumpets, snorts, and screams. Using these sounds, elephants tell their group members about how and where they are. They can send information about food and safety, as well as about playfulness and pleasure.

Making Sounds

The elephant has a huge head and body. This means that sounds an elephant makes can vibrate and echo through its body. Every part of the elephant helps it make sounds. The way an elephant holds its head, trunk, or throat can change how a sound is made.

By changing the length of its trunk, an elephant is able to change the sound it makes.

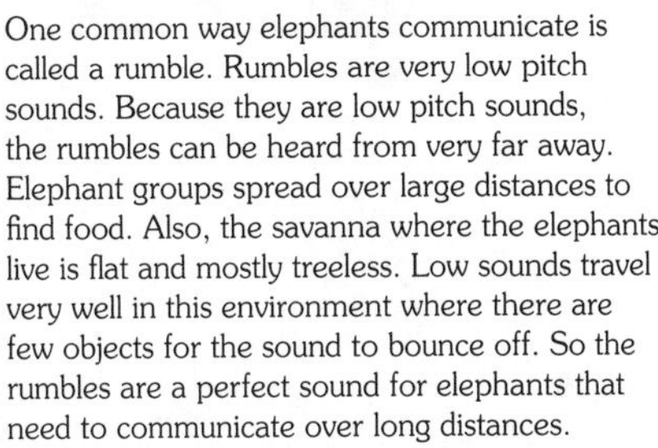

Types of Sounds

One common way elephants communicate is called a rumble. Rumbles are very low pitch sounds. Because they are low pitch sounds, the rumbles can be heard from very far away. Elephant groups spread over large distances to find food. Also, the savanna where the elephants live is flat and mostly treeless. Low sounds travel very well in this environment where there are few objects for the sound to bounce off. So the rumbles are a perfect sound for elephants that need to communicate over long distances.

Until recently, scientists did not know that elephants made these low rumbles. That is because the rumbles are too low for people to hear. Scientists observed elephants and how they seemed to communicate with one another. They thought the only way elephants could be communicating was through sound, but they were not hearing anything. These observations made scientists think they needed to listen to the elephants more carefully. Scientists needed tools that would allow them to listen to lower sounds.

◀ *2 class periods**

3.5 Read

How Are Elephants Adapted for Communication?

15 min.

Introduce and discuss information about how elephants communicate.

△ Guide

Begin by letting students know that they will now learn more about how elephants communicate to help them explain some of the communication behaviors they have observed. Let students know that they will have a chance to revise their explanations of elephant communication using what they have learned.

*A class period is considered to be one 40 to 50 minute class.

TEACHER TALK

"You have observed and explained some ways elephants communicate, but our observations from the video did not provide us with a lot of information about elephant communication. Scientists have done numerous lengthy observations and measurements concerning elephant communication and have learned many things about the ways elephants communicate. One thing found through research is that elephants make a number of different sounds. Did any of you pay attention to the different sounds that elephants made in the video you watched? What were the sounds that you heard? How many sounds do you think elephants usually make?

Then let students know that elephants can make at least 25 different sounds including rumbles, trumpets, snorts, and screams."

META NOTES

Consider showing the video again and let students pick out the number of different sounds they hear.

Making Sounds

5 min.

well as about playfulness and pleasure.

Making Sounds

The elephant has a huge head and body. This means that sounds an elephant makes can vibrate and echo through its body. Every part of the elephant helps it make sounds. The way an elephant holds its head, trunk, or throat can change how a sound is made.

By changing the length of its trunk, an elephant is able to change the sound it makes.

○ Engage

Ask students whether they think the elephant in the picture is making a sound. If so, what sound do they think it's making? What are their reasons?

△ Guide

Then discuss how the elephant's body helps it to make sounds. Let students know that elephants, like humans, produce sounds using their lungs, larynx and vocal track. Discuss how the sounds an elephant makes vibrate through its whole body, and it can change the sound it makes by changing the position of its head, trunk, throat, or even ears. You can explain how by changing the length of its trunk as it is producing a sound, the elephant can alter the sound it is making.

to chang... ...roat can change ... ...nade.
it makes.

Types of Sounds

One common way elephants communicate is called a rumble. Rumbles are very low pitch sounds. Because they are low pitch sounds, the rumbles can be heard from very far away. Elephant groups spread over large distances to find food. Also, the savanna where the elephants live is flat and mostly treeless. Low sounds travel very well in this environment where there are few objects for the sound to bounce off. So the rumbles are a perfect sound for elephants that need to communicate over long distances.

Until recently, scientists did not know that elephants made these low rumbles. That is because the rumbles are too low for people to hear. Scientists observed elephants and how they seemed to communicate with one another. They thought the only way elephants could be communicating was through sound, but they were not hearing anything. These observations made scientists think they needed to listen to the elephants more carefully. Scientists needed tools that would allow them to listen to lower sounds.

AIA 120

Project-Based Inquiry Science

By using microphones that can pick up sounds below human hearing, they were able to hear the rumbles of elephants.

Using these tools, scientists found out that elephants also make other noises. Elephants have a tremendous range of sounds. They can make low sounds and high sounds. They can make quiet sounds as well as loud sounds.

Making a wide range of sounds does not really help an elephant if it cannot hear the sounds. An elephant's hearing is also amazing. Not only can elephants hear rumbles over long distances (10 km, 6.2 mi), they can also determine the direction of the sound. Just by listening, elephants are able to know where their group members are, even if the other group members are far away.

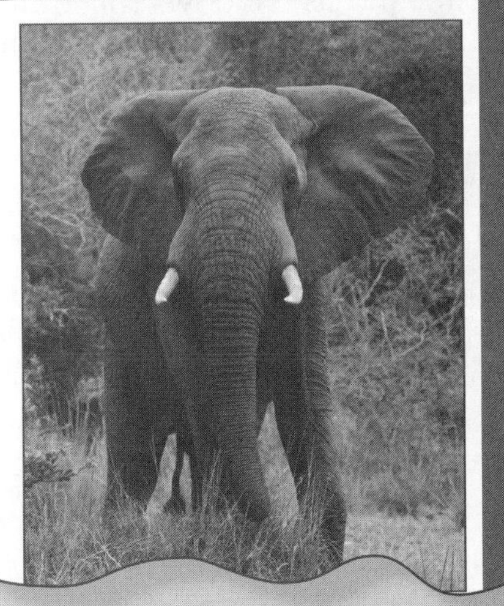

Types of Sounds

5 min.

△ Guide

Discuss the sounds that elephants make to communicate and what these sounds mean.

TEACHER TALK

"Elephants make many different sounds such as rumbles, trumpets, snorts, and screams. Scientists have made many observations of these. Most of the sounds elephants make are rumbles. There are distinctive rumbles that have been interpreted by scientists to mean "let's go." Elephants use another contact rumble to let one another know where they are. Males use a rumble called *musth* when they want to mate, but it has also been observed that they use these calls when they are angry, and when they feel challenged.

So far, over seventy different call types have been observed. It has been observed that much of elephant communication is in a frequency range that humans cannot hear."

META NOTES

Sound attenuates (or diminishes) as it travels through substances (such as air, water, or solids) and tends to diminish depending on the absorption factors of the material it is traveling through and the square of the frequency. The lower pitched frequencies do not attenuate as rapidly as the higher frequencies do.

Using the student text, describe how rumbles are very low sounds that can travel long distances—especially in the open savanna, where there are few trees or other objects that the sound can bounce off of or get absorbed by. Give students an example of how lower pitch (or lower frequency) sounds travel further. One example is a car driving by with the music on, the bass turned up, and its windows rolled up. You may not hear the singing voices but you can feel the bass rumbling through your body.

Next, explain that many of the rumbles are too low for humans to hear, and that scientists need to use special microphones to learn more about elephants' rumbles. Using these microphones, scientists have discovered a range of sounds elephants make.

Point out the picture of the elephant in the student text with its ears outward. Students may notice that it seems to have its ears directed toward something. Explain that elephants are also very good at determining the direction of sound. This allows them to know where their group members are, even when they are very far away.

By using microphones that can pick up sounds below human hearing, they were able to hear the rumbles of elephants.

Using these tools, scientists found out that elephants also make other noises. Elephants have a tremendous range of sounds. They can make low sounds and high sounds. They can make quiet sounds as well as loud sounds.

Making a wide range of sounds does not really help an elephant if it cannot hear the sounds. An elephant's hearing is also amazing. Not only can elephants hear rumbles over long distances (10 km, 6.2 mi), they can also determine the direction of the sound. Just by listening, elephants are able to know where their group members are, even if the other group members are far away.

An elephant has amazing hearing that can pick up sound over great distances.

What Are Some Other Ways Elephants Communicate?

Think back to when you were working on the puzzle. Some students used only gestures to help their partner assemble the puzzle. These groups found that they could be effective with gestures, but that it required being able to see each other. Elephants also use different parts of their bodies to communicate visually. In a threatening situation, an elephant will stand very tall with its head held high. Sometimes it will even stand on a log or rock to appear taller. When an elephant flaps its ears rapidly, it means that it is excited. A large male elephant may kneel as an invitation to a smaller elephant to play. Some elephants have even been observed smiling.

Touching is another way elephants communicate with their bodies. Elephants use their trunks to greet one another. Female elephants will entwine their trunks in a way that is similar to a human handshake. They also use their trunks to caress one another. When mother elephants brush their **calves** with their trunks, they seem to be comforting them. Elephants can also be aggressive. They can use their trunks to uproot and throw things, or use their tusks to push one another.

calf (plural, calves): the young of certain kinds of animals, such as an elephant or cow.

One way elephants communicate is not easily observable. Scientists know about chemical communication from many different types of animals, including humans. Elephants have a keen sense of smell. Other than sound,

What Are Some Other Ways Elephants Communicate?

5 min.

△ Guide

Remind students of the different ways to communicate that they discovered during the puzzle activity and how the constraints imposed on them determined how they communicated. Ask students about some of the gestures they came up with.

Discuss the information in the student text, describing the other ways elephants communicate by gestures, touch, and chemicals (odor).

META NOTES

Decide how you want students to use the student text. They could take turns reading aloud, read in groups, summarize paragraphs, pick out key topics or main ideas, or you could read aloud to them. The class should discuss the information in the text.

Describe how elephants may use their eyes and gestures to communicate.

TEACHER TALK

"Elephants use different parts of their bodies to communicate. Sometimes elephants communicate visually, using gestures as students did in the nonverbal groups. Elephants may stand tall in threatening situations, or may flap their ears rapidly when they are excited. Elephants may kneel to invite smaller elephants to play. Elephants can communicate with their eyes, and they can even smile."

Describe how elephants may use touch or odor (chemicals) to communicate.

META NOTES

Male and female African elephants have tusks, but only the male Asian elephants have tusks that protrude beyond the lip. African elephants have much larger ears than Asian elephants.

TEACHER TALK

"Sometimes elephants communicate by touch. Females may entwine their trunks in greeting each other, much like a hand shake. Females may touch their calves with their trunks to comfort them. Elephants may also push one another with their tusks or throw things with their trunks. Elephants also use chemical communication, following trails marked by urine and other odors that they pick up with their very keen sense of smell."

An elephant calf closely follows its mother. To comfort the calf, the mother may touch it with her trunk.

elephants probably use chemical communication more than any other type of communication. Elephants are often seen using the tips of their trunks to follow trails marked by urine or other waste material.

Stop and Think

1. What are some ways elephants communicate with each other?
2. People communicate about a variety of things. What are some things elephants must communicate about?
3. How does an elephant's environment affect how it communicates?

Revise Your Explanation

You have just read more about how elephants communicate. With your group, look back at your *Observing and Interpreting Animal Behavior* pages. Look at your original interpretations of elephant communication. Now, with the new science knowledge from what you have read, reinterpret the behavior you saw in the video.

Go back to your explanation on your *Create Your Explanation* page. First, add the science you just learned to the science knowledge box. Then, check to make sure your claim is still accurate. If your claim does not match the science you have read, you should revise it. Next, support your claim with the science knowledge you just learned.

Rewrite your explanation to make it more complete. Remember that an explanation is a statement that connects a claim to evidence and science knowledge in a logical way. Write your explanation so that it tells why your claim is accurate. Be sure your explanation matches the science you just

Stop and Think

10 min.

Have a class discussion about elephant communication.

○ Get Going

Let students know that they should answer the questions with their group members and that a class discussion will follow.

△ Guide and Assess

Have a class discussion on students' responses. Guide and assess students' answers based on the information provided below.

1. Elephants can communicate with each other by making and detecting (hearing) sounds, by seeing and making gestures, by touch, and with chemicals (through odors).
2. Elephants must communicate for survival. Elephants communicate to each other in threatening situations and for reproduction. Although it is not in the reading, elephants live on average about 70 years. They do not begin mating until they are 20. Many of their behaviors are learned and they have a highly intricate social system. Animal behavior that is learned rather than instinctive is very important.
3. Elephants live in various environments. It may be assumed that the low frequencies elephants primarily use are an adaptation to their environments.

META NOTES

African elephants live in forests, savannas, river valleys, and deserts. Asian elephants live in open grasslands, marshes, and forests.

Revise Your Explanation

10 min.

Have students revise their explanations.

Revise Your Explanation

You have just read more about how elephants communicate. With your group, look back at your *Observing and Interpreting Animal Behavior* pages. Look at your original interpretations of elephant communication. Now, with the new science knowledge from what you have read, reinterpret the behavior you saw in the video.

Go back to your explanation on your *Create Your Explanation* page. First, add the science you just learned to the science knowledge box. Then, check to make sure your claim is still accurate. If your claim does not match the science you have read, you should revise it. Next, support your claim with the science knowledge you just learned.

Rewrite your explanation to make it more complete. Remember that an explanation is a statement that connects a claim to evidence and science knowledge in a logical way. Write your explanation so that it tells why your claim is accurate. Be sure your explanation matches the science you just

read. Make sure your claim now matches what you have learned. If it does not, revise your explanation. Use the information from your reading about elephant communication to support your revised explanation. You might need to write an explanation that has a few sentences rather than just one long sentence. This is fine. The goal is to to tie everything together and help others understand why elephants communicate the way they do.

△ Guide

Let students know that they will now have the opportunity to revise their explanations about what affects elephant communication based on all the information they now have. Emphasize that the claims they have made must be supported by the science knowledge introduced in this section, or they will need to revise and/or construct new claims.

TEACHER TALK

"Now you know more about elephant communication than when you wrote your explanations about what affects elephant communication. When you look at your explanations, you may see that information from your reading might be helpful in supporting your explanation or it may go against your explanation.

First, check to make sure your claim fits with the new science knowledge. If your claim is inconsistent with what you learned, change it.

Then, put any science knowledge you can use in the science knowledge box and rewrite your explanation to make it more complete."

○ Get Going

Distribute new *Create Your Explanation* pages and have students revise their explanations to include the new science knowledge.

◇ Evaluate

As groups are revising their explanations, monitor students' progress. Check their explanations to make sure they are using the science knowledge they have just learned to support their claims. If it looks like their claims are inconsistent with what they have learned, note this as something to discuss when the class discusses the groups' presentations.

Students should now be able to construct explanations with valid claims in which the evidence supports the claims. If groups are still using unfounded claims or using opinions for evidence, stop the class and review what they should know about constructing explanations.

Communicate

15 min.

Have a class discussion on the groups' presentations of explanations.

read. Make sure your claim now matches what you have learned. If it does not, revise your explanation. Use the information from your reading about elephant communication to support your revised explanation. You might need to write an explanation that has a few sentences rather than just one long sentence. This is fine. The goal is to to tie everything together and help others understand why elephants communicate the way they do.

Communicate

Share Your Explanation

Share your new explanation with the class. When you share your explanation, tell the class what makes this revised explanation better than your earlier one. As each group shares their explanation, pay special attention to how the other groups have supported their claims with science knowledge. Ask questions or make suggestions if you think a group's claim is not as accurate as it could be or if the group has not supported their claim well enough with observations and science knowledge.

As a class, create your best explanation of what affects how elephants communicate.

Update the *Project Board*

As you read more about how elephants communicate, you may have thought of new questions you want to ask. Add what you learned from your reading to the third column of the *Project Board*. Make sure you add evidence to support any new science you learned. Also, think about how what you are learning will help you answer the *Big Question, How do scientists answer big questions and solve big problems?* As your teacher records this information on the big *Project Board*, add the information to your own *Project Board* page.

What's the Point?

Animals communicate in a variety of ways. An elephant uses gestures, sound, and smell to communicate with others. This is similar to how humans and other mammals communicate. How each type of animal communicates depends on its physical characteristics. The elephant uses its trunk, head and body size, and ears to help with communication. Also, the elephant's environment affects how it communicates. The sounds an elephant makes are tailored to meet the needs of the savanna where it lives. Animals, including humans and elephants, communicate to keep track of family members, to alert others to danger, to identify a food source, to play, and to take care of their young.

META NOTES

If students ask you questions about other groups' explanations, redirect the question to the group presenting. In general, encourage students to discuss their explanations with each other and the presenting groups.

△ Guide

Let the class know that each group will be presenting their explanation and that it is important for each presenting group to clarify how they have supported their claims with observations and science knowledge. Emphasize that it is also important that the audience take notes and ask questions if it is not clear. Then let them know that after hearing everyone's explanations, the class will come up with an explanation together about what affects how elephants communicate.

Have a discussion after each presentation. Model for students how they should seek clarification or point out areas they may not understand.

TEACHER TALK

"I don't see how that fact backs up your claim. Could you clarify that for me?

I'm not sure I understand your claim. Can you explain it to me?"

Also, encourage students to point out where they think the group could have used some of the science knowledge they just learned to support their claims.

◇ Evaluate

Evaluate students' use of the reading to revise their explanations.

△ Guide

After all groups have presented, ask students to point out the similarities and differences between the explanations presented.

Then ask the class to pick out the claims they think they should include in their class explanation. Record these claims on the board and then ask for the supporting evidence and science knowledge.

After the class has agreed on an explanation, have students copy them on *Create Your Explanation* pages.

Update the *Project Board*

As you read more about how elephants communicate, you may have thought of new questions you want to ask. Add what you learned from your reading to the third column of the *Project Board*. Make sure you add evidence to support any new science you learned. Also, think about how what you are learning will help you answer the *Big Question, How do scientists answer big questions and solve big problems?* As your teacher records this information on the big *Project Board*, add the information to your own *Project Board* page.

Update the Project Board

5 min.

Have a class discussion focused on updating the class Project Board.

△ Guide

Ask students if they have any new questions about how animals communicate and record these in the second column *(What do we need to investigate?)* of the *Project Board*.

Then, ask students what they are learning and what is their evidence. Record their answers in the third and fourth columns of the *Project Board*. Remember to link the claims (column 3) and evidence (column 4) together on the *Project Board*.

◇ Evaluate

Make sure that the claims from students' explanations are in the third column and that information from the reading is in the fourth column before moving on.

Students' claims should include claims specific to elephant communication. Why elephants communicate, what they communicate, the types of sounds they make, and how their bodies and their environment affects their communications are all good claims.

What's the Point?

5 min.

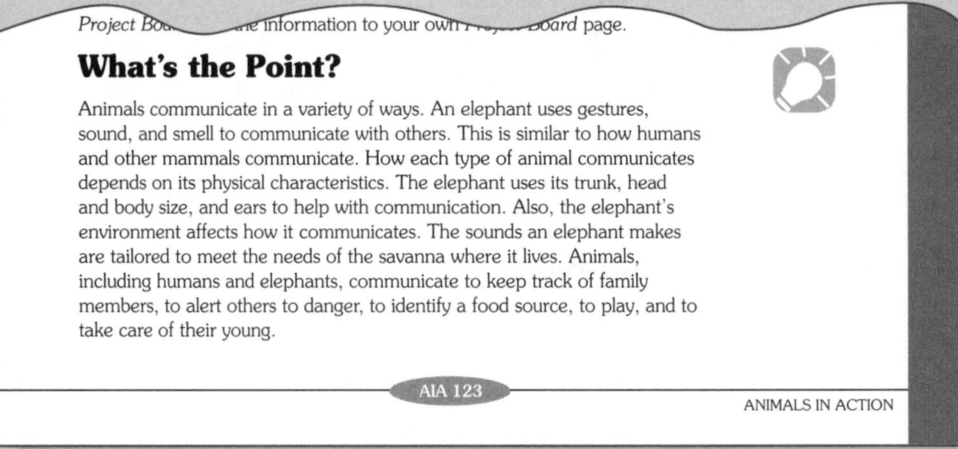
Project Bo... information to your own ... Board page.

What's the Point?

Animals communicate in a variety of ways. An elephant uses gestures, sound, and smell to communicate with others. This is similar to how humans and other mammals communicate. How each type of animal communicates depends on its physical characteristics. The elephant uses its trunk, head and body size, and ears to help with communication. Also, the elephant's environment affects how it communicates. The sounds an elephant makes are tailored to meet the needs of the savanna where it lives. Animals, including humans and elephants, communicate to keep track of family members, to alert others to danger, to identify a food source, to play, and to take care of their young.

AIA 123

ANIMALS IN ACTION

Emphasize that the way elephants communicate is determined by their physical characteristics and their environments. For example, an elephant can make various sounds by changing the length of its trunk. Sound is an effective way for elephants to communicate across the open spaces of the savanna.

More to Learn

What Is Sound?

Sound is one way animals can communicate with each other. You have read about how elephants communicate using sound. But what exactly is sound?

Matter is made of small particles, called molecules. Sound is created by the back-and-forth motion, or **vibration**, of the molecules in matter. Because sound is the movement of molecules, sound can move through gases, liquids, and solids. Sound cannot travel where there are no molecules to move.

Sound as Waves

When you make a sound with your voice, your vocal chords vibrate very quickly. The vibration causes molecules in the surrounding air to be pushed together in a pattern. Between each of the places where the air molecules are pushed together is a part where the air molecules are spread out. This pattern is repeated over and over again very rapidly. Scientists call this movement of molecules a **sound wave**.

As the vibration in your vocal chords begins, the air nearest your vocal chords is pushed together. This group of air molecules that has been pushed together is called a **compression**. Next, the vocal chords move back, giving the air molecules room to spread apart. The space where molecules are spread apart is called a **rarefaction**.

The diagram on the next page shows a vibrating tuning fork. All sound waves happen in the same way, regardless of how they are made. They all have compressions and rarefactions. The vibrations that create sound are repeated very rapidly, so rapidly that you cannot see them. The sound waves then move through the matter that surrounds the source of the sound. Often, the matter is air, but sound can also move through liquids and solids.

You have probably noticed that the farther you are from the source of a sound, the softer the sound is. When you are very close to the source of a loud sound, the sound may even hurt your ears. But for any sound, no matter how loud, from very far away, you cannot hear it at all. This is because sound waves change and lose strength as they move away from their source. For example, as you speak, the sound waves you make move away from you. As they move away, each wave spreads out, which decreases the strength of the wave at any point.

vibration: the back-and-forth movement of molecules.

sound wave: the movement of molecules in a pattern, repeated over and over again, very rapidly.

compression: part of a sound wave where molecules are pushed together.

rarefaction: part of a sound wave where molecules are spread apart.

More to Learn

40 min.

Discuss sound, its characteristics, and how humans hear sound.

Engage

Let students know that many animals communicate by making and hearing sounds. Ask students what they think creates a sound and record their responses.

Guide

Then explain sound using the information in the student text. Describe how sound is created (by the vibration or back-and-forth motion of atoms or molecules). Emphasize that sound cannot travel where there are no atoms or molecules (such as in outer space), but it can travel in solids, liquids, and gases.

META NOTES

You may want to point out that electromagnetic waves (light) can travel in places where there are no atoms, but sound cannot.

META NOTES

Decide how you want to read the *More to Learn* segment. Consider having students take turns reading aloud, reading in groups, summarizing, or pulling out key ideas. It is important to discuss the ideas in this segment.

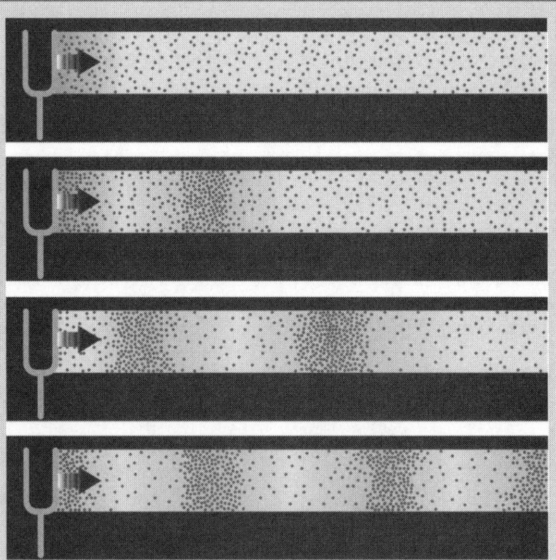

Air molecules are pushed together as the tuning fork arms vibrate quickly. The compressions move away from the tuning fork as sound waves.

Also, when sound waves run into another object, they either bounce off it or are absorbed by it. When waves bounce off an object, some of the wave strength passes into the object. The remaining wave is weaker.

When an object absorbs waves, the sound cannot move any farther. So sounds in wide-open spaces can move farther than sounds in closed areas, or crowded areas, such as areas with many trees. The objects in closed-in or crowded areas absorb sound waves, which decreases the amount of sound traveling through them.

Sound waves travel at different speeds in different kinds of matter. Molecules in gases are very spread out. Molecules in liquids are packed together more tightly. Molecules in solids are packed together even more tightly than in liquids. Because molecules in gases are more spread out than those in liquids, sound moves more slowly in gas than in liquid. Because molecules in solids are packed even more closely together, sound waves move fastest through solids.

wavelength: measured from one compression to another compression, or from one rarefaction to another rarefaction.

frequency (of waves): the number of waves that pass a point in a second.

pitch: a measure of the frequency of the vibration of the source of a sound.

Characteristics of Sound Waves

The diagrams shown on the next page can help you imagine sound waves. If there were no sound, all of the molecules would spread out evenly. A vibrating object pushes molecules together in some places. You can see that on the diagrams there are places, shown as dark areas, where the molecules are pushed together. These places are the compressions. There are other places, which are lighter, where the molecules are spread out. These places are the rarefactions. A single wave is made up of one compression and one rarefaction.

Scientists measure certain characteristics of waves. All waves—light waves, radio waves, and sound waves—share three characteristics: wavelength, frequency, and amplitude.

Wavelength is the length of the wave. Scientists determine wavelength by measuring individual waves from one compression to another compression, or from one rarefaction to another rarefaction. Diagrams A and B have the same wavelength. Diagram C has a longer wavelength than A or B.

Imagine the waves shown in the diagrams moving across the page. You can imagine that, as the wave in Diagram A or B moves, more waves will pass any one point on the page than when the wave in Diagram C moves. The number of waves that pass a point each second is called the wave **frequency**. Sound waves vibrate very fast. The faster something vibrates, the more waves it produces each second. Wavelength and frequency are related to each other. Waves with a shorter wavelength have a higher frequency than waves with a longer wavelength. Something that vibrates fast produces sound waves of a higher frequency than something that vibrates more slowly. **Pitch** is determined by the frequency of the sound wave. Sounds with a high frequency have a high pitch. Sounds that have a low frequency have a low pitch. You hear sound waves with higher frequencies, like those from a whistle, as higher pitched sounds. Other sounds, like sound waves from a bass drum, you hear as lower pitched sounds.

The diagrams are models that help you imagine sound waves, but sound waves cannot be seen. Also, most sound waves in air have a longer wavelength than those shown. The waves in the diagrams have relatively short wavelengths, which are close to the wavelengths of the highest pitch sounds humans can hear.

Characteristics of Sound Waves

15 min.

Let students know that scientists describe and measure waves using certain characteristics such as its wavelength, frequency, and amplitude. Then describe what these characteristics are using the information in the student text.

Describe the wavelength of a sound wave as the distance between one compression to another using the images on page AIA 127 in the student text. Explain that the dark bands in the images are compressions and the light areas are rarefactions. Point out that the wavelengths are labeled on each of the diagrams.

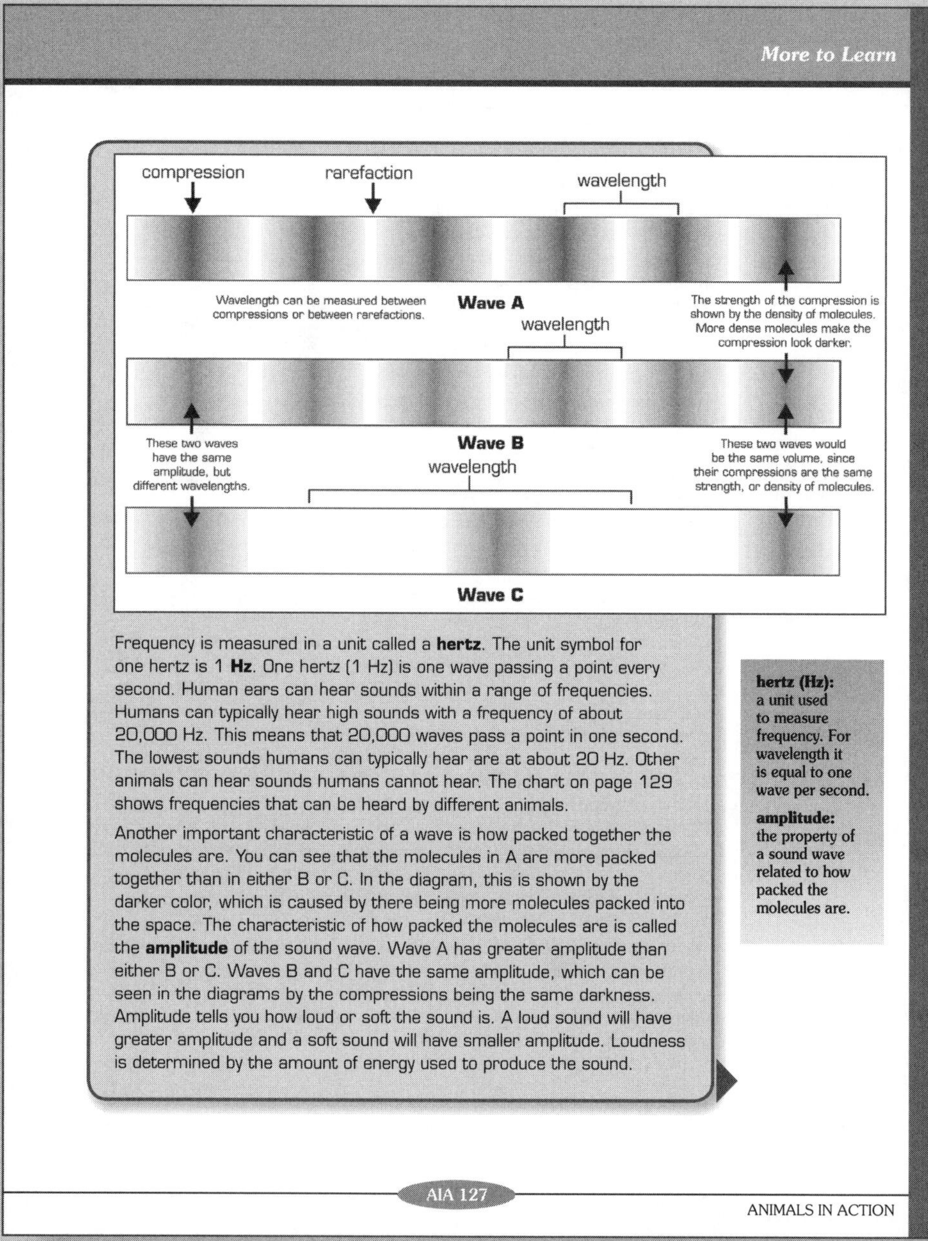

More to Learn

compression
rarefaction
wavelength

Wavelength can be measured between compressions or between rarefactions.

Wave A

The strength of the compression is shown by the density of molecules. More dense molecules make the compression look darker.

wavelength

These two waves have the same amplitude, but different wavelengths.

Wave B

These two waves would be the same volume, since their compressions are the same strength, or density of molecules.

wavelength

Wave C

Frequency is measured in a unit called a **hertz**. The unit symbol for one hertz is 1 **Hz**. One hertz (1 Hz) is one wave passing a point every second. Human ears can hear sounds within a range of frequencies. Humans can typically hear high sounds with a frequency of about 20,000 Hz. This means that 20,000 waves pass a point in one second. The lowest sounds humans can typically hear are at about 20 Hz. Other animals can hear sounds humans cannot hear. The chart on page 129 shows frequencies that can be heard by different animals.

Another important characteristic of a wave is how packed together the molecules are. You can see that the molecules in A are more packed together than in either B or C. In the diagram, this is shown by the darker color, which is caused by there being more molecules packed into the space. The characteristic of how packed the molecules are is called the **amplitude** of the sound wave. Wave A has greater amplitude than either B or C. Waves B and C have the same amplitude, which can be seen in the diagrams by the compressions being the same darkness. Amplitude tells you how loud or soft the sound is. A loud sound will have greater amplitude and a soft sound will have smaller amplitude. Loudness is determined by the amount of energy used to produce the sound.

hertz (Hz): a unit used to measure frequency. For wavelength it is equal to one wave per second.

amplitude: the property of a sound wave related to how packed the molecules are.

AIA 127

ANIMALS IN ACTION

Explain that waves such as the ones depicted in the diagrams move rapidly away from their source. Students can imagine the waves in the diagrams moving across the page. The frequency of a sound is the number of waves that pass a point each second. Sounds with longer wavelengths have lower frequencies, while sounds with shorter wavelengths have higher frequencies. Pitch, often described as how high or low a sound is, is a measure of the frequency of the vibrations of the source of a sound. Frequency is measured in hertz (Hz).

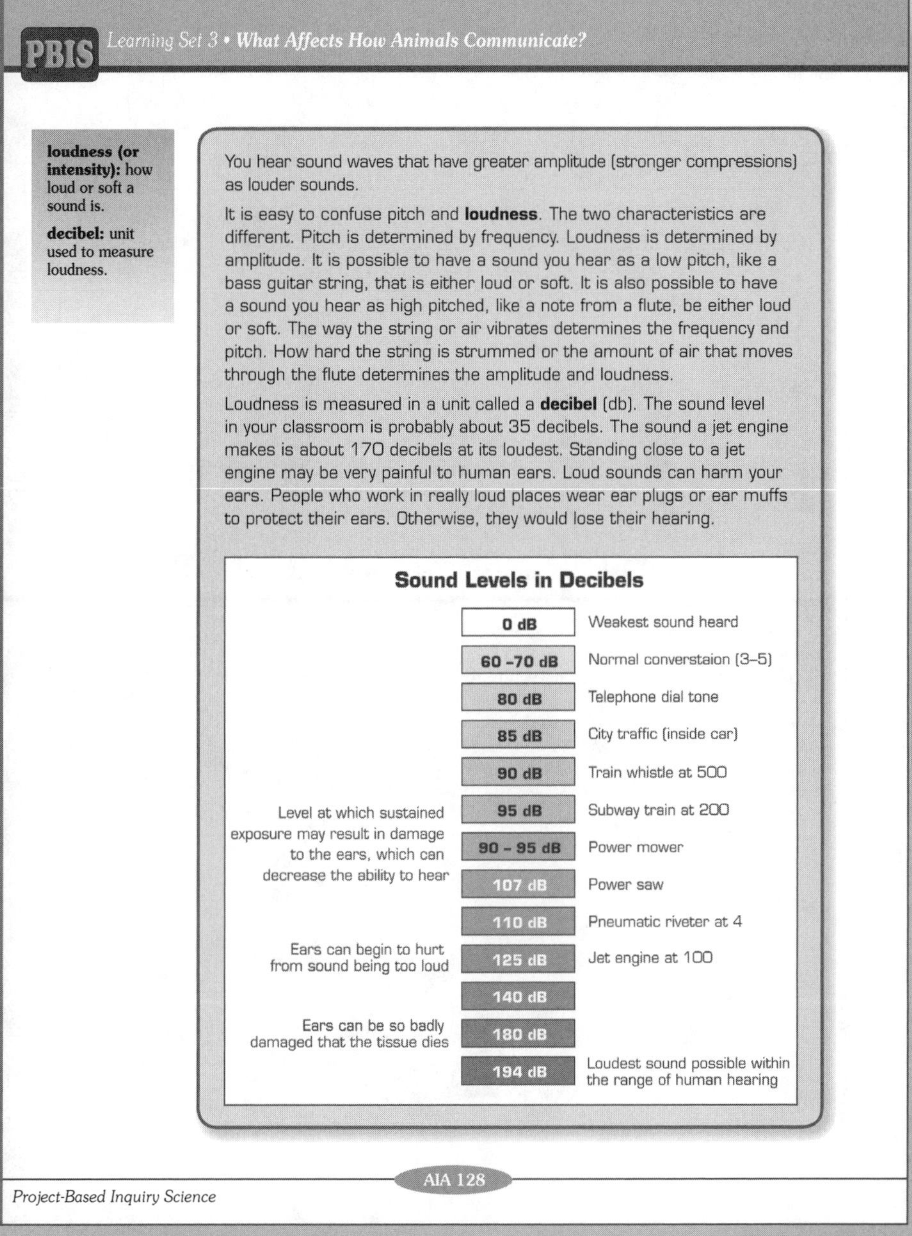

loudness (or intensity): how loud or soft a sound is.

decibel: unit used to measure loudness.

You hear sound waves that have greater amplitude (stronger compressions) as louder sounds.

It is easy to confuse pitch and **loudness**. The two characteristics are different. Pitch is determined by frequency. Loudness is determined by amplitude. It is possible to have a sound you hear as a low pitch, like a bass guitar string, that is either loud or soft. It is also possible to have a sound you hear as high pitched, like a note from a flute, be either loud or soft. The way the string or air vibrates determines the frequency and pitch. How hard the string is strummed or the amount of air that moves through the flute determines the amplitude and loudness.

Loudness is measured in a unit called a **decibel** (db). The sound level in your classroom is probably about 35 decibels. The sound a jet engine makes is about 170 decibels at its loudest. Standing close to a jet engine may be very painful to human ears. Loud sounds can harm your ears. People who work in really loud places wear ear plugs or ear muffs to protect their ears. Otherwise, they would lose their hearing.

Sound Levels in Decibels

	0 dB	Weakest sound heard
	60 -70 dB	Normal converstaion (3–5)
	80 dB	Telephone dial tone
	85 dB	City traffic (inside car)
	90 dB	Train whistle at 500
Level at which sustained exposure may result in damage to the ears, which can decrease the ability to hear	95 dB	Subway train at 200
	90 - 95 dB	Power mower
	107 dB	Power saw
	110 dB	Pneumatic riveter at 4
Ears can begin to hurt from sound being too loud	125 dB	Jet engine at 100
	140 dB	
Ears can be so badly damaged that the tissue dies	180 dB	
	194 dB	Loudest sound possible within the range of human hearing

Explain that the amplitude of a sound wave is a measure of how compressed the molecules are. If they are highly compressed, the sound has high amplitude, which means that it will be loud. If they are less compressed, the sound has low amplitude, which means that the sound will be quiet.

Next, discuss how the loudness of a sound wave is measured. Point out that a sound wave is measured in decibels and discuss the diagram depicting sound levels on page AIA 128 in the student text.

How Humans Hear

10 min.

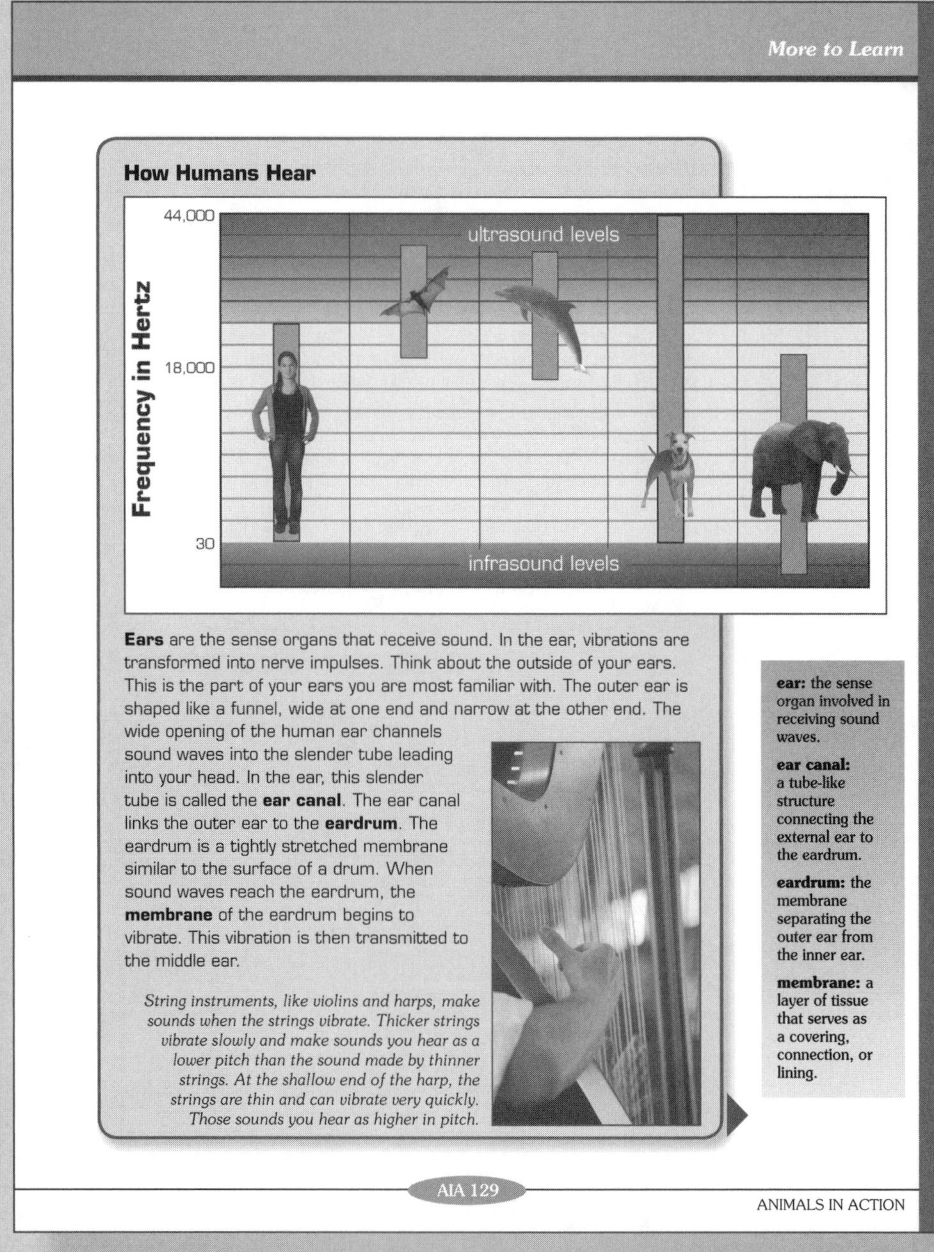

More to Learn
How Humans Hear
44,000
18,000
30
Frequency in Hertz
ultrasound levels
infrasound levels
Ears are the sense organs that receive sound. In the ear, vibrations are transformed into nerve impulses. Think about the outside of your ears. This is the part of your ears you are most familiar with. The outer ear is shaped like a funnel, wide at one end and narrow at the other end. The wide opening of the human ear channels sound waves into the slender tube leading into your head. In the ear, this slender tube is called the ear canal. The ear canal links the outer ear to the eardrum. The eardrum is a tightly stretched membrane similar to the surface of a drum. When sound waves reach the eardrum, the membrane of the eardrum begins to vibrate. This vibration is then transmitted to the middle ear.
String instruments, like violins and harps, make sounds when the strings vibrate. Thicker strings vibrate slowly and make sounds you hear as a lower pitch than the sound made by thinner strings. At the shallow end of the harp, the strings are thin and can vibrate very quickly. Those sounds you hear as higher in pitch.
ear: the sense organ involved in receiving sound waves.
ear canal: a tube-like structure connecting the external ear to the eardrum.
eardrum: the membrane separating the outer ear from the inner ear.
membrane: a layer of tissue that serves as a covering, connection, or lining.
AIA 129
ANIMALS IN ACTION

Discuss the diagram on page AIA 129 that shows the frequency at which different animals hear. Ask students what frequency is. If they are unsure, remind them that it is a measure of how many cycles of the wave go by in a second. Point out that different animals can hear sounds at varied frequencies Elephants can hear many of the frequencies humans can hear, although they can also hear some frequencies that are too low for humans to hear, and they cannot hear some of the higher frequencies that humans can hear.

Moving along in the student text, discuss the structure and function of the human ear using the information on page AIA 130. Point out that the ear is a sense organ that receives and processes sound. Explain that sound is

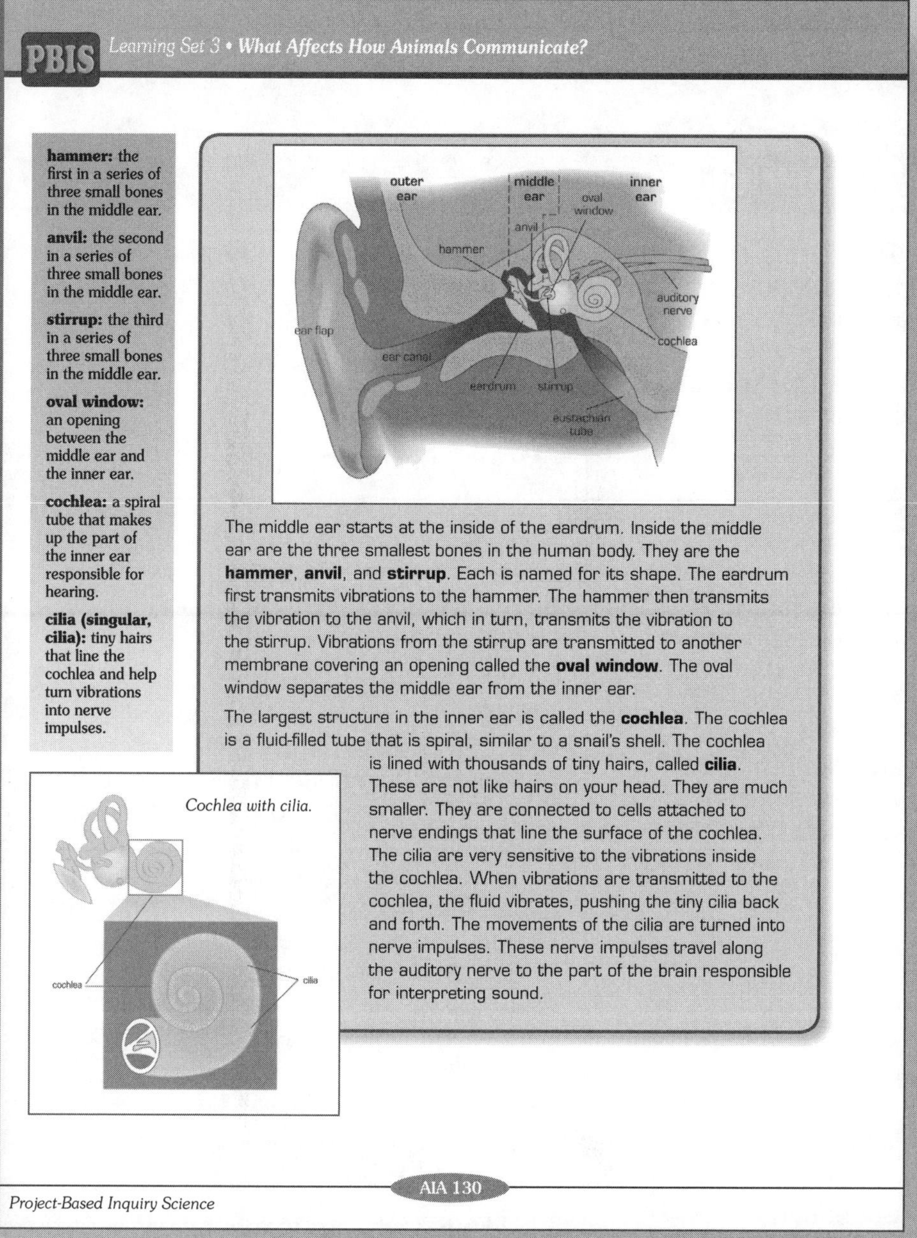

PBIS Learning Set 3 • *What Affects How Animals Communicate?*

hammer: the first in a series of three small bones in the middle ear.

anvil: the second in a series of three small bones in the middle ear.

stirrup: the third in a series of three small bones in the middle ear.

oval window: an opening between the middle ear and the inner ear.

cochlea: a spiral tube that makes up the part of the inner ear responsible for hearing.

cilia (singular, cilia): tiny hairs that line the cochlea and help turn vibrations into nerve impulses.

The middle ear starts at the inside of the eardrum. Inside the middle ear are the three smallest bones in the human body. They are the **hammer**, **anvil**, and **stirrup**. Each is named for its shape. The eardrum first transmits vibrations to the hammer. The hammer then transmits the vibration to the anvil, which in turn, transmits the vibration to the stirrup. Vibrations from the stirrup are transmitted to another membrane covering an opening called the **oval window**. The oval window separates the middle ear from the inner ear.

The largest structure in the inner ear is called the **cochlea**. The cochlea is a fluid-filled tube that is spiral, similar to a snail's shell. The cochlea is lined with thousands of tiny hairs, called **cilia**. These are not like hairs on your head. They are much smaller. They are connected to cells attached to nerve endings that line the surface of the cochlea. The cilia are very sensitive to the vibrations inside the cochlea. When vibrations are transmitted to the cochlea, the fluid vibrates, pushing the tiny cilia back and forth. The movements of the cilia are turned into nerve impulses. These nerve impulses travel along the auditory nerve to the part of the brain responsible for interpreting sound.

Cochlea with cilia.

Project-Based Inquiry Science AIA 130

channeled into the ear canal by the outer ear flap and then passes through the ear canal to the eardrum, a membrane similar to the surface of a drum. The membrane vibrates, and the vibration is transmitted to the hammer, anvil, and stirrup (three tiny bones in the middle ear). These bones transmit the vibration to a membrane covering the oval window, which separates the middle ear from the inner ear. The vibrations then pass to the fluid in the cochlea in the inner ear. The cochlea is lined with cilia, tiny hairs that are sensitive to vibrations and are connected to cells attached to nerve endings. The vibrations of the cilia are converted to nerve impulses, which travel along the auditory nerve to the brain.

Assessment Options

Targeted Concepts, Skills, and Nature of Science	How do I know if students got it?
The structure and function of animals' bodies are complementary and affect animal behavior.	**ASK:** What features of elephants' bodies affect how they communicate? **LISTEN:** Students should identify the parts of the elephants' bodies that are used to produce sound, their ears, their sense of smell, and the parts of their bodies that they use in visual communication.
Vibrations of molecules produce sound. Sound is a compression wave that can be described by its amplitude, frequency, and wavelength. Sound moves differently through different matter.	**ASK:** What creates sound? What are the characteristics of a sound wave? How does sound travel differently through solids, liquids, and gases? Does sound need to travel through something? **LISTEN:** Sound is created by a vibration that compresses and rarefies atoms and molecules. The characteristics of a sound wave are its wavelength (distance between compressions or between points where it repeats itself), frequency (number of cycles per second), and amplitude (amount of compression). Sound travels most quickly through solids, then liquids, and finally gases. Sound must travel through something that contains matter (atoms/molecules).
Animals' ears are adapted to hearing sounds in their environment. Some animals use sound that is out of the range of human hearing.	**ASK:** How do elephants' ears affect their communication? **LISTEN:** Elephants can hear sounds that are too low for humans to hear and they communicate in this range. They also can hear over long distances—as much as 10 km—so they are able to communicate with members of their group who are very far away.

Targeted Concepts, Skills, and Nature of Science	How do I know if students got it?
Scientists must keep clear, accurate, and descriptive records of what they do so they can share their work with others, consider what they did, why they did it, and what they want to do next.	**ASK:** How will the *Project Board* help to design an enclosure that encourages communication? **LISTEN:** Students should recognize that to design enclosures, they will need the relevant information available and organized in a useful way.
Scientists often work together and then share their findings. Sharing findings makes new information available and helps scientists refine their ideas and build on others' ideas. When another person's or group's idea is used, credit needs to be given.	**ASK:** How did sharing your explanations help you to develop valid claims? **LISTEN:** Students should have been able to help each other see where their evidence did not support their claims.
Scientific knowledge is developed through observations, recording and analysis of data, and development of explanations based on evidence.	**ASK:** What were the important steps you took to develop your explanations? How are these steps similar to what scientists do? **LISTEN:** Students should note that they observed elephants in the video, recorded observations, analyzed and interpreted their observations, and developed explanations. Similarly, scientists make observations, analyze and interpret their data, and develop explanations.
Explanations are claims supported by evidence. Evidence can be experimental results, observational data, and other accepted scientific knowledge.	**ASK:** How did you use your new science knowledge to revise your explanations? **LISTEN:** Students should have used the new science knowledge to support their claims, or to determine how their claims needed to be modified.

Teacher Reflection Questions

- What concepts in this section did students have difficulties with? What ideas do you have to assist their understanding these concepts?
- What difficulties did students have using the new science knowledge to revise their explanations? What ideas do you have to assist students with this?
- How did you manage the various reading segments in this section? What ideas do you have for next time?

NOTES

SECTION 3.6 INTRODUCTION

3.6 Explore

What Affects How Marine Mammals Communicate?

◀ *2 class periods**

*A class period is considered to be one 40 to 50 minute class.

Overview

Students observe dolphins' communication behavior and develop explanations of how dolphins communicate. Students begin by making predictions of how dolphins communicate by considering the physical characteristics and needs of dolphins. Then students observe dolphin behavior in a video, develop observation plans, and make detailed observations during their next viewing of the video. Working with their groups, they analyze their observations and interpretations, and present their results to the class. After all observations and interpretations are shared, groups develop explanations of how dolphins communicate, building on each other's work as scientists do. Groups then share their explanations.

Targeted Concepts, Skills, and Nature of Science	Performance Expectations
Scientists often work together and then share their findings. Sharing findings makes new information available and helps scientists refine their ideas and build on others' ideas. When another person's or group's idea is used, credit needs to be given.	Student groups should collaborate during an *Investigation Expo* to share their observations and interpretations with the class. Based on everyone's observations and interpretations, they should come up with an explanation for the communication behaviors of dolphins and share these with the class.
Scientists must keep clear, accurate, and descriptive records of what they do so they can share their work with others, consider what they did, why they did it, and what they want to do next.	Students should use *Observing and Interpreting Animal Behavior* pages to analyze their observations. Then they create and record explanations of dolphins' communication behaviors.
Tables are an effective way to communicate results of a scientific investigation.	Students use tables where appropriate in their *Investigation Expos* and to share their observations and interpretations of dolphin communication behaviors.

Targeted Concepts, Skills, and Nature of Science	Performance Expectations
Scientists differentiate between observations and interpretations. They use their observations and interpretations to explain animal behavior.	Students should clearly distinguish observations from interpretations they make after observing dolphin communication. Students should be able to describe the differences between observations and interpretations.
Scientists make claims (conclusions) based on evidence obtained (trends in data) from reliable investigations.	Students should construct interpretations and draw conclusions based on trends they find in their observations.
Explanations are claims supported by evidence. Evidence can be experimental results, observational data, and other accepted scientific knowledge.	Students should be able to support their interpretations based on their observations and science knowledge. These explanations should be revised as students gather more information.
Behavior is a type of response to internal or external stimuli. Behavior is determined by experience, physical characteristics, and environment.	Students should consider how stimuli in the environment affect dolphins' communication behavior.
The structure and function of animals' bodies are complementary and affect animal behavior.	Students should recognize that dolphins' physical features allow for the kinds of communication they observed.
Animals' ears are adapted to hearing sounds in their environment. Some animals use sound that is out of the range of human hearing.	Students should be able to describe sounds dolphins use to communicate.

Materials	
2 per student and 1 per group	*Observing and Interpreting Animal Behavior* pages
1 per student	*Create Your Explanation* pages
1 per group	Sticky notepad
10-15 per group	3" x 5" Index cards
1 per class	*Animals in Action* DVD *(How Do Dolphins Communicate?)*, and a way to view the video

Activity Setup and Preparation

Set up the DVD equipment before class. View the *How Do Dolphins Communicate?* video and make sure students will be able to see and hear the video from all viewing locations.

Homework Options

Reflection

- **Science Process:** How was your plan for observing dolphins similar to your plan for observing elephants? How was it different? What were some reasons for observing each the way you did? How well did your plans work? *(This question is meant to get students thinking about the important factors involved in creating an observation plan.)*
- **Science Process:** What were some differences between your explanations and other groups' explanations? What claims do you think you can support with the evidence presented by groups in the class? Would you need more information to support any of the claims presented by groups in the class? *(This question should encourage students to evaluate the explanations presented in class and to think about how to create scientific explanations.)*
- **Science Content:** Describe the similarities and differences between dolphin, elephant, bee, and human communications. (Students should describe similarities, like the needs to communicate and differences based on environmental and physical constraints.)

Preparation for 3.7

- **Science Content:** If you could ask a dolphin expert any questions about dolphin behavior to help explain your observations, what would they be? *(This question should engage students in thinking about what they have to learn.)*

NOTES

SECTION 3.6 IMPLEMENTATION

3.6 Explore

What Affects How Marine Mammals Communicate?

Animals communicate through sound, sight, touch, and smell. The puzzle activity you did, using verbal and non-verbal communication, showed the advantage humans have in using spoken language. However, you also observed some fascinating examples of how other animals can exchange important information using their own forms of communication.

The means of communication used by elephants and bees depends largely on their environment. Elephants can send sounds over long distances to keep in contact with other members of their extended family. Scientists think bees perform a waggle dance on the surface of their hive to communicate the distance and direction of a food source. There are many animals, however, that live in a different type of environment. They are **marine mammals,** or mammals that live in seawater. Sometimes their world is a dark and murky underwater environment. At other times, they are on the surface of the water where there is plenty of sunlight.

marine mammal: a mammal that lives in the sea and/or gets its food from the sea.

pod: a social group of whales or dolphins. Members of a pod may protect one another.

Marine mammals, like dolphins and whales, live in groups, called **pods**. They swim long distances through the ocean in these groups. They rise to the surface of the water to breathe oxygen, and then dive deep into the water to feed. Like elephants, bees, and people, marine mammals need to communicate to their group members their location and the location of food.

Dolphins live in groups, called pods.

AIA 131

ANIMALS IN ACTION

◀ *2 class periods**

3.6 Explore

What Affects How Marine Mammals Communicate?

5 min.

Discuss characteristics of dolphins with the class.

○ Engage

Begin by eliciting students' ideas about what a mammal is and if there are any mammals that live in water. Record students' ideas.

Describe to students that a mammal is an animal with a vertebrate, has fur or hair, usually gives live birth, and that mother mammals produce milk for their newborns.

Explain that marine mammals are mammals that live in seawater and dive deep below the ocean's surface, traveling far under water and only coming up for oxygen. If students did not mention whales and dolphins as marine mammals, do so now.

*A class period is considered to be one 40 to 50 minute class.

PBIS Learning Set 3 • What Affects How Animals Communicate?

Predict

5 min.

Have groups make predictions about how dolphins communicate.

Predict

Working by yourself, think about where marine mammals live and the reasons why they need to communicate. Make a list of marine mammals' communication needs. Then get together with your group and develop a group list of marine-mammal communication needs.

Dolphins must be able to communicate on the surface and underwater.

Work together with your group to predict how you think dolphins might communicate with other dolphins. Remember that environment plays an important role in communication. Identify the constraints the dolphins' bodies and environment place on dolphins' forms of communication. Consider the fact that sometimes dolphins swim at the top of the water, where there is plenty of sunlight, and sometimes they swim much deeper in the ocean, where it can be very dark and hard to see.

Observe Dolphin Communication

A baby dolphin relies on touch and sound to remain close to its mother.

Once again, you will watch a video to observe animal behavior. You will watch a video showing dolphins communicating. As before, you will watch the video twice. The first time, you will be watching it to come up with an observation plan. The second time, you will watch more carefully, using your plan, and then work with your group to identify all the different forms of dolphin communication.

Watch the video. Pay attention to the behavior of the dolphins. Try to figure out what they are doing and what behaviors might be communication. Notice if all the dolphins are behaving the same way.

Next, elicit students' ideas about how marine mammals communicate. Emphasize that they live in groups (pods) and need to communicate with the members of their pods. Record students' ideas.

△ Guide

Let students know that they should list as many reasons that they can think of why marine mammals might need to communicate. Then they will meet with their groups to discuss their ideas and construct their group prediction of how dolphins might communicate.

Get Going

Let students know how much time they have and then have groups meet for discussion.

Guide and Assess

Monitor groups and check to see if students are considering the constraints of the dolphins' bodies and environment.

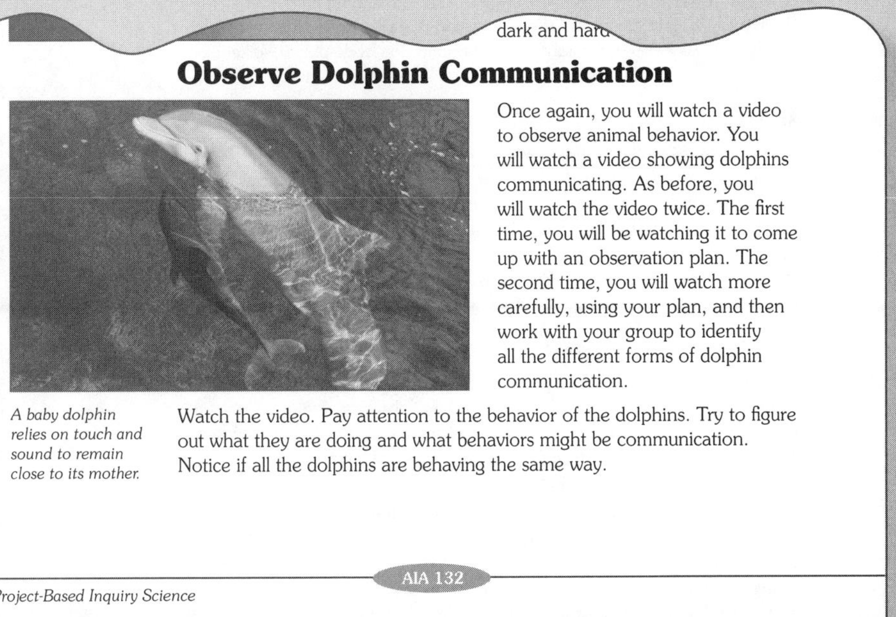

dark and hard

Observe Dolphin Communication

Once again, you will watch a video to observe animal behavior. You will watch a video showing dolphins communicating. As before, you will watch the video twice. The first time, you will be watching it to come up with an observation plan. The second time, you will watch more carefully, using your plan, and then work with your group to identify all the different forms of dolphin communication.

A baby dolphin relies on touch and sound to remain close to its mother.

Watch the video. Pay attention to the behavior of the dolphins. Try to figure out what they are doing and what behaviors might be communication. Notice if all the dolphins are behaving the same way.

AIA 132

Project-Based Inquiry Science

Observe Dolphin Communication

5 min.

Have students view the How Do Dolphins Communicate? *video.*

Guide

Let students know that they will observe dolphins by watching a video of dolphins in their natural habitat. Emphasize that they should not be writing down detailed observations yet. They should be thinking about what they should observe so that they can construct an observation plan that they will follow the next time they view the video.

Then show the dolphin video.

Plan

10 min.

Have groups construct observation plans.

Plan

You watched a video of dolphins communicating. When you watch the video again, you will need to make detailed observations about how dolphins communicate. Make an observation plan with your group. When you observed elephant communication, you also developed a plan. You might use a similar plan this time. Create your plan, keeping in mind that you need to observe as much of the scene as possible, and that your observations will determine how well you understand the communication of dolphins. Make sure that part of your plan includes using a page to record your notes.

Observe

Once you have your plan written and all the members of your group understand the plan, watch the video again. Remember to take notes about the dolphins' behavior and habitat. Pay attention to the dolphins' bodies. Think about how their body structure helps them to communicate. How does a dolphin's habitat affect how it communicates? Remember to follow the plan your group decided on. Record your observations so you will be able to share them with your group and your class.

Analyze Your Data

The video you watched showed a group of dolphins. How did you observe them communicating? You may have predicted and observed that dolphins make different sounds. These sounds communicate a variety of things to other dolphins. This is one way other mammals, such as people, communicate. You may also have seen the dolphins communicate in other ways.

Meet with your group to share your observations. Each member of the group should have a chance to share all their observations. Remember that, depending on your observation plan, you may not have been watching what the other group members were watching. Other group members may have a lot of information to share with you. Listen carefully for observations you might not have seen. Create a group list of observations that everyone agrees on. Make another list of observations that were not agreed on. Save the observations you could not agree on for the *Investigation Expo*.

Using the same procedure you used while analyzing the behavior of elephants (in *Section 3.4*), create a sticky note for each observation you all agreed on. Organize your observations into groups based on the type of communication behavior you observed. For example, you might have seen dolphins touch. If you think that some behaviors might fall into two groups, you can list those behaviors twice.

META NOTES

Students will use their observations to create an explanation of dolphins' communication behavior based on the form and function of the dolphins' bodies and on their habitat.

△ Guide

Briefly discuss some things students should think about as they develop their observation plans such as what observation methods were effective in previous sections, what questions they want to answer with their observations, and how they want to work with the other members of their groups to make detailed observations. Emphasize that they should observe dolphins' characteristics and environment as well as their behaviors.

Remind students that they will be presenting their observations and interpretations to the class when they have completed them.

⬡ Get Going

Let groups know how much time they have to develop their observation plans. Remind students that their goal is to gather information to answer the question and to record their data with as much detail as possible.

☐ Assess

Monitor groups and check what they plan to observe. They should plan to observe dolphins' physical characteristics and the environmental factors that might affect how they communicate. Look for gaps in students' plans and help them identify ways in which they could tighten up their plans so they can collect accurate, detailed data.

Observe

Once you have your plan written and all the members of your group understand the plan, watch the video again. Remember to take notes about the dolphins' behavior and habitat. Pay attention to the dolphins' bodies. Think about how their body structure helps them to communicate. How does a dolphin's habitat affect how it communicates? Remember to follow the plan your group decided on. Record your observations so you will be able to share them with your group and your class.

Observe

5 min.

Have students observe dolphins' communication behavior.

⬡ Get Going

Remind students to follow their observation plans and show the video again. If needed, show the video more than once.

Analyze Your Data

15 min.

Have groups discuss, analyze, and categorize their observations and interpretations.

the pla... on. Record y... you wi... able to sha... with your group and your cla...

Analyze Your Data

The video you watched showed a group of dolphins. How did you observe them communicating? You may have predicted and observed that dolphins make different sounds. These sounds communicate a variety of things to other dolphins. This is one way other mammals, such as people, communicate. You may also have seen the dolphins communicate in other ways.

Meet with your group to share your observations. Each member of the group should have a chance to share all their observations. Remember that, depending on your observation plan, you may not have been watching what the other group members were watching. Other group members may have a lot of information to share with you. Listen carefully for observations you might not have seen. Create a group list of observations that everyone agrees on. Make another list of observations that were not agreed on. Save the observations you could not agree on for the *Investigation Expo*.

Using the same procedure you used while analyzing the behavior of elephants (in *Section 3.4*), create a sticky note for each observation you all agreed on. Organize your observations into groups based on the type of communication behavior you observed. For example, you might have seen dolphins touch. If you think that some behaviors might fall into two groups, you can list those behaviors twice.

AIA 133 ANIMALS IN ACTION

PBIS Learning Set 3 • *What Affects How Animals Communicate?*

It is time to think about what your observations mean. Think about when dolphins make different sounds, and what the different sounds might mean. Think about the different ways dolphins use their bodies to send information to each other, and what those body movements might mean. These interpretations will help you as you begin to develop explanations. Using an *Observing and Interpreting Animal Behavior* page, record the observations your group agreed on in the left column. Include one behavior in each row of the chart. Then add to the middle column what you know about what allows these behaviors, and record your interpretations of these behaviors in the right column. Collaborate with the members of your group to create one chart. Be prepared to share your observations and interpretations with the class.

△ Guide

Let students know that while they share their observations and interpretations with their group members, they should be thinking about how they knew the dolphins were communicating.

Then describe for students how they will analyze the data using the information in the student text.

TEACHER TALK

"To analyze your data, you will first be sharing your observations with your group members. Remember that you may not have observed the same things.

Then, you should create a list of all the observations your group agrees on. Using this list, create a sticky note for each observation and group the observations into categories. One category might be how the dolphins use sound. These categories should identify types of communication.

With your group, record the observations and interpretations you agree upon in an *Observing and Interpreting Animal Behavior* page. Be prepared to share these with the class."

○ Get Going

Distribute sticky notepads, index cards, and *Observing and Interpreting Animal Behavior* pages and let students know how much time they have for their analysis.

△ Guide and Assess

Monitor students and identify places where students have confused observations and interpretations. Note where they disagree about observations, or where they have many observations in common, which may indicate that they have created and implemented a good observation plan.

Discuss the groups' lists and help students identify interpretations versus observations, faults in the plan that may have affected the data collection, and difficulties they had with making their observations.

As groups are categorizing their observations, note the different categories groups are using. Check whether groups are identifying any inconsistencies in their results and problems with their procedures. Keep track of the differences in groups' interpretations and categories so that they can be discussed during the *Investigation Expo*.

META NOTES

During this discussion, students may require clarification of what they have seen. They may request to watch the video again, especially the parts they were uncertain about. Provide time for students to check their observations and interpretations in this way.

Communicate

35 min.

Have a discussion on the groups' presentations of their observations and interpretations.

It is time to think about what your observations mean. Think about when dolphins make different sounds, and what the different sounds might mean. Think about the different ways dolphins use their bodies to send information to each other, and what those body movements might mean. These interpretations will help you as you begin to develop explanations. Using an *Observing and Interpreting Animal Behavior* page, record the observations your group agreed on in the left column. Include one behavior in each row of the chart. Then add to the middle column what you know about what allows these behaviors, and record your interpretations of these behaviors in the right column. Collaborate with the members of your group to create one chart. Be prepared to share your observations and interpretations with the class.

Communicate

Investigation Expo

You will share your observations and data analysis in an *Investigation Expo*. In your presentation, you will share what you have seen with the class. Remember that each group created their own plan for making their observations and analyzing their data. Some groups' observations and interpretations may be different from yours.

For your presentation, create a poster that describes each of the following:

- the questions you were trying to answer in your observations
- your observation procedure and how it helped you make observations
- your analysis of your observations and how confident you are about the analysis

Add to the bottom of your poster the list of observations your group did not agree on.

Before the presentations, read these questions. Plan your group's presentation, making sure you will answer all of the questions. Then as each group is presenting, listen for the answers to these questions. If a group does not answer all the questions, ask them to help you understand what they observed and why they interpreted behaviors the way they did. When you ask your questions, focus on better understanding the observations the group made.

- What was the group trying to find out?
- What procedure did they use to collect their data?
- Were they able to make clear observations that were detailed?

META NOTES

It is important for the group and the class to notice differences in observations, interpretations, and procedures. The questions for the *Investigation Expo* will help focus students on these points.

- How did they group their observations? What did their groupings allow you to see?
- What conclusions do their results suggest?
- Do you trust their observations? Why or why not?

△ Guide

Let students know that they will be presenting their results in an *Investigation Expo* and that each group will need to create a poster of their results and present it to the class. Emphasize that students posters and presentations should address the first set of bulleted items in the student text. Also, let students know that they should list the observations they did not agree on at the bottom of their poster.

⬡ Get Going

Distribute poster materials and have groups make posters for the *Investigation Expo*. Let students know how much time they have.

△ Guide

As groups are working on their posters and preparing for their presentations, assist them as needed. Some issues that may arise are:

- Students might include opinions or interpretations in their observations. Ask them if anyone could argue with their observations. If any of the observations are debatable, they probably include opinions or interpretations.
- Determining confidence levels may still be difficult for students. Several issues can affect their confidence. They may think they were unable to record their observations quickly enough or with enough detail. They may also have confused observation and interpretation. This can either raise or lower their confidence. Finally, students may suggest that their confidence is high, even though it is clear that the observation plan was not effective or was not followed. Be cautious of students' stating more confidence than they should have in their results. Ask students if the observations on their posters give a clear picture of what they observed and how they observed it.

◇ Evaluate

Make sure groups have included responses for the bulleted points in the student text on their posters.

△ Guide

Let students know that before anyone presents, they will have the chance to familiarize themselves with each group's work and formulate questions if needed. Emphasize to the class that they will need to be able to answer each of the bulleted items in the second set of bullets for each group's results.

⬡ Get Going

Have students display their posters around the room and let students visit each poster for no more than a minute.

△ Guide

Again, emphasize to the class that they will need to be able to answer each of the bulleted items in the second set of bullets for each presentation. Then, let the groups present. Have a brief discussion after each presentation Encourage students to lead the discussion.

□ Assess

As you listen to groups' presentations, focus on the way students have used their observations to support their claims (interpretations). Identify areas of strength and weakness in students' observational plans. Note any interpretations mixed in with the observations and look for interpretations that are supported by observations.

△ Guide

After all the groups have presented, have a class discussion looking for similarities in the observations and interpretations from all the presentations. Also discuss major differences. Ask students to consider what these similarities and differences might indicate.

META NOTES

When students are presenting their posters, they should be presenting to each other, not to the teacher. By moving away from the presenters or sitting with the audience, you will encourage students to speak to the entire class.

Provide sufficient time for students to ask questions about the data analysis and conclusions (interpretations). If students are struggling, use some model questions to support thinking about the posters in a different way. By modeling questions about content, you will also assist students in asking better questions.

META NOTES

If needed, show the video again to resolve disputes in observations and/or interpretations.

- How did they group their observations? What did their groupings allow you to see?
- What conclusions do their results suggest?
- Do you trust their observations? Why or why not?

Explain

Once again, using your observations and interpretations, create an initial explanation of what affects dolphin communication. Your explanation of dolphin communication behavior must bring together the evidence (observations), the interpretations the class agreed on, and your science knowledge. Using a *Create Your Explanation* page, develop your explanation of why dolphins communicate as they do. Remember that the best explanations help others understand what makes a claim valid. At this point, your science knowledge may be limited. In the next section, you will read more about dolphins, and you will have an opportunity to improve your explanations.

Communicate

Share Your Explanation

Share your explanation with the class. As each group shares their explanation, pay special attention to how the other groups have supported their claims with science knowledge. Ask questions or make suggestions if you think a group's claim is not as accurate as it could be or if the group has not supported their claim well enough with observations and science knowledge. Remember that the best explanations help others understand what is true about the world that makes the claim trustworthy.

What's the Point?

Marine mammals, such as dolphins, communicate in very different ways than other animals. Dolphins live in a different type of environment than humans, bees, and elephants. The ocean can be a dark and murky place deep underwater or a sunny place near the surface. Dolphin communication must adapt to the type of environment.

While observing dolphins communicating, once again it is important to make observations first then interpret the observations. This is how scientists study animal behavior. They are aware of the difference between observations and interpretations.

Explain

10 min.

Have students create and discuss explanations of the behaviors they observed.

△ Guide

Now let students know that they will be constructing another explanation. This explanation will be on what affects dolphin communication. As in previous explanations, they should construct a claim (which may be one of their interpretations or a combination of their interpretations) that is supported by their observations and science knowledge. Emphasize that explanations are comprised of a claim (a conclusion or interpretation) supported by evidence (observations), and science knowledge (ideas accepted in the scientific community) in a logical way.

△ Guide and Assess

As groups are working on their explanations, go around the room and ask students what their claims are and what evidence they are using to support their claims. Their claims should be valid (based on the evidence) and their evidence should not include any opinions.

Example:

Dolphins communicate by making a variety of sounds. They do this because their bodies are adapted to making sounds under water. Our evidence is that dolphins make sound continuously until they leap out of the water. Once they are in the air, they are silent.

□ Assess

As groups are working on their explanations, ask students what their claims are and what evidence they are using to support their claims. At this point, they do not have much evidence (observations) to support their claims and they do not have a lot of science knowledge they can use. Note any difficulties students are having with this.

Communicate

10 min.

Have students share their explanations.

dolphins, and you will have an opportunity to improve your explanations.

Communicate

Share Your Explanation

Share your explanation with the class. As each group shares their explanation, pay special attention to how the other groups have supported their claims with science knowledge. Ask questions or make suggestions if you think a group's claim is not as accurate as it could be or if the group has not supported their claim well enough with observations and science knowledge. Remember that the best explanations help others understand what is true about the world that makes the claim trustworthy.

⬡ Get Going

Remind groups that they should be prepared to defend their explanations, and to discuss how they revised them after hearing everyone's explanations.

Have groups present their explanations to the class about dolphin communication. Ask the class to help pick out their claims, their evidence, and their reasons.

△ Guide

As groups present, students should ask questions to clarify and point out where something seems to be missing from an explanation. Model how you expect students to ask questions by asking about how the evidence in an explanation supports the claim.

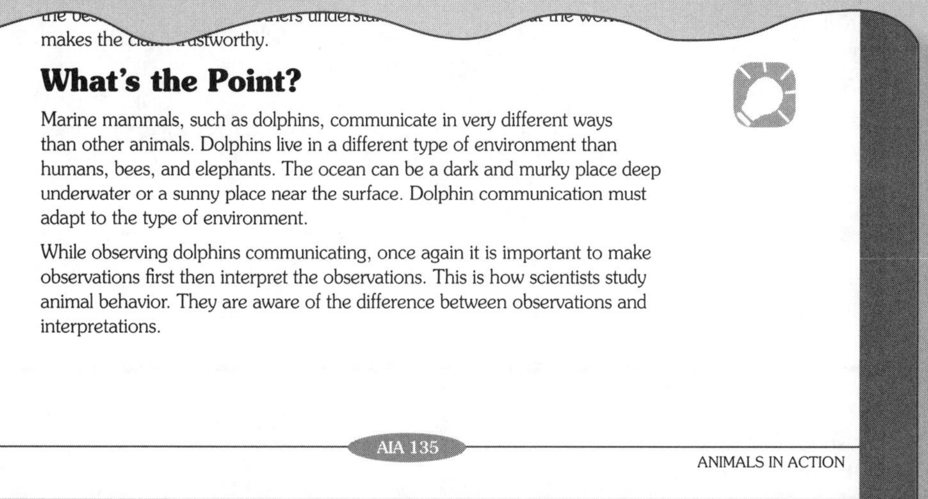
makes the claim trustworthy.

What's the Point?

Marine mammals, such as dolphins, communicate in very different ways than other animals. Dolphins live in a different type of environment than humans, bees, and elephants. The ocean can be a dark and murky place deep underwater or a sunny place near the surface. Dolphin communication must adapt to the type of environment.

While observing dolphins communicating, once again it is important to make observations first then interpret the observations. This is how scientists study animal behavior. They are aware of the difference between observations and interpretations.

AIA 135

ANIMALS IN ACTION

What's the Point?

5 min.

Point out to students that they carefully made observations first and then they interpreted their observations. This is how biologists study and interpret animal behavior. They also collaborated with their groups and the class as they made observations, interpreted them, and created explanations, just as scientists collaborate to develop explanations.

Assessment Options

Targeted Concepts, Skills, and Nature of Science	How do I know if students got it?
Behavior is a type of response to internal or external stimulus. Behavior is determined by experience, physical characteristics, and environment.	**ASK:** Dolphin communication might be a response to what things in the environment? **LISTEN:** Students should recognize that communication in all animals is often a response to danger or to the opportunity for food in the environment.

Targeted Concepts, Skills, and Nature of Science	How do I know if students got it?
The structure and function of animals' bodies are complementary and affect animal behavior.	**ASK:** What features of dolphins' bodies affect how they communicate? **LISTEN:** Students should identify the parts of the dolphins' bodies that are used to produce sound, such as the trunk, and they should also identify the ears, which allow dolphins to hear but which can also be moved to communicate.
Animals' ears are adapted to hearing sounds in their environment. Some animals use sound that is out of the range of human hearing.	**ASK:** What features of dolphins' bodies affect how they communicate? **LISTEN:** Students should identify dolphins' hearing as an important adaptation that allows them to communicate.
Scientists must keep clear, accurate, and descriptive records of what they do so they can share their work with others, consider what they did, why they did it, and what they want to do next.	**ASK:** How did you use your observations to create explanations? **LISTEN:** Students should have used their observations as evidence to support their claims.
Scientists often work together and then share their findings. Sharing findings makes new information available and helps scientists refine their ideas and build on others' ideas. When another person's or group's idea is used, credit needs to be given.	**ASK:** How did sharing your ideas in *Investigation Expos* help you create explanations? **LISTEN:** Students should have used the observations and interpretations of the class to create their explanations.

Targeted Concepts, Skills, and Nature of Science	How do I know if students got it?
Tables are an effective way to communicate results of a scientific investigation.	**ASK:** How did, or could you have used tables in your presentations? What advantages do tables have? **LISTEN:** Students should describe how tables make it easy to see relationships between the ideas or observations presented.
Scientists differentiate between observations and interpretations. They use their observations and interpretations to explain animal behavior.	**ASK:** Did you discover that there were interpretations mixed in your observations? How did you identify them? **LISTEN:** Students should have recognized that anything that was debatable was probably an interpretation.
Scientific knowledge is developed through observations, recording and analysis of data, and development of explanations based on evidence.	**ASK:** What were the important steps you took to develop your explanations? How are these steps like what scientists do? **LISTEN:** Students observed dolphins in the video, recorded observations, analyzed and interpreted their observations, and developed explanations. Similarly, scientists make observations, analyze and interpret their data, and develop explanations.
Explanations are claims supported by evidence. Evidence can be experimental results, observational data, and other accepted scientific knowledge.	**ASK:** How did you support the claims in your explanations? **LISTEN:** Students should have used evidence from their observations to support their claims.

Teacher Reflection Questions

- What difficulty did students have observing dolphin behavior?
- What difficulties did students have in creating explanations? What can you do to help students with this?
- How was managing the explanatory presentations and discussions different from managing the *Investigation Expo?* What might you do differently next time?

NOTES

SECTION 3.7 INTRODUCTION

3.7 Read

How Do Dolphins Communicate?

◀ *1 class period**

*A class period is considered to be one 40 to 50 minute class.

Overview

Students read and discuss the ways dolphins communicate and how their communications are determined by their physical characteristics and environment. They read about how dolphins use echolocation and have signature whistles that allow them to identify one another and find one another. They also read that dolphins communicate visually and by touch, as well as with sound. Using the science knowledge introduced in this section, students revise their explanations. Students then share their explanations with the class, create a class explanation about how dolphins communicate, and then update the *Project Board*.

Targeted Concepts, Skills, and Nature of Science	Performance Expectations
Scientists often work together and then share their findings. Sharing findings makes new information available and helps scientists refine their ideas and build on others' ideas. When another person's or group's idea is used, credit needs to be given.	Students should work in small groups to revise their explanations. They then share these with the class and with the class create a final explanation of how dolphins' physical features allow them to communicate.
Scientists must keep clear, accurate, and descriptive records of what they do so they can share their work with others, consider what they did, why they did it, and what they want to do next.	Students should use their previous records to assist them in revising their explanations. They should also keep accurate and detailed records to refer back to when they design their zoo enclosure or critique others'.
Scientific knowledge is developed through observations, recording and analysis of data, and development of explanations based on evidence.	Students should recognize that the things they learned and recorded on the *Project Board* were developed through observation, reading, analysis, and creating explanations.

Targeted Concepts, Skills, and Nature of Science	Performance Expectations
Explanations are claims supported by evidence. Evidence can be experimental results, observational data, and other accepted scientific knowledge.	Students should revise their explanations about what affects dolphins' communication behaviors.
The structure and function of animals' bodies are complementary and affect animal behavior.	Students should recognize that dolphins' physical features allow them to communicate using sound, touch, and sight.
Vibrations of molecules produce sound. Sound is a compression wave that can be described by its amplitude, frequency, and wavelength. Sound moves differently through different matter.	Students should be able to describe that sound needs a medium to travel through and that it travels more quickly through water than air.
Animals communicate with other animals using sound. The sounds they can make and the sounds they can hear are adaptations to their environment.	Students should describe how dolphins make sounds to communicate.
Animals' ears are adapted to hearing sounds in their environment. Some animals use sound that is out of the range of human hearing.	Students should describe how dolphins have sensitive hearing that they use to help them locate their young, find food, gather information about objects through echolocation, and communicate with each other.
Animals' sense of sight is adapted to their environment. Some animals see things that humans cannot see.	Students should describe how echolocation uses sound waves to map out where things are located or to "see" where objects are. Students should be able to describe how these are used.

Materials	
2 per student	*Create Your Explanation* pages
1 per class	*Project Board*
1 per student	*Project Board* pages
1 per class	*Animals in Action* DVD *(How Do Dolphins Communicate?)* and a way to view the video

Activity Setup and Preparation

Consider doing an Internet search on dolphin communication. It may be helpful to look for information on dolphin research center Web sites.

Homework Options

Reflection

- **Science Content:** Why is echolocation an effective way for dolphins to find things in their environment? *(In parts of the ocean where there isn't enough light to see well, dolphins need an alternative way to sense things in their environment. They have excellent hearing, and are able to judge how near or far something is and what its shape and speed are based on how long it takes sound to bounce off of it and return.)*

Preparation for BBC

- **Science Content:** What have you learned about dolphin communication that you could apply to designing an enclosure that would encourage dolphins to communicate? *(This question should help students connect what they're learning to the* Big Challenge.)

NOTES

SECTION 3.7 IMPLEMENTATION

3.7 Read

How Do Dolphins Communicate?

Sometimes, when people cannot see each other but know that someone is close enough to hear, they use their voices. The sound of a human, or an elephant, travels well through the air. When marine mammals communicate, the sounds they make must be able to travel well through the water. Marine mammals have developed many different ways to communicate in a water environment.

How Do Dolphins Use Sound to Communicate?

To survive, dolphins must be able to keep in touch with other members of their pod, identify and avoid obstacles and predators in the ocean, and find food. Their environment requires that they communicate in ways other than just visual communication. The ocean can be dark and murky, and finding food, ocean hazards, or other dolphins through sight alone is not always possible.

Dolphins must be able to communicate in dark, murky areas of the ocean.

Project-Based Inquiry Science AIA 136

◀ *1 class period**

3.7 Read

How Do Dolphins Communicate?

10 to 15 min.

Guide students through the science of dolphin communication.

META NOTES

Using different reading strategies may keep students focused. Consider having students read aloud, reading in groups, summarizing paragraphs, pulling out key ideas, or reading to them. Keep in mind the reading levels, needs of your students, and class temperament when picking a reading strategy.

△ Guide

Begin by letting students know that they will now learn more about how dolphins' bodies allow them to communicate and that this will help them explain some of the communication behaviors they have observed. Let students know that they will have a chance to revise their explanations using what they have learned about dolphin communication.

*A class period is considered to be one 40 to 50 minute class.

TEACHER TALK

"You have observed and explained some ways dolphins communicate, but our observations from the video did not provide us with a lot of information about dolphin communication. Scientists have done numerous and lengthy observations and measurements concerning dolphin communication and have learned many things about the ways dolphins communicate. They have learned that dolphins make a number of different sounds. Did any of you pay attention to the different sounds that dolphins made in the video you watched? What were the sounds that you heard? How many sounds do you think dolphins make?"

META NOTES

Consider showing the video again and let students pick out the number of different sounds they hear.

Then, let students know that bottlenose dolphins make sounds described as clicks, moans, trills, grunts, squeaks, creaks, and whistles, and they make these sounds at any time and at considerable depths. The sounds vary in volume, wavelength, frequency, and pattern.

How Do Dolphins Use Sound to Communicate?

5 to 10 min.

Discuss how dolphins use echolocation and signature whistles.

How Do Dolphins Use Sound to Communicate?

To survive, dolphins must be able to keep in touch with other members of their pod, identify and avoid obstacles and predators in the ocean, and find food. Their environment requires that they communicate in ways other than just visual communication. The ocean can be dark and murky, and finding food, ocean hazards, or other dolphins through sight alone is not always possible.

○ Engage

Ask students why they think dolphins made clicks and whistles. Record their ideas.

△ Guide

Discuss how dolphins' survival depends on communicating to keep track of other members of their pod, to identify and avoid obstacles and predators, and to find food. Then let students know that dolphins have very good eyesight, but as they go deeper in the ocean, there is not much light. Dolphins use sound not only to communicate, but also to "see."

Then introduce echolocation using the information in the student text. Emphasize that one of the sounds dolphins make, known as the sonar click, is used for echolocation. Let students know that sonar stands for SOund NAvigation and Ranging. Then describe how a dolphin emits a clicking sound, the sound projects through the water, hits objects in the water, and is reflected back to the dolphin. The dolphin hears the reflected sound and is able to learn about the size, shape, speed, distance, and direction of objects in the water from hearing the returning sound wave and taking into account the time it takes the sound wave to return.

Dolphins use clicks, whistles, squeaks, and trills to communicate. These kinds of sounds travel well through water. One specific type of click is called a "*sonar* click." (Sonar stands for "SOund NAvigation and Ranging." You will read more about sonar later in this section.) Sonar clicks allow dolphins to communicate with **echolocation**. Echolocation works just like an echo. When a dolphin makes a sound in the water, the sound waves move through the water and hit an object. The sound waves then bounce off the object and travel back to the dolphin. The dolphin hears the returning sound wave. The time it takes for the sound wave to travel out and come back gives the dolphin information about the size, shape, speed, distance, and direction of objects in the water. Echolocation is very accurate. Dolphins are so good at using echolocation that they can even use it to tell the difference between types of fish.

echolocation: a method used by dolphins and some other animals such as bats, to locate objects. The animal sends out sound waves that bounce off the object. The returning sound (echo) is interpreted to determine the shape and the location of the object.

In water, the molecules are closer than they are in air, therefore sound travels almost five times faster in water than it does in air. This helps dolphins receive information more quickly than if they were sending the sounds through the air.

Marine mammals have excellent hearing. The ocean is a noisy place. Many creatures make sounds. People have also added many sounds to the ocean. Dolphins can distinguish those sounds from one another. They can hear the echoes of their own clicks, and they can hear the clicks of other dolphins. They can use the clicks of other dolphins to find them. By using their excellent hearing, dolphins can find food, avoid obstacles, find other dolphins, and avoid some dangers.

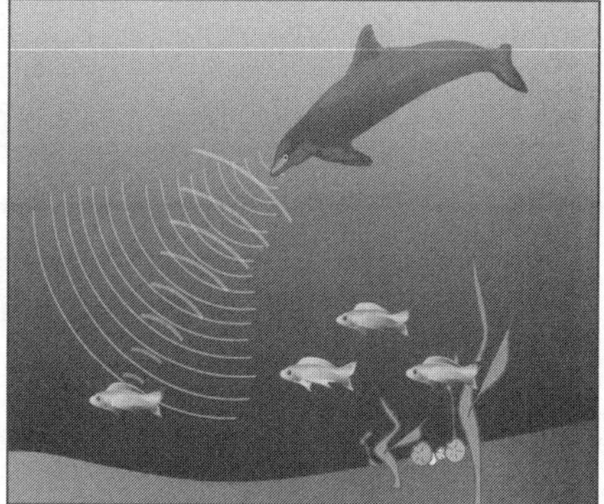

Dolphins use echolocation to find fish.

Dolphins use sound in other ways, too. Each individual dolphin has its own "signature whistle." A mother dolphin will repeat this signature whistle over and over to her newborn. The calf becomes able to identify its mother's special sound. When a dolphin mother is separated from her calf, she whistles her sound. Because the calf can recognize the sound, it knows where to find its mother.

Remind students that sound travels faster in liquids than gas and so it travels faster in the ocean than in the air. Describe how this helps dolphins locate objects, such as food, quickly. Describe the signature whistle and, if needed, show the video again.

META NOTES

How humans use echolocation is discussed later in this section.

TEACHER TALK

"Dolphins have many learned behaviors. One behavior they learn early is to listen to their mother's special whistle, called a *signature whistle.* The mother dolphin will repeat her signature whistle to her newborn until the newborn recognizes it. If the mother and baby are separated she uses the whistle so that her calf will be able to find her."

What Are Some Other Ways Dolphins Communicate?

5 min.

Discuss how dolphins use sight and touch to communicate.

What Are Some Other Ways Dolphins Communicate?

Dolphins are mammals and must breathe air to survive. To get air, they must return to the water surface. In the brighter ocean near the surface, dolphins often use sight and touch, as well as sound, to communicate with one another. Dolphins can be seen leaping high into the air. They also slap their flippers or tails on the surface of the water. Using this body language, dolphins can alert others to danger and possible food sources, or can tell others they want to play.

*Dolphins often jump high above the water surface and fall back with a loud splash. This is called **breaching**. Scientists interpret this behavior as a form of communication.*

Touch is an important part of communication for dolphins. Scientists have observed dolphins rubbing, petting, or hitting one another. Often, they maintain contact with one another while they are swimming. When dolphins meet, they may rub fins.

breaching: a behavior seen in marine mammals and some fish, where the animal jumps high above the water surface and falls back with a splash.

sonar: a technique that uses sound to provide images of objects that are under water.

oceanographer: a scientist who studies the ocean.

Stop and Think

1. What are some ways dolphins communicate with one another?
2. People communicate about a variety of things. What do you think are some things dolphins communicate about?
3. How does a dolphin's environment affect how it communicates?

Humans also Use Echolocation

Echolocation is a very important tool for marine mammals and some other mammals, such as bats. Scientists also use echolocation, called **sonar**. **Oceanographers** are scientists who study the ocean.

△ Guide

Using the student text, discuss how dolphins use visual and tactile information in communication. Describe how dolphins use gestures and vision to communicate. Breaching, when a dolphin jumps high above the water's surface and falls back with a splash, is interpreted as communication by scientists. They also slap their flippers or tails on the water's surface. They can use this kind of communication to alert other dolphins to danger and food sources or to invite others to play.

Then describe some ways dolphins use touch such as rubbing, petting, or hitting one another to communicate. Point out that dolphins often maintain contact with each other while swimming and sometimes rub fins when they meet.

marine mam... and some fish, where the animal jumps high above the water surface and falls back with a splash.

sonar: a technique that uses sound to provide images of

Stop and Think

1. What are some ways dolphins communicate with one another?
2. People communicate about a variety of things. What do you think are some things dolphins communicate about?
3. How does a dolphin's environment affect how it communicates?

Stop and Think

10 min.

Have a class discussion of the Stop and Think *questions.*

Get Going

Let groups know that they should work together to answer the questions and that if they cannot agree on an answer, they should list all their answers. Then, let students know they will be sharing their answers during a class discussion. Let students know how much time they have to answer the questions.

Guide and Assess

Once groups have answered the questions, initiate a discussion of some of the ways dolphins communicate, how dolphins' needs, and their environments affect their communication. Ask groups what their answers to the *Stop and Think* questions were. Listen for the following responses:

1. Students should identify various ways dolphins communicate with sounds such as clicks and whistles. Students should describe how dolphins use gestures and sight to communicate (breaching, slapping flippers, or tails on the surface of the water) to indicate danger, food, or play. Students should also describe how dolphins use touch such as rubbing flippers, petting, and hitting to communicate. An example is when dolphins rub fins to greet each other.

2. Students should describe dolphins' need to keep track of other members of their pods, especially their young. Dolphins also need to communicate to help each other locate food and avoid danger.

3. Students should describe how the difficulty of seeing deep in the ocean hinders visual communication. Since sound travels well in water, it makes a good way for dolphins to communicate. In addition, dolphins are able to use sound in echolocation to determine what is in their environment where it might be difficult to see.

Humans also Use Echolocation

5 min.

Discuss how echolocation is used by humans.

uses sound ... provide images of objects that are under water.

oceanographer: a scientist who studies the ocean.

Humans also Use Echolocation

Echolocation is a very important tool for marine mammals and some other mammals, such as bats. Scientists also use echolocation, called **sonar**. **Oceanographers** are scientists who study the ocean.

AIA 138

Project-Based Inquiry Science

Discuss how echolocation is used by humans using the information in the student text.

TEACHER TALK

"Sonar *(Sound Navigation and Ranging)* is a kind of echolocation, which allows scientists to make images of things deep in the sea, such as the ocean floor. They send sound waves from the surface of the water and measure the time it takes for each sound wave to return. When the measurements are put together, they create an image of what is on the ocean floor. This is very similar to the way dolphins emit clicks and use the time it takes a click to return to determine what is in their environment. Humans have been using sonar since World War I to intercept enemy vessels, image other ships, and locate obstacles in the water.

Sonar has been linked to the deaths of some marine mammals. There have been several cases of whales beaching themselves after being exposed to sonar."

3.7 **Read**

Because parts of the ocean are very deep and dark, these scientists cannot explore it themselves. However, they can make accurate images of what the ocean floor looks like using sonar. This allows them to study underwater mountains and ridges. Sonar use also allows oceanographers to find sunken ships.

Scientists use sonar to map the floor of the ocean.

As mentioned, sonar stands for "SOund NAvigation and Ranging." Scientists send out sound waves from the water and measure the time it takes for each one to return. Each measurement is recorded. When the measurements are put together, they create an image of what is on the ocean floor. Sonar has expanded scientists' ability to see where they have never seen before.

Sonar can also be used by ships to see other ships, find obstacles in the water, and even find fish. Since World War I, in 1915, nations have used sonar on naval ships to find and intercept enemy vessels. Unfortunately, these uses of sonar have been linked to deaths of marine mammals around the world.

Although scientists do not understand why, it is clear that some types of sonar used by these ships interfere with the echolocation of marine mammals. Scientists have found several cases of whales having beached themselves after having been exposed to the sonar of ships.

One of 12 sperm whales that ***beached*** *and died on Karekare Beach, West Auckland, New Zealand, in late 2003.*

beached (beaching, to beach): when a marine mammal that cannot live out of the water strands itself on land, usually a beach.

AIA 139

ANIMALS IN ACTION

Revise Your Explanation

10 min.

Have students revise their explanations.

Revise Your Explanation

You have just read more about how dolphins communicate and what scientists know about dolphin communication. With your group, look back at your *Observing and Interpreting Animal Behavior* pages. Look at your original interpretations of dolphin communication. Now, with your new science knowledge, reinterpret the behavior you saw in the video.

Go back to your explanation on your *Create Your Explanation* page. First, add the science you just learned to the science knowledge box. Then, check to make sure your claim is still accurate. If your claim does not match the science you have read, revise it. Next, support your claim with your new science knowledge.

Rewrite your explanation to make it more complete. Include in your explanation how the dolphin's body structure and environment influence its communication. The dolphin's environment includes the dark area under water as well as the sunlit area on the surface of the water. Be sure to include both of these areas in your explanation. Remember that an explanation is a statement that connects a claim to evidence and science knowledge in a logical way. Write your explanation so that it tells why your claim is accurate. Be sure your explanation matches the science you just read. Make sure your claim now matches what you have learned. If it does not, revise your explanation. Use the information from your reading about dolphin communication to support your revised explanation. You might need to write an explanation that has a few sentences rather than just one long sentence. The goal is to tie everything together and help others understand why your claim is true.

Communicate

Share Your Explanation

Share your new explanation with the class. When you share your explanation, tell the class what makes this revised explanation more accurate than your earlier one. As each group shares their explanation, pay special attention to how the other groups have supported their claims with science knowledge. Ask questions or make suggestions if you think a group's claim is not as accurate as it could be or if the group has not supported their claim well enough with observations and science knowledge.

As a class, work together to develop your best explanation of what affects how dolphins communicate.

△ Guide

Let students know that they will now have the opportunity to revise their explanations about what affects dolphin communication based on all the information they now have. Emphasize that the claims they have made must be supported by the science knowledge introduced in this section, or they will need to revise and/or construct new claims.

TEACHER TALK

"You know more now about dolphin communication than when you wrote your explanations. When you look at them now, you may see that information from your reading might be helpful in supporting your explanation.

First, check to make sure your claim fits with the new science knowledge. If your claim is inconsistent with what you learned, revise it.

Then, put any science knowledge you can use in the science knowledge box and rewrite your explanation to make it more complete."

○ Get Going

Distribute new *Create Your Explanation* pages and have students revise their explanations to include the new science knowledge.

◇ Evaluate

As groups are revising their explanations, monitor students' progress. Check their explanations to make sure that they are using the science knowledge they just learned to support their claims. If it looks like their claims are inconsistent with what they just learned, note this as something to discuss when the class discusses the groups' presentations.

If groups are still using unfounded claims or opinions for evidence, stop the class and review what they should know about constructing explanations. They should make valid claims in which the evidence supports the claims.

Communicate

15 min.

Have a class discussion on the groups' presentations of their explanations.

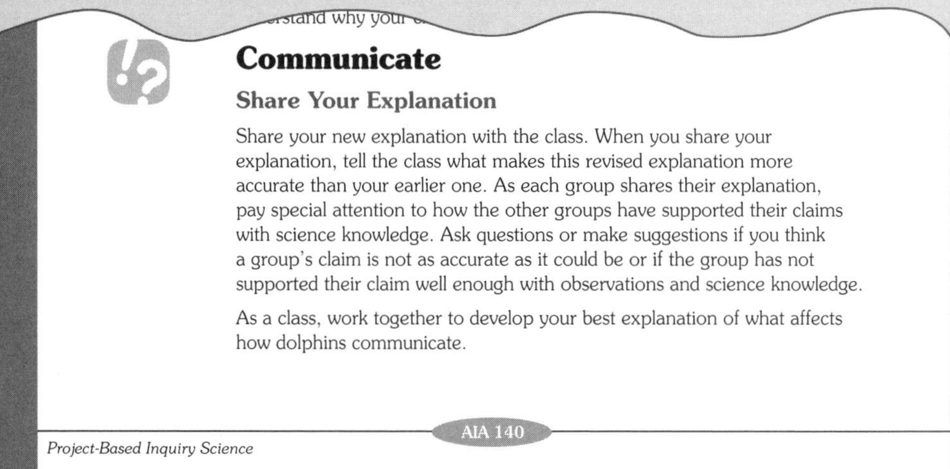

Communicate

Share Your Explanation

Share your new explanation with the class. When you share your explanation, tell the class what makes this revised explanation more accurate than your earlier one. As each group shares their explanation, pay special attention to how the other groups have supported their claims with science knowledge. Ask questions or make suggestions if you think a group's claim is not as accurate as it could be or if the group has not supported their claim well enough with observations and science knowledge.

As a class, work together to develop your best explanation of what affects how dolphins communicate.

Project-Based Inquiry Science AIA 140

△ Guide

Let the class know that each group will be presenting their explanation and that it is important for each presenting group to make clear how they have supported their claims with observations and science knowledge. Emphasize that it is also important that the audience take notes and ask questions if anything is not clear. Then let them know that after hearing everyone's explanations, the class will come up with an explanation together about what affects how dolphins communicate.

Have a discussion after each presentation. Model for students how they should seek clarification or point out areas they may not understand.

META NOTES

If students ask you questions about other groups' explanations, redirect the question to the group presenting. In general, encourage students to discuss their explanations with each other and the presenting groups.

TEACHER TALK

"I don't see how that fact backs up your claim. Could you clarify that for me?

I'm not sure I understand your claim. Can you explain it to me?"

Also, encourage students to point out where they think the group could have used some of the science knowledge they just learned to support their claims.

◇ Evaluate

Evaluate students' use of the reading to revise their explanations.

△ Guide

After all groups have presented, ask students to point out the similarities and differences between the explanations presented.

Then, ask the class to pick out the claims they think they should include in their class explanation. Record these claims on the board and then ask for the supporting evidence and science knowledge.

After the class has agreed upon an explanation, have students copy them on *Create Your Explanation* pages.

NOTES

Update the Project Board

5 min.

Have a class discussion focused on updating the class Project Board.

Update the *Project Board*

As you read more about how dolphins communicate, you may have thought of new questions you want to ask. Add what you learned from your reading to the *Project Board*. Make sure you add evidence to support any new science you learned. Also, think about how this learning will help you answer the *Big Question, How do scientists answer big questions and solve big problems?* As your teacher records this information on the big *Project Board*, add the information to your own *Project Board* page.

What's the Point?

Echolocation and sonar both rely on sound waves being sent out, bounced off objects, and returned. The way the sound comes back helps determine the shape and location of an object. Dolphins use echolocation. Sets of sonar clicks make it possible for a dolphin to find food, other dolphins, and hazards in the water. Echolocation is very important for dolphins because ocean water can be murky and dark. Other forms of communication are not as effective in that environment.

Sonar is also a tool used by scientists to "see" into the deep and dark parts of the ocean. Sonar uses sound waves to create an image of the ocean floor, fish, and objects in the ocean, such as sunken ships. Naval vessels use sonar to find and intercept other ships. Scientists think there are times when the sonar from these ships interferes with the echolocation of marine mammals.

This dolphin is using echolocation to identify an object inside a box.

△ Guide

Ask students if they have new questions about how animals communicate and record these in the second column *(What do we need to investigate?)* of the *Project Board.*

Then, ask students what they are learning and what is their evidence. Record their answers in the third and fourth columns of the *Project Board.* Remember to link the claims (column 3) and evidence (column 4) together on the *Project Board.*

◇ Evaluate

Make sure that the claims from students' explanations are in the third column and that information from the reading is in the fourth column before moving on.

Students' claims should include claims specific to dolphin communication, like why dolphins communicate, what they communicate, the types of sounds they make, and how their bodies and environment affect their communications.

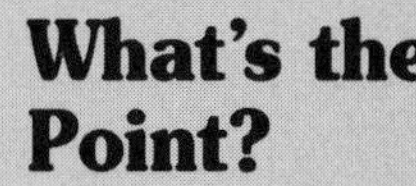

What's the Point?

5 min.

Project Bo... information to your ow... *Board* page.

What's the Point?

Echolocation and sonar both rely on sound waves being sent out, bounced off objects, and returned. The way the sound comes back helps determine the shape and location of an object. Dolphins use echolocation. Sets of sonar clicks make it possible for a dolphin to find food, other dolphins, and hazards in the water. Echolocation is very important for dolphins because ocean water can be murky and dark. Other forms of communication are not as effective in that environment.

Sonar is also a tool used by scientists to "see" into the deep and dark parts of the ocean. Sonar uses sound waves to create an image of the ocean floor, fish, and objects in the ocean, such as sunken ships. Naval vessels use sonar to find and intercept other ships. Scientists think there are times when the sonar from these ships interferes with the echolocation of marine mammals.

◇ Evaluate

Make sure students recognize the parts of an explanation and how to construct one.

Students should be able to describe dolphin communication, the need to communicate for survival, and how their physical characteristics and environment affect their communication.

Assessment Options

Targeted Concepts, Skills, and Nature of Science	How do I know if students got it?
The structure and function of animals' bodies are complementary and affect animal behavior.	**ASK:** What features of dolphins' bodies affect how they communicate? **LISTEN:** Students should recognize that dolphins' have physical features necessary to make sounds and their excellent hearing allows them to find each other using signature whistles and allows them to use echolocation.
Scientists must keep clear, accurate, and descriptive records of what they do so they can share their work with others, consider what they did, why they did it, and what they want to do next.	**ASK:** How will the *Project Board* help to design an enclosure that encourages communication? **LISTEN:** Students should recognize that to design enclosures, they will need the relevant information available and organized in a useful way.
Scientists often work together and then share their findings. Sharing findings makes new information available and helps scientists refine their ideas and build on others' ideas. When another person's or group's idea is used, credit needs to be given.	**ASK:** How did sharing your explanations help you to develop valid claims? **LISTEN:** Students should have been able to help each other see where their evidence did not support their claims.

Targeted Concepts, Skills, and Nature of Science	How do I know if students got it?
Scientific knowledge is developed through observations, recording and analysis of data, and development of explanations based on evidence.	**ASK:** What were the important steps you took to develop your explanations? How are these steps like what scientists do? **LISTEN:** Students should describe how they observed dolphins in the video, recorded observations, analyzed and interpreted their observations, and developed explanations. Similarly, scientists make observations, analyze and interpret their data, and develop explanations.
Explanations are claims supported by evidence. Evidence can be experimental results, observational data, and other accepted scientific knowledge.	**ASK:** How did you use your new science knowledge to revise your explanations? **LISTEN:** Students should have used the new science knowledge to support their claims, or to determine how their claims needed to be modified.

Teacher Reflection Questions

- What concepts in this section were difficult for students? What can you do to help make these concepts clear?
- What difficulties remain for students in constructing explanations and what ideas do you have to assist them?
- How did you encourage collaboration between group and class members? What ideas do you have for next time?

NOTES

BACK TO THE BIG CHALLENGE INTRODUCTION

Learning Set 3

Back to the Big Challenge

◀ ***1 class period****

*A class period is considered to be one 40 to 50 minute class.

Overview

Students use what they have learned about animal communication behavior to make recommendations for designing animal enclosures that encourage communication. They work with their groups to develop several recommendations. Then they share their recommendations with the class and get feedback and ideas from the class. Finally, the class records the recommendations on the *Project Board,* where they can reference them when they *Address the Big Challenge.*

Targeted Concepts, Skills, and Nature of Science	Performance Expectations
Scientists often work together and then share their findings. Sharing findings makes new information available and helps scientists refine their ideas and build on others' ideas. When another person's or group's idea is used, credit needs to be given.	Students work with their groups to develop recommendations. Then they share their recommendations with the class and get feedback from the class.
Scientists must keep clear, accurate, and descriptive records of what they do so they can share their work with others, consider what they did, why they did it, and what they want to do next.	Students should use their records from *Learning Set 1* and *Learning Set 3* to construct recommendations for an animal enclosure design.
Explanations are claims supported by evidence. Evidence can be experimental results, observational data, and other accepted scientific knowledge.	Students include observations and scientific knowledge in their explanations, to support their recommendations for designing an enclosure.

Targeted Concepts, Skills, and Nature of Science	Performance Expectations
Criteria and constraints are important in determining effective scientific procedures and answering scientific questions.	Students' recommendations for designs should address all the criteria and constraints.
Behavior is a type of response to internal or external stimulus. Behavior is determined by experience, physical characteristics, and environment.	Students' recommendations should be sensitive to how the animal's physical characteristics and environment affect its communication behaviors.
The structure and function of animals' bodies are complementary and affect animal behavior.	Students' recommendations should include how the animal's structure affects its communication behavior and the design of the enclosure.

Materials	
1 per class	*Project Board*
1 per student	*Project Board* pages
4 per student	*Create Your Explanation* pages
1 per group	Large sheets of paper for *Solution Briefing* posters

Homework Options

Reflection

- **Science Process and Content:** What ideas from other groups' recommendations can you use for your final animal enclosure design plan? *(Students should look for ideas they can use in other groups' recommendations.)*

BACK TO THE BIG CHALLENGE IMPLEMENTATION

◀ *1 class period**

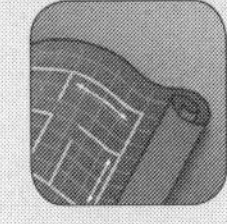

Learning Set 3

Back to the Big Challenge

You have been developing an answer to the question, *What affects how animals communicate?* Answering this question will help you complete the challenge of the Unit, to design an enclosure for an animal. Your enclosure needs to be designed so that your animal can communicate as naturally as possible. The enclosure needs to be as similar as possible to the animal's habitat.

To develop your recommendations about the design of an animal enclosure, you will need to think about some of the big ideas that affect how animals communicate. Several big ideas were introduced in this *Learning Set*, and you need to think about how those ideas are going to be built into your enclosure.

You are going to focus on one of the animals you learned about in this section. You need to use the explanations regarding that animal to help you develop recommendations about designing your enclosure. You can also use information from the *Project Board*, especially the third and fourth columns where you have included information about each animal. You will then share your recommendations with the class.

Develop Recommendations

A recommendation is a type of claim. You will support each of your recommendations with evidence from your observations and reading. Some of this evidence may come from your previous explanations. Some will come from your observations and interpretations. To prepare for presenting to the class, and so that you can be sure that your recommendations match the evidence you have collected and what science tells you, use a *Create Your Explanation* page for each recommendation your group makes. Your recommendation will be your claim. Add evidence and science knowledge that supports it. Then develop a logical statement linking your recommendation to the evidence and science knowledge.

You will write a set of recommendations for how to provide the best environment for some animal so it can communicate in the most natural way possible. Think about starting your recommendations with "If," "When,"

AIA 142

Project-Based Inquiry Science

Learning Set 3

Back to the Big Challenge

5 min.

Introduce the section.

△ Guide

Remind students that they have focused on answering the question: *What Affects How Animals Communicate?* in this *Learning Set,* and that this is one of the main issues to consider in their challenge to design an animal enclosure. Let students know that they will be presenting their ideas on how to design an animal enclosure based on the communication behaviors of one of the animals presented in this *Learning Set*.

*A class period is considered to be one 40 to 50 minute class.

TEACHER TALK

"During this *Learning Set,* you have focused on what affects how various animals communicate. What affects how animals communicate? This is important in the design of an animal enclosure because you want the animals to communicate as closely to how they would in their natural environment. Today, your group will construct recommendations for an enclosure based solely on what affects how an animal communicates."

Develop Recommendations

10 min.

Have groups write recommendations for designing animal enclosures that encourage communication or feeding.

...ommendations with ...

Develop Recommendations

A recommendation is a type of claim. You will support each of your recommendations with evidence from your observations and reading. Some of this evidence may come from your previous explanations. Some will come from your observations and interpretations. To prepare for presenting to the class, and so that you can be sure that your recommendations match the evidence you have collected and what science tells you, use a *Create Your Explanation* page for each recommendation your group makes. Your recommendation will be your claim. Add evidence and science knowledge that supports it. Then develop a logical statement linking your recommendation to the evidence and science knowledge.

You will write a set of recommendations for how to provide the best environment for some animal so it can communicate in the most natural way possible. Think about starting your recommendations with "If," "When,"

Project-Based Inquiry Science AIA 142

Back to the Big Challenge

or "Because." For example, you might begin your recommendation by writing, "If the dolphins were going to communicate in the most natural way possible. . . Even better would be a recommendation of the form, "Because dolphins need to. . . " Then you need to complete the statement.

It will be necessary to create more than one recommendation. You have learned a lot in this *Learning Set*, and you want to show what you have learned. You will need to write as many recommendations as you can about how an enclosure will ensure that your animal is able to communicate naturally.

△ Guide

Give students a brief overview of what they will be doing: Developing recommendations for the design of an enclosure for one of the animals they learned about, presenting and discussing their recommendations with the class, and updating the *Project Board*.

Then, remind the class what a recommendation is and that they will need to develop more than one recommendation for their enclosure design.

TEACHER TALK

"In our class, when we talk about a recommendation we mean a claim that suggests what to do in a situation based on evidence and science knowledge. This is the information we use when we make explanations and what we keep track of in columns 3 and 4 of our *Project Board.*

Recommendations may be of the forms:

- **If... then... because; or**
- **When ... occurs, then do, try, or expect ...**
- **Because ..., ...should....**

You will need more than one recommendation to describe what is important in the design of your enclosure."

Let students know that they should use a *Create Your Explanation* page for each recommendation their group makes and that they will be presenting their recommendations to the class.

Next, provide some examples of recommendations such as those in the student text in the *Making Recommendations* information box and those given below. Again, emphasize that recommendations require evidence and science knowledge to back them up.

⬡ Get Going

Distribute the *Create Your Explanation* pages to groups and let students know how much time they have.

△ Guide and Assess

Monitor students' progress and determine if students understand what is involved in making a recommendation by reviewing what they are doing. Guide them as needed by asking them to point out their claim, evidence, science knowledge, and how the evidence and science knowledge support their claim.

Here's an example of a recommendation a group might write:

Because dolphins need to communicate with other members of their pods, they need to be together in one large enclosure. We have observed dolphins swimming and communicating with members of their pods. We also know from our reading that dolphins in pods use whistles and other sounds to keep track of one another.

NOTES

or "Because." For example, you might begin your recommendation by writing, "If the dolphins were going to communicate in the most natural way possible. . . Even better would be a recommendation of the form, "Because dolphins need to. . . " Then you need to complete the statement.

It will be necessary to create more than one recommendation. You have learned a lot in this *Learning Set*, and you want to show what you have learned. You will need to write as many recommendations as you can about how an enclosure will ensure that your animal is able to communicate naturally.

Communicate Your Solution

Solution Briefing

After you have developed your recommendations, you will communicate your recommendations to one another in a *Solution Briefing*. In a *Solution Briefing*, you present your solution, or recommendations, in a way that will allow others to evaluate how well it achieves the criteria and to make suggestions about how you might improve it.

As you prepare for this briefing, make sure you revisit the criteria and constraints you identified in the beginning of the Unit. Use the following questions to plan your presentation.

- How is your enclosure addressing the communication needs of your selected animal?
- How does the enclosure meet the criteria?
- How did the constraints affect your recommendations?
- What information did you use to help you make each recommendation?
- What other ideas did you think about along the way, and why did you not recommend them?
- What questions do you still have?

As you listen to the presentations, make sure you understand the answers to these questions. If you do not understand something, or if something is not presented clearly enough, ask questions.

You can use the questions above as a guide. When you think something can be improved, make sure to contribute your ideas. Be careful to ask your questions and make your suggestions respectfully. As you listen, record notes on a *Solution-Briefing Notes* page.

Communicate Your Solution: *Solution Briefing*

20 min.

Discuss Solution Briefings, *then have groups present their recommendations.*

META NOTES

A *Solution Briefing* is a common pedagogical tool used in *PBIS*. In a *Solution Briefing*, students present their ideas and hear others' ideas so that they can build on each other's ideas. In this *Solution Briefing*, students are sharing and building on each other's ideas about how to design an enclosure that will accommodate communication or feeding. In this and many other class presentations, students will learn how to communicate their ideas, ask questions, and sharpen their critical thinking skills.

NOTES

△ Guide

Remind students what a *Solution Briefing* is and how it works. Explain that they will be presenting their recommendations to share their ideas and gather advice. The goal is for the larger group (the class) to help each smaller group to make their solution better.

Then, describe how students should prepare for this *Solution Briefing* using the information in the student text. Emphasize that their presentations should address the bulleted questions on page AIA 143 in the student text.

Also explain that the audience should ask clarifying questions and offer suggestions. Everyone should voice their questions and ideas in a polite and considerate manner, using language such as, "I agree with ... because..." or "I disagree with... because."

TEACHER TALK

"It is important that you all think about what you are going to say when you present your recommendations. In the student text, there is a list of questions that will help you think about what to talk about during your presentation. When you are in the audience or not presenting you will need to listen to make sure that you understand each design. You should also voice your opinions and ideas. Remember to always be polite and considerate whether you are presenting or are in the audience."

⬡ Get Going

Point out where you have displayed the class's Project Board and their list of criteria and constraints for the enclosure.

Give groups five minutes to prepare their presentation.

Then, have student groups present.

△ Guide

Describe the *Solution-Briefing Notes* page to students as a place to keep track of the presented design recommendations and the ideas they found useful. Remind students that each group will be presenting their recommendations and that the audience will be taking notes, asking questions and giving advice. Emphasize the importance of listening for reasoning that supports the recommendations and to prepare their questions and advice for the presenting group.

Distribute the *Solution-Briefing Notes* page and begin the presentations.

While groups are presenting, you and the class should be listening for how recommendations are supported, if the criteria and constraints are met, and if the questions in the student text have been addressed.

If groups get off track or stuck, you can prompt them with one or two of the questions listed in the student text. As the *Solution Briefing* progresses, encourage students in the audience to participate more in asking questions to help the presenters clarify their design recommendations.

Listen for the following types of responses to the student text questions:

- Students' recommendations should describe the communication needs of the animal based on information provided in this *Learning Set* and address these in their recommended design.
- The recommendations for the enclosure must meet the class list of criteria.
- Students should describe how their recommendations address the constraints of the enclosure, such as building the enclosure locally and dealing with the local weather.
- Students' recommendations should be supported by their observations and science knowledge in the student text and they should be able to describe how these support their recommendations.
- Students should describe ideas they decided against and their reasoning for choosing or not choosing an idea.
- Students should list questions they still have.

After all groups have presented, wrap up the *Solution Briefing* by asking students if they heard any interesting ideas or have thought of any new ideas they would like to use to improve their own zoo animal enclosure design. Guide students to think about how sharing ideas will help them improve their own ideas.

Finally, compare this session with what scientists and designers do to improve their solutions. Reflect on the usefulness of iterations and record keeping. Tell students that scientists are always building on each other's ideas. Bring up the idea of giving credit to others when you build from their ideas. This is not copying. Emphasize that the difference between this and copying is that copying doesn't give credit to the people who thought of the ideas you used.

NOTES

Update the *Project Board*

5 min.

Record the ideas on the class Project Board.

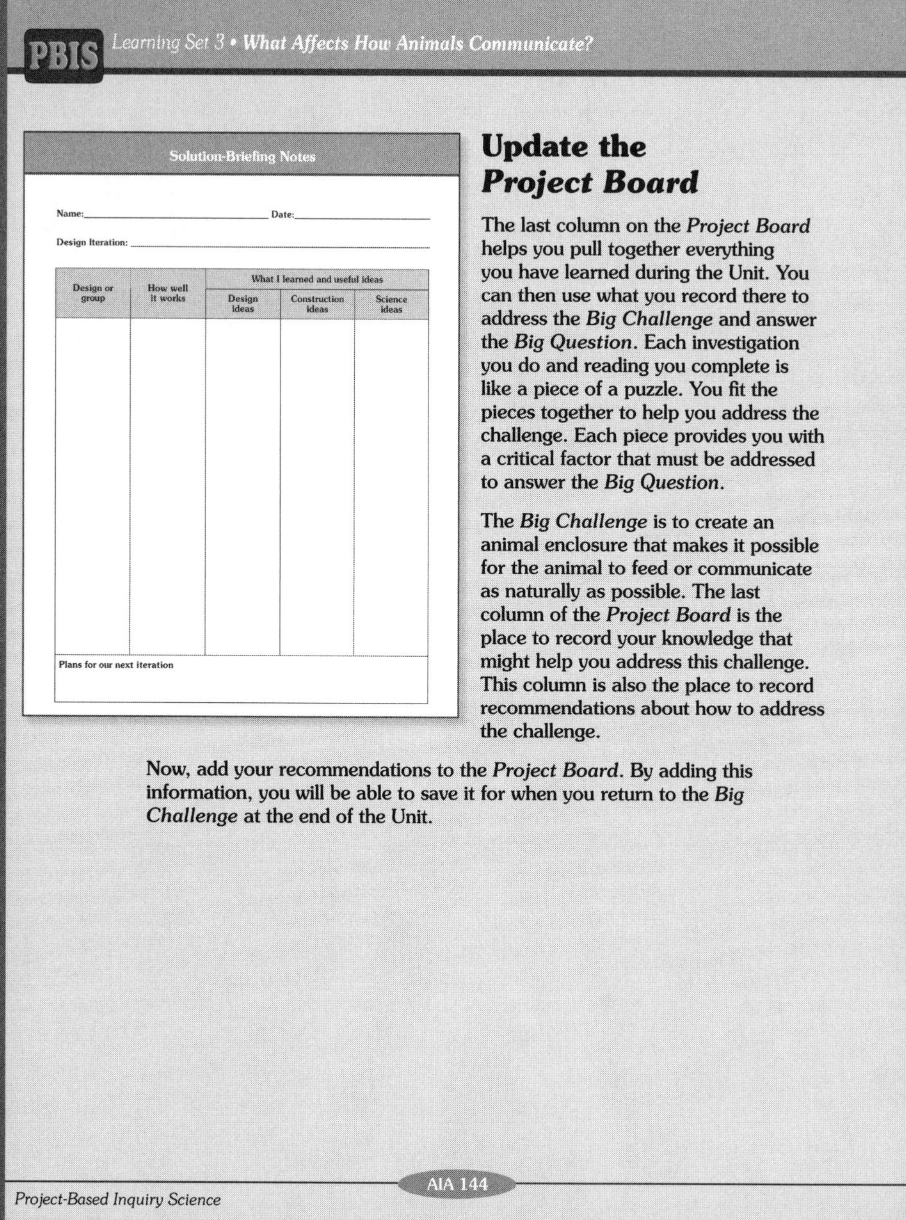

Solution-Briefing Notes

Name: ______ Date: ______

Design Iteration: ______

Design or group	How well it works	What I learned and useful ideas		
		Design ideas	Construction ideas	Science ideas

Plans for our next iteration

Update the *Project Board*

The last column on the *Project Board* helps you pull together everything you have learned during the Unit. You can then use what you record there to address the *Big Challenge* and answer the *Big Question*. Each investigation you do and reading you complete is like a piece of a puzzle. You fit the pieces together to help you address the challenge. Each piece provides you with a critical factor that must be addressed to answer the *Big Question*.

The *Big Challenge* is to create an animal enclosure that makes it possible for the animal to feed or communicate as naturally as possible. The last column of the *Project Board* is the place to record your knowledge that might help you address this challenge. This column is also the place to record recommendations about how to address the challenge.

Now, add your recommendations to the *Project Board*. By adding this information, you will be able to save it for when you return to the *Big Challenge* at the end of the Unit.

Have a class discussion on what to add to the *Project Board,* focusing on the last column—*What does it mean for the challenge or question?* Record students' ideas on the class *Project Board,* and have the students record their ideas on their own *Project Board* pages.

TEACHER TALK

“This *Learning Set* has been focused on investigating what affects how animals communicate and today you considered how that might affect the challenge by constructing recommendations for designing an enclosure that allows animals to communicate as they would in their natural environment. What do we know about what affects how animals communicate now and how it will affect the enclosure designs? Think about all of the recommendations you heard and our class discussion on them to help you decide what to put up on the class *Project Board*.”

Example Recommendation:

Dolphins live in large groups called pods and use special signature whistles so their young can find them. Because of this, it is a good idea to have a large water filled tank for the dolphin enclosure that allows a number of dolphins to be together and allows mother and calf to be together.

◇ Evaluate

Make sure the class *Project Board* contains recommendations for enclosures only pertaining to animals discussed in *Learning Set 3* and focuses on communication.

Teacher Reflection Questions

- What difficulties did students have in making recommendations? What ideas do you have to help them?
- What evidence do you have that students understand that the sharing of ideas and building on other's ideas are practices of scientists? How can you further their understanding of the importance of these practices?
- What questions or comments did students ask or say that might be rude or inappropriate? (For example, “She/He copied from us!” or talking out of turn.) What will help your students to learn how to question or comment respectfully?

NOTES

ADDRESS THE BIG CHALLENGE INTRODUCTION

Address the Big Challenge

Design an Enclosure for a Zoo Animal that Will Allow it to Feed or Communicate as in the Wild.

◀ ***2 class periods****

*A class period is considered to be one 40 to 50 minute class.

Overview

Students apply the concepts of this Unit as they construct design plans for a zoo enclosure. Students begin by identifying the criteria and constraints of the challenge. Using the criteria and constraints as guides, groups then plan their designs for animal enclosures. They share their designs with the class in *Plan Briefings* and get feedback from the class, working collaboratively, as scientists do. Using the feedback from their peers, they revise their design plans. They present their revised designs again, get feedback, and revise their plans. Students gain understanding of the value of iteration in design. They then present their final plans to the class and update the *Project Board*.

Materials	
3 per group	Large sheets of paper for *Plan Briefing* posters and *Solution Showcase* posters
1 per class	*Project Board*
1 per class	Class list of criteria and constraints

Homework Options

Reflection

- **Science Process:** How did you use explanations from this Unit to plan your design? *(Students should have used the claims from class explanations to determine what their animal needs to communicate or feed as it does in the wild.)*
- **Science Process:** How do scientists answer big questions and solve big problems? *(Students responses should contain the information in the* Answer the Big Question *section that follows.)*

NOTES

ADDRESS THE BIG CHALLENGE IMPLEMENTATION

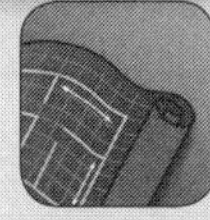

Address the Big Challenge

Design an Enclosure for a Zoo Animal that Will Allow it to Feed or Communicate as in the Wild

Your challenge for this Unit is to design a zoo enclosure that will accommodate the feeding or communication of one of the animals you studied in the Unit. Your goal will be to design the zoo environment so it is similar enough to the natural environment of the animal to allow the animal to feed or communicate effectively. The enclosure will also have to allow visitors and scientists to observe the animals clearly.

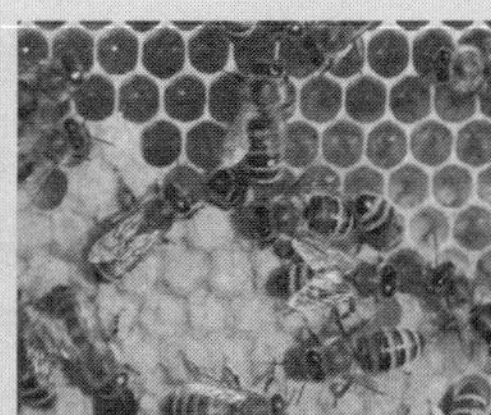

Animals you observed in the Unit: chimpanzees, bees, elephants, and dolphins.

You now know enough to address the challenge. You observed the behavior of several animals in the Unit. You analyzed their behavior, identified why they were behaving the way they were, and recorded your data on *Observing and Interpreting Animal Behavior* pages. These pages will be useful in designing your enclosure. They contain the behavior you saw and how you interpreted that behavior. The details you entered may help you identify how your animal's environment affects its behavior. You also

◀ 2 class periods*

Address the Big Challenge

Design an Enclosure for a Zoo Animal that Will Allow it to Feed or Communicate as in the Wild.

5 min.

Introduce the Section.

△ Guide

Remind students of the *Big Challenge: Design an Enclosure for a Zoo Animal that Will Allow it to Feed or Communicate as in the Wild* that will accommodate the feeding or communication of one of the animals studied in the Unit. Emphasize that students will need to refer to their records from the entire Unit when planning their design and when helping other groups plan designs.

*A class period is considered to be one 40 to 50 minute class.

Identify Criteria and Constraints

10 min.

Have students identify the criteria and constraints of the challenge.

developed explanations of why the different animals behave the way they do. As a class, you developed recommendations about designing enclosures for each of the animals. You recorded your claims and recommendations on *Create Your Explanation* pages and included evidence and science knowledge to support your claims. All of this will be useful to you as you address the challenge.

Identify Criteria and Constraints

The first thing designers do when they are asked to address a challenge is to identify the criteria and constraints. You began the Unit by identifying some of the criteria and constraints for this challenge. But you now know a lot more about animal behavior than you did at the beginning of the Unit. You also now know which animal you will be planning an enclosure for and whether you will be focusing on the animal's feeding or communication.

As you develop a more complete set of criteria and constraints for your design, keep the questions in the table in mind.

Questions for those addressing feeding needs	Questions for those addressing communication needs
How large is your animal?	How large is your animal?
What size group does it live with?	What size group does it live with?
What kind of climate does it live in?	What kind of climate does it live in?
What kinds of plants and animals live in its habitat?	What kinds of plants and animals live in its habitat?
Are there other special features of its habitat that are important to its feeding behavior?	Are there other special features of its habitat that are important to its communication behavior?
What foods does it eat?	For what purposes does it communicate?
Where does it find its food?	How does it send messages?
How does it obtain its food?	How does it receive messages?

As you answer the questions in the table and identify criteria and constraints, listen to each other's ideas. It is important for everyone's answers and ideas to be heard.

△ Guide

Let students know that they will begin by identifying the criteria and constraints of this challenge. Initiate a discussion of what students need to do to design an enclosure that encourages communication or feeding. Remind them that two important features of a challenge are criteria and constraints.

> ***Address the Big Challenge***
>
> Each animal has different needs. You must keep those needs in mind as you plan the enclosure for your animal. Your next step is to identify the criteria and constraints you will need to keep in mind as you design an enclosure for your animal. For example, if your animal eats leaves from the tops of tall trees, one of your criteria will be to include those kinds of trees in the enclosure. If your animal lives in large groups, one of your criteria will be that the enclosure has to be large enough for a natural group.

Some criteria and constraints the class should have identified are:

- The enclosure has to be similar enough to the animals' natural habitat that it will allow the animals to communicate and feed effectively through instinctual behavior.
- We must be able to build the enclosure at the zoo.
- The enclosure must take into account the local weather conditions.

Let students know that they will be meeting in their groups to discuss specific criteria and constraints for the animal and behavior they are choosing to design an enclosure for. Discuss the table in the student text that students should use as a guide as they construct their criteria and constraints. Emphasize that each group member should answer each question and the group should discuss and agree upon the final answers.

⬡ Get Going

Have groups begin working and let them know how much time they have.

◇ Evaluate

As groups develop their sets of criteria and constraints, monitor their progress and make sure that groups are identifying specific criteria and constraints that will help them start designing their enclosures. Check to see that groups' are discussing each member's response to the questions and that all members are participating politely.

META NOTES

Student groups are often changed at the beginning of a Unit. It may be useful to begin considering groups for the next Unit.

Plan Your Design

10 min.

Have groups design animal enclosures.

Each animal has different needs. You must keep those needs in mind as you plan the enclosure for your animal. Your next step is to identify the criteria and constraints you will need to keep in mind as you design an enclosure for your animal. For example, if your animal eats leaves from the tops of tall trees, one of your criteria will be to include those kinds of trees in the enclosure. If your animal lives in large groups, one of your criteria will be that the enclosure has to be large enough for a natural group.

Plan Your Design

You have learned about the ways each animal's body and habitat affect its ways of finding food and communicating. As you design your enclosure, keep these constraints in mind. Remember to use the *Project Board* and your interpretations, explanations, and recommendations as resources as you design your enclosure.

Make decisions about the design of the enclosure together with your group. For example, how much space will you need and why? How will you make sure your animal will be able to use its body the way it does in the wild? What plants and trees will the habitat need, and why? What other animals will the habitat need to include, and why? How will your animal be able to get the food it needs? How will you make sure it can communicate the way it needs to? For each decision you make, discuss alternatives with your group. Know why you are making each decision, and record the evidence and science knowledge you are using.

When you are finished, you will have a chance to share your plan with other groups in a *Plan Briefing*. Others in the class might be able to help you with any difficult decisions you need to make as you work on your plan. You will get a chance to iterate on your plan and improve it based on others' feedback.

Communicate Your Design Plan

Plan Briefing

As you are finishing your plan, begin to draw a poster for presentation of your plan to the class. Include in that plan a drawing of your enclosure. Be sure to label the details of the enclosure. Be prepared to describe all the parts of your plan to others and to support each of the parts of your plan with evidence and science knowledge.

△ Guide

Have groups plan their enclosure designs. Emphasize that groups should use the recommendations the class put on the class *Project Board*. It is also important that they meet the criteria and constraints they identified.

Make sure the class *Project Board* and the class list of criteria and constraints are on display.

⬡ Get Going

Let groups know how much time they have and let them begin.

☐ Assess

As groups are working on their enclosure designs, ask students what ideas they have. Also ask them how they are using the recommendations of the class and what decisions they have had to make.

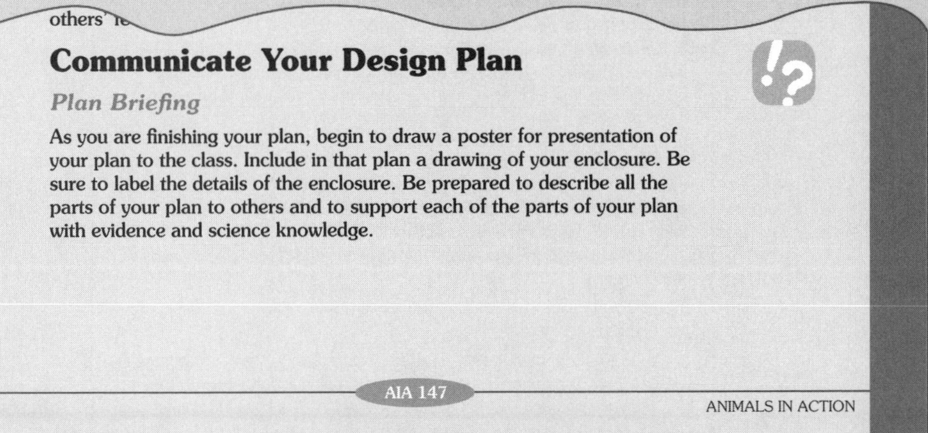

others' re

Communicate Your Design Plan

Plan Briefing

As you are finishing your plan, begin to draw a poster for presentation of your plan to the class. Include in that plan a drawing of your enclosure. Be sure to label the details of the enclosure. Be prepared to describe all the parts of your plan to others and to support each of the parts of your plan with evidence and science knowledge.

AIA 147

ANIMALS IN ACTION

Communicate Your Design Plan

30 min.

Have a class discussion on groups' enclosure design plans.

△ Guide

After the allotted time for planning their designs, introduce the class to *Plan Briefings*. These are similar to *Solution Briefings,* but in *Plan Briefings* groups present details of their design plans rather than their solutions.

TEACHER TALK

"You've already shared your recommendations using a *Solution Briefing*. You will be doing something similar now called a *Plan Briefing.* In a *Plan Briefing* you share your design plans. You will be sharing and discussing your enclosure design plans and your reasoning behind your design choices. You'll then be able to revise your plan. To describe your plans to the class, you're going to make a *Plan Briefing* poster and then you will present it."

Let students know that they will make posters with detailed drawings of their plans and with all parts labeled, descriptions of any special features, explanations and recommendations that support the design, and credit to anyone else's ideas used.

Let students know that the class will discuss what criteria the features of the design achieve, what they expect the animals to do in the enclosure, and any possible problems. Emphasize that *Plan Briefings* should be specific and contain the reasoning behind design choices. Students should use the information in the student text under *Introducing a Plan Briefing* for preparing their poster and their presentation for the *Plan Briefings*.

Be a Scientist

Introducing a *Plan Briefing*
Preparing a *Plan-Briefing* Poster

A *Plan Briefing* is much like the other briefings you have engaged in. In a *Plan Briefing*, you present the plan you are developing. You must present it well enough so that your classmates can understand your ideas and why you made each of your decisions. They should be able to identify if you have made any mistakes in your reasoning. Then they can provide you with advice that you can use to improve your plan. As a presenter, you will learn the most from a *Plan Briefing* if you can be very specific about your plan and about why you made your plan decisions. You will probably want to draw pictures, maybe providing several views. You certainly want everyone to know why you expect your plan to achieve the challenge.

The following guidelines will help you as you decide what to present on your poster.

- Your poster should have a detailed sketch of your plan with at least one view. You might consider sketching multiple views so that the audience can see your plan from different angles. It is important that the audience can picture your design.
- Parts of the plan and any special features should all be labeled. The labels should describe how and why you made each of your plan decisions. Show the explanations and recommendations that support your decisions. Convincing others that your plan choices are quality ones means convincing them that you are making informed decisions backed by scientific evidence.
- Make sure to give credit to groups or students who gave you information or ideas for your plan. If another group provided an Explanation or a Recommendation you are using, you should credit them with their assistance in developing your final plan.

Participating in a *Plan Briefing*

As in other presentation activities, groups will take turns making presentations. After each presentation, the presenting group will take comments and answer questions from the class.

When presenting, be very specific about your plan and what evidence helped you make your plan decisions.

○ Get Going

Distribute materials and let students know how much time they have.

◇ Evaluate

As groups work on their *Plan Briefing* posters, walk around the room and look at groups' posters to see if their drawings are clear and detailed enough so that the class will be able to discuss what criteria the plan will achieve and any possible problems with the plan. Also make sure that students list the explanations and recommendations that support their design plan.

○ Get Going

Have groups present and follow up with class discussions on presentations.

△ Guide

After groups have finished their posters let them know where they should display them. Then allow the groups to visit each poster for about a minute.

When groups have finished viewing everyone's posters, begin the class presentations.

Remind students that during the presentations they should check if all the criteria and constraints have been met, they should ask clarifying questions, and they should give their advice. Then have groups begin their presentations

During the presentations, model the participation you expect by asking questions of the presenting group when anything is not clear and offer suggestions to improve their design.

Point out the presenting group can ask for help with a specific aspect of their design from the class.

After each presentation, you may need to ask a question or two to begin the discussion. Then ask a student to ask a question. Questions should ask how they meet the criteria and constraints, and why they think their design choices are best. Students should also give ideas on how to improve designs. With the class, summarize the suggested improvements for each group so they can record ideas.

META NOTES

Assess students' skill in sharing ideas, asking questions, and responding to peers. Notice if students present their design choices based on what they listed in their *Project Board*, and how these choices affect their design.

META NOTES

Keep track of revisions that groups should make when they revise their plans so you may check for them later.

Update Criteria and Constraints

10 min.

Have groups revise their plans.

Your presentation should answer the following questions:

- What are the critical features of the plan?
- For each feature, what criterion will it achieve? Why is this way the right way to achieve that criterion?
- What issues are you still thinking about?
- How did you use the explanations and recommendations the class developed to help you with your plan?
- In what ways do you need the help of other groups? What issues can they help you solve?

As a listener, you will provide the best help if you ask probing questions about the things you don't understand. Be respectful when you point out errors in the reasoning of others. These kinds of conversations will also allow listeners to learn.

For each presentation, if you don't think you understand the answers to the questions above, be sure to ask questions. When you ask others to clarify what they are telling you, you can learn more. They can learn, too, by trying to be more precise.

Update Criteria and Constraints

Revise Your Plan

You will now have some time to work further on your plan based on suggestions your classmates made in the recent briefing. As you listened to the presentations of others, you may also have thought about other things you want to add to your plan. Each change and new plan is called an iteration.

You may have received some good advice during the *Plan Briefing*. Others may have made suggestions about ways to make your plan better. You may also have discovered new criteria and constraints that you were not aware of earlier. You will have a chance to update your criteria and constraints and to revise your plan. As you do that, be sure to update the pages you are using to record your plan and to justify your decisions. Be sure to revise your sketches and the labels on them.

As you think about which of your classmates' suggestions to include in your new plan, think about whether each suggestion is valid, given what you know about your animal.

As you update your criteria and constraints and then your plan, keep in mind one further issue. Observers need to be able to make good observations of your animal in the enclosure. Consider the type of enclosure and the type of observations the observers might want to make. To create an enclosure that will allow good viewing, you will need to answer the following questions.

- Where in your enclosure would the observer be able to observe feeding and communication?
- What type of data would an observer like to collect?
- Where would an observer sit or stand to get the best view?
- Would all the observer's observations be made with her or his eyes?
- How would the observer use her or his ears to make observations?
- What special instruments might the observer use to gather data?
- Where might those instruments be located in the enclosure?

⬡ Get Going

Ask students to revise their design plans based on the feedback they received from the class. Let students know how much time they have. Point out that they should also consider the bulleted questions in the student text as they make their revisions. Emphasize that they need to record their new plans and justify their revisions.

☐ Assess

As groups are revising their plans, walk around the room and check to see what revisions groups are making. They should be using the feedback they got from the class and the bulleted items in the student text.

Communicate Your Plan: Plan Briefing

15 min.

Have groups present their revised plans.

As you think about which of your classmates' suggestions to include in your new plan, think about whether each suggestion is valid, given what you know about your animal.

As you update your criteria and constraints and then your plan, keep in mind one further issue. Observers need to be able to make good observations of your animal in the enclosure. Consider the type of enclosure and the type of observations the observers might want to make. To create an enclosure that will allow good viewing, you will need to answer the following questions.

- Where in your enclosure would the observer be able to observe feeding and communication?
- What type of data would an observer like to collect?
- Where would an observer sit or stand to get the best view?
- Would all the observer's observations be made with her or his eyes?
- How would the observer use her or his ears to make observations?
- What special instruments might the observer use to gather data?
- Where might those instruments be located in the enclosure?

Communicate Your Plan

Plan Briefing

In this *Plan Briefing*, you will focus on the revisions you made to your plan to allow visitors to observe your animals. You may present a revised poster, or you may need to make a new one.

When presenting, be very specific about your revised enclosure plan and what evidence helped you make your new decisions. Also, make sure you give credit to the groups that helped you think about how you might plan your enclosure. If another group provided an explanation or evidence you are using, credit them with their assistance in developing your plan.

Your presentation should answer the following questions.

- What are the critical features of the revised plan?
- What criterion of the challenge will each feature achieve? What makes the revisions you made to your plan a better way to achieve that criterion?

△ Guide

When groups have finished revising their plans, let them know they will present *Plan Briefings* to the class. Have them edit their *Plan Briefing* posters or, if necessary, create new ones. Emphasize that their presentations should answer the bulleted questions on page 150 of the student text.

Have groups present their revised designs, exactly as they presented before. Once again, make sure that students are asking questions and contributing ideas. Emphasize that it is important for students in the audience to share their ideas, and for presenting groups to listen to the feedback they get from the class.

Address the Big Challenge

- What issues are you still thinking about?
- In what ways do you still need the help of others groups? What issues can they help you solve?

For each presentation, if you do not think you understand the answers to the above questions, be sure to question your classmates. When you ask them to clarify what they are telling you, you can learn more. They can learn, too, by trying to be more precise.

Revise Your Plan

With your group, take into account the advice of your classmates, and revise your plan one last time.

Be a Scientist

Copying versus Crediting

When you build on someone else's idea, it is important to give them credit. Why isn't this "copying"? Copying means taking the work of someone else and claiming it as your own. If you simply sketch what some other group sketched, that is copying. But if you add to another group's idea and acknowledge from where you got your idea, you are doing what scientists do. When you explain how you used their ideas and made them better, you are adding your contribution to theirs.

This means you need to keep good records of where you get your ideas. When you use someone else's ideas, always record from whom you borrowed the idea. Record how you included it in your design and why you did it that way. Then make sure to give credit to the other person or group in your presentations.

Communicate Your Solution

Solution Showcase

After every group has a chance to iterate on their plans, it will be time to complete this challenge. You will present your final design in a *Solution Showcase*.

The goal of a *Solution Showcase* is to have everyone better understand how each group approached the challenge. You will see several solutions for the challenge, each designed to accommodate a specific animal's needs. Be sure to discuss how you included in your final design the explanations

Revise Your Plan

Have students revise their plans again.

△ Guide

Once students have presented their revised plans to the class, have them use the suggestions they got from the class to revise their plans again.

Communicate Your Solution: Solution Showcase

20 min.

Introduce a Solution Showcase *and have students present their* Solution Showcases *to the class.*

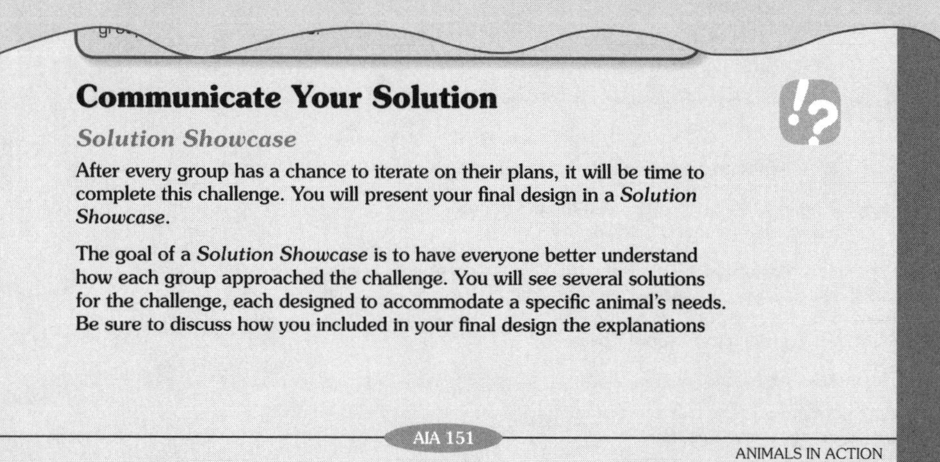

△ Guide

Once students have finished iteratively designing their enclosures, let them know that they will present their final designs to the class in a *Solution Showcase.* Introduce *Solution Showcases,* highlighting the information in the information box on page 152, *Introducing a Solution Showcase.* Emphasize that a *Solution Showcase* should include the original design plan, the history of the group's revisions, the way the group used the explanations and recommendations of the class, and the final design. Emphasize that it should also detail the changes the group made in each iteration and the reasons for the changes.

⬡ Get Going

Let students know how much time they will have for their presentations and how much time they have to prepare their presentations.

△ Guide

Have each group present their *Solution Showcase.* As each group presents, encourage students to ask questions and model the kinds of questions students should be asking. These should be about what techniques groups have tried, what criteria the designs achieve, how the constraints were or were not accounted for in the design, and whether there are any problems with the design.

Some of the questions they might ask are at the bottom of the *Introducing a Solution Showcase* textbox.

PBIS

and recommendations that the class generated. Explain why you think your solution is a good one.

Your presentation should show a picture of your animal enclosure. The picture should include as much detail as possible, and all the critical parts should be labeled. Be prepared to carefully describe how the features of your enclosure will help meet the needs of your animal for feeding or communication. Be sure to give credit to others who helped you improve your plan.

Your presentation of your enclosure design should show

- how your animal will feed or communicate naturally
- how you accounted for criteria and constraints of the challenge
- how you took into account any special needs your animal might have
- how observers will view the behavior of your animal

Be a Scientist

Introducing a Solution Showcase

The goal of a *Solution Showcase* is to present a completed solution and help everyone understand how each group arrived at their solution. You have the opportunity to see how each group took into account the advice given earlier.

Your presentation during a *Solution Showcase* should include the history of your plan. Review your original plan. Then tell the class why and how you revised it. Make sure to present the reasons you made the changes you did. Do this for the whole set of iterations you did. Make sure too, that the class understands what your final plan is. As you prepare, you will need to organize your thoughts so you can present the history of your plan quickly and completely.

As you listen, it will be important to look at each plan carefully. You should ask questions about how each plan meets the criteria of the challenge. Be prepared to ask (and answer) questions such as these.

- What approaches were tried and how were they done?
- How well does the solution meet the goals of the challenge?
- How did the challenge constraints affect your solution decisions?
- What problems remain?

Update the *Project Board*

5 min.

Have the class update their Project Board.

Update the *Project Board*

Now that you have completed your challenge, you will go back to the *Project Board* for one final edit. You will focus mainly on the middle column, *What are we learning?* and the last column, *What does it mean for the challenge or question?* Record what you have learned about what animals need to feed and communicate, and how this applies to designing a natural enclosure.

What's the Point?

To help in designing zoo enclosures, ethologists study animals and how animals behave. Addressing this challenge requires a lot of thinking. The scientists who design zoo enclosures must think about the behavioral and environmental needs of the animal. In many cases, they must also think about how other scientists can observe the animal in its enclosure. They aren't always able to fully design a proper enclosure the first time.

By observing animals, listening to the advice and ideas of other scientists, and using that information to improve upon their original design, scientists can build the best enclosure for an animal.

△ Guide

Draw the students' attention to their *Project Board,* and ask them what they can add to the columns *What are we learning?* and *What is our evidence?*

TEACHER TALK

"During the design of your enclosures and the class discussions, you may have realized that something else should be included on the class *Project Board* in the columns *What are we learning?* and *What is our evidence?* Does anyone have anything they think should be added?"

Discuss what should be included in the final column—*What does it mean for the challenge or question?* Remind students that this column contains the recommendations they made at the end of the second and third *Learning Sets*.

TEACHER TALK

"Think about all of the recommendations the class came up with and all the designs the class made. Is there anything else we should put in the last column —*What does it mean for the challenge or question?*"

Remind students to record the new answers on their own *Project Board* pages.

a natural enclosure.

What's the Point?

To help in designing zoo enclosures, ethologists study animals and how animals behave. Addressing this challenge requires a lot of thinking. The scientists who design zoo enclosures must think about the behavioral and environmental needs of the animal. In many cases, they must also think about how other scientists can observe the animal in its enclosure. They aren't always able to fully design a proper enclosure the first time.

By observing animals, listening to the advice and ideas of other scientists, and using that information to improve upon their original design, scientists can build the best enclosure for an animal.

What's the Point?

5 min.

Have students consider the different aspects they needed to consider to focus on the zoo enclosure.

ANSWER THE BIG QUESTION IMPLEMENTATION

Answer the Big Question

How Do Scientists Answer Big Questions and Solve Big Problems?

You addressed the *Big Challenge* by planning, sketching, and sharing with others an enclosure for an animal. The Unit also includes a *Big Question*. The *Big Question* is *How do scientists answer big questions and solve big problems?* Like scientists, you planned, recorded, analyzed, and explained your animal observations. You also read about several ethologists who study animal behavior. The following questions will help you remember what you now know about how scientists answer big questions. Your answers to these questions will help you answer the *Big Question*. Write an answer to each question. Use examples from class to justify your answers. Be prepared to discuss your answers in class.

1. **Teamwork—**Scientists often work in teams. Think about your teamwork during the Unit. Record the ways you helped your team. What were you able to do better as a team than you could have done alone? What things made working together difficult? What did you learn about working as a team?
2. **Learning from other groups—**What did you learn from other groups? What things did you help other groups learn? What is required to learn from another group or help another group learn? How can you make *Plan* and *Solution Briefings* work better?
3. **Iteration—**You used iteration to improve your explanations and your solutions to the challenge. Iteration is more than simply trying again. What is iteration helpful for? How did iteration help you with observations and explanations? How did it help you design your enclosure?
4. **Meeting criteria and dealing with constraints—**What are criteria? What are constraints? How is specifying criteria and constraints useful for solving problems and addressing design challenges?

Answer the Big Question

How Do Scientists Answer Big Questions and Solve Big Problems?

10 min.

Discuss the answer to the Big Question.

△ Guide

Remind students of the *Big Question: How do scientists Answer Big Questions and Solve Big Problems?* and ask them if they have an answer for this. Let them know that they have been answering big questions and solving big problems during this Unit just as scientists do. Then have students answer the questions in the student text.

△ Guide and Assess

Have a brief class discussion on students' responses. Their responses should contain the answer to the *Big Question: How do scientists answer big questions and solve big problems?*

5. **Using cases to learn—**It is interesting to learn how scientists do their work. In the Unit, you learned about Jane Goodall, Karl von Frisch, and Alfred Wenner. They each showed you a little about how scientists work. You worked as an ethologist in the Unit in ways similar to these ethologists. What are the benefits of learning how other scientists work?
6. **Interpretation—**Scientists observe and then interpret. What is the difference between observation and interpretation? Why is it important to observe before you interpret? Why is it important not to confuse observation and interpretation?
7. **Explaining—**All scientists attempt to explain the way the world works. Ethologists explain animal behavior. What are the parts of a valid explanation? What makes some explanations better than others?
8. **Supporting decisions with evidence—**Scientists seek evidence so they can explain the way the world works. Ethologists' evidence often comes from their observations of animal behavior. Why is evidence so important for making good recommendations?

Look for the following in students' answers:

1. Look for any collaborative behaviors you observed, as well as any difficulties students had working together. Note students' ideas about what worked and what didn't work. These may help you to guide future activities.
2. In addition to using the feedback they got from the class, students should have built on the ideas they got from other groups.

3. Students should have been able to make changes more effectively when they made incremental changes, rather than changing many things at once.

4. Criteria are the goals that must be achieved; constraints are the limitations on how we can achieve those goals. Identifying criteria and constraints at the beginning allows designers to focus on what is to be accomplished and how.

5. When you learn about how other scientists work, you can use aspects of their solutions to solve similar problems that you face. You can see what worked and what didn't work in their experiences, and you can use that knowledge to guide your efforts.

6. An observation states what was seen or heard without explaining why. Two observers at the same event should have little, if any, trouble agreeing on an observation. Interpretations include ideas about why something happened, or why something is the way it is. It is important not to include interpretations in your observations, because interpretations must be supported with reasoning and evidence before they can be accepted as valid claims.

7. An explanation includes a claim, evidence, and science knowledge. An explanation is good if the claim is supported by accurate evidence and science knowledge in a logical way.

8. Evidence is important because it makes a recommendation persuasive, but it is also important because recommendations should not be based on theoretical ideas about the world, they need to be based on observations and experimental measurements.

Teacher Reflection Questions

- What difficulties did students have with the concept of iteration and the application of iteration in creating their designs?
- How did students' answers to the questions in the *Answer the Big Question* section help them to answer the *Big Question?*
- How well were you able to assess how students were giving and receiving feedback and building on each other's ideas? Were there any indicators that you could look out for next time?

Blackline Masters

Animals In Action Blackline Masters

Name of BLM	Location *
Observing and Interpreting Animal Behavior	1.3.1 / 1.BBQ.1 / 2.3.1 / 2.5.1 / 2.7.1 / 3.3.1/ 3.4.1 / 3.6.1
Create Your Explanation	1.4.1 / 2.3.2 / 2.7.2 / 2.BBC.1 / 3.6.2
Flower Dissection Observations	2.MTL.1
Solution-Briefing Notes	2.BBC.2 / 3.BBC.1

* Number indicates Learning Set.section.sequence within section

Observing and Interpreting Animal Behavior

1.3.1 / 1.BBQ.1 / 2.3.1 / 2.5.1
2.7.1 / 3.3.1 / 3.4.1 / 3.6.1

Name: ______________________________ **Date:** ____________________

Animal I am observing: ______________________________

Observations	What about the environment and animal allows that behavior?	Interpretations

Name: ______________________________ **Date:** ______________

Use this page to explain the lesson of your recent investigations.

Write a brief summary of the results from your investigation. You will use this summary to help you write your Explanation.

Claim – a statement of what you understand or a conclusion that you have reached from an investigation or a set of investigations.

Evidence – data collected during investigations and trends in that data.

Science knowledge – knowledge about how things work. You may have learned this through reading, talking to an expert, discussion, or other experiences.

Write your Explanation using the ***Claim***, ***Evidence*** and ***Science knowledge***.

Flower Dissection Observations

Name: ______________________ **Date:** ______________

Use the space below to sketch, tape and describe the parts of your flower.

My flower has __________ sepals.

The color of the sepals is __________

A sepal from my flower

My flower has __________ petals.

The color of the petals is __________

A petal from my flower

Use the boxes below to sketch the inside of your flower and tape a sample of pollen.

This is the inside of my flower.

This is a sample of pollen from my flower.

Solution-Briefing Notes

Name: ______________________ Date: ______________

Design or group	How well it works	What I learned and useful ideas		
		Design ideas	Construction ideas	Science ideas

Plans for our next iteration

NOTES

NOTES